RIDE
of the
Second Horseman

RIDE

of the

Second Horseman

THE BIRTH AND DEATH OF WAR

Robert L. O'Connell

New York   Oxford
OXFORD UNIVERSITY PRESS
1995

Oxford University Press

Oxford   New York
Athens   Auckland   Bangkok   Bombay
Calcutta   Cape Town   Dar es Salaam   Delhi
Florence   Hong Kong   Istanbul   Karachi
Kuala Lampur   Madras   Madrid   Melbourne
Mexico City   Nairobi   Paris   Singapore
Taipei   Tokyo   Toronto

and associated companies in
Berlin   Ibadan

Published by Oxford University Press, Inc.
198 Madison Avenue, New York, New York 10016

Oxford is a registered trademark of Oxford University Press

Library of Congress Cataloging-in-Publication Data
O'Connell, Robert L.
Ride of the second horseman : the birth and death of war
/ Robert L. O'Connell.
p. cm.   Includes index.
ISBN 0-19-506460-7
1. War—History.   2. Military art and science—History.
I. Title.
U21.2.028   1995
355.02—dc20   95-9858

1 3 5 7 9 8 6 4 2

Printed in the United States of America
on acid-free paper

To H. B.

PREFACE

This book springs from a long conversation with Robert Cowley in the summer of 1988, during which I expressed my surprise at how little had been written on war's origins among humans. He in turn encouraged me to write an article on the subject. The process, however, raised more questions than it answered, and before long I found myself facing what amounted to a panorama of early human history. For warfare simply could not be separated from the other basic activities involved in the great transformation of our species from a life of hunting and gathering to a much more organized existence. I came to believe that the question "Why warfare?" could not be answered without addressing "Why civilization?" To put it mildly, these are not issues which lend themselves to easy resolution. What follows is the product of five years of intensive thought and writing. I am under no illusions that it will be the last word on the subject. But I do think it makes a good start at broadening the context to the point that ultimate answers are at least conceivable. In the meantime, I hope *Ride of the Second Horseman* will stimulate discussion and serve as a catalyst for further exploration.

Bringing a book to life is more like parenting a child than most realize, for control is more imagined than real, and there comes a time, when the issue leaves the nest, that only hope remains. No child is raised, nor any book written, without the substantial help of others. This is particularly true in this case, for the subject led me onto ground far afield from the normal province of a historian. For much of the time I managed to navigate a course by myself, reading, interpreting, and learning as I went. But there were also numerous instances when I could proceed no further without the direct advice of those primarily grounded in the fields of archaeology, anthropology, biology, and political theory, among others. Virtually without exception I was greeted with graciousness, generosity, and, above all, patience. Dorcas Brown, Dean Falk, R. Brian Ferguson, Helen E. Fisher, Jonathan Haas, Marvin Harris, William McNeill, Roger D. Masters, Olga Soffer, and Edward O. Wilson were kind enough to either talk or correspond with me on a variety of subjects. Also, I particularly want to thank David W. Anthony for spending what amounted to many hours with me on the phone, explaining issue after issue and in general providing a sympathetic ear. This book literally would not have been possible without his help. I can say much the same for my friends Alex Roland and Gordon Bowen, whose interest in the project and detailed written comments proved invaluable.

We live in a time when literally everyone seems busy. Therefore, I am very grateful to Carl Brandt, John Casey, Robert Cowley, Ronald Dimberg, Norman A. Graebner, David Lee, Peter Kracht, and William Suggs for having found time to read the manuscript, either in its entirety or in part. I want to thank you all for your suggestions.

The research for this book was conducted primarily at the Alderman Library of the University of Virginia, and the staff was, as always, extremely helpful and cooperative. At Oxford University Press my friend and editor, Nancy Lane, deserves a great deal of credit. She supported the project from its inception, gently shepherded me through the research and writing, and at a critical juncture provided wise counsel. So did my agent, Carl Brandt, whose wisdom and sagacity has left me much in his debt.

Finally, I want to thank my family—my wife, Benjie, and my daughters, Jessica and Lucy—for their forbearance during the last half-decade. I have stolen much time from you as a husband and parent. But rest assured you are always in my thoughts and in my heart.

Charlottesville, Va.   R. L. O.
April 1995

CONTENTS

RIDE

of the
Second Horseman

And he was clothed with a vesture dipped in blood; and his name is called The Word of God. And the armies which were in heaven followed him upon white horses, clothed in fine linen, white and clean. And out of his mouth goeth a sharp sword, that with it he should smite the nations; and he shall rule them with a rod of iron; and he treadeth the wine press of the fierceness and wrath of Almighty God.

——REVELATION 19:13–15

{1}

INTRODUCTION

IN SEARCH OF A BEGINNING

1

The attack had been utterly unexpected, and the few survivors who crept back at dawn were far from understanding what had befallen them. Certainly, they were unprepared for the scene of devastation unveiled by the morning light. As they staggered through the smoking ruins of their village they found, one by one, the mutilated corpses of their friends and relatives. Not only men, but women and children—the oldest and the youngest—all of them slaughtered. Some had died clutching figurines representing the Great Mother and the Bull. But nothing had protected them from the hatred of the marauders, an anger incomprehensible to these Neolithic farmers. It was apparent the attackers had searched for and stolen valuables, and eaten their fill. But the fact that they had taken care to find and burn the village's grain supplies stored in underground pits was an act of wantonness the survivors simply could not comprehend. It was as if they had been visited by demons, for these attackers seemed no more familiar. They had come out of nowhere and disappeared without a trace. More certain but perhaps strangest of all was their terrible enthusiasm for killing. The villagers had known trouble before, sporadic raids on the hamlet's animal pens by the shepherds who lived in the hills beyond their fields. But this was entirely different, or so they thought.

Seventy years later a much larger settlement would sit atop the wreckage, built by the survivors of this and four other localities. Now the entire living area would be surrounded by a massive wall, twelve feet high, constructed of boulders laboriously hauled from a nearby streambed. In its way this stone shield was a monument to the marauders, one more tangible than any they themselves would leave behind.

3

And one day in the not-too-distant future, a body of armed men would march out of this settlement, or one still larger, on their way to a similarly walled town eight miles to the north. Here they would do battle with an equivalent force, striving to conquer and so come to rule them. This act, too, was a memorial to the marauders, the fruit of an idea planted on that day of desolation, now barely remembered but destined never to be forgotten.

This reconstruction compresses a sequence of events that would have begun sometime after 5500 B.C. at a location marked by one of the innumerable *tells*, or mounds of ruins, dotting southwestern Asia. Perhaps it was at the very northern edge of the Mesopotamian basin, on the plains below the Taurus Mountains in Anatolia, or in the foothills of the Zagros range at the edge of the Iranian plateau— we will never know for certain. But such an attack would have marked a highly significant development in human violence. For the aggressors are assumed here to have been pastoralists, and their foray is hypothesized as representing a fundamental discordance between their own way of life and that of the farmers, a true clash of subsistence patterns. Had this not been the case—had the raiders not represented another way of life—then it seems entirely plausible that the course of war and ultimately history would have taken a considerably different path.

II

Even so, the act of pinning the origins of human warfare to any specific time frame must be an arbitrary gesture. For example, there is clear evidence of a stone wall having been built around at least one settlement as much as three thousand years before the date of our hypothetical attack.[1] But this was an isolated occurrence in apparent response to localized violence, not a manifestation of a broader, more lasting trend. By the same token, it is also apparent that war began among men at least one other time—in an entirely separate New World context—and that its evolution took a considerably different path. While this sequence serves as an invaluable basis for comparison—a legitimate, if alternative, origin of war—it took place thousands of years later at a time when organized combat was already well established throughout Eurasia.[2] So it seems logical to treat it as something of a secondary phenomenon.

As the topic is addressed, it becomes increasingly apparent that locating the origins of war depends ultimately on what is meant by warfare. And that, in turn, becomes a matter of perspective. For instance, there is every indication that human beings and their evolutionary predecessors engaged in acts of violence, both individually and in groups, long before the proposed sequence of events. Therefore, it is not so much violence per se but the quality and implications of that violence that are critical here. Although this study will deal largely with information derived from the fields of biology, anthropology, psychology, and particularly archaeology, it is intended as history. Consequently, the concept of warfare will be viewed and framed in a manner that is at least analogous to a contemporary definition of the term.

What, then, is meant by war? Herein lies a primary reason why the origins of war are not better understood and have not been more coherently pursued. All too frequently, definitions of war, particularly those employed by anthropologists, have sorely lacked precision, relying instead on catchphrases not much more descriptive than "organized violence" or "fighting between territorial units."[3] This has led to problems in distinguishing blood feuds and other acts of violence motivated by issues of reproduction and revenge from a concept of warfare relevant to our own world. War and mayhem blend together in a manner that not only makes the former appear logically eternal but actually detracts from the understanding of both. Of course, definitional specificity can be overdone, establishing preconditions so rigorous and arbitrary that they exclude interactions that common sense tells us are obviously warlike. Bearing this in mind, it is probably better to proceed in terms of general characteristics rather than to attempt construction of an all-inclusive, internally consistent definition.

Thus warfare—organized fighting (for those who absolutely demand a bare-bones formulation)—should be understood to encompass most, though perhaps not all, of the following characteristics. There must be an element of premeditation and planning; it is not simply a random emotion-driven act. Nor is it concerned primarily with the individual or those closely related but instead focuses on societal issues, with the intent of resolving them by force, using the resources of the group. Similarly, war implies direction by some form of governmental structure, and a military organization determined at least in part by that structure. It is assumed that combatants are willing (though perhaps not enthusiastic) and able to conduct a somewhat protracted campaign aimed at palpable economic and/or political goals, though these may be as simple as defense and survival. Participants also are presumed ready to apply lethal violence and risk injury and death in pursuit of these objectives and in accordance with the dictates of the command authority. Finally, some understanding can be attributed to the parties involved that the results of war, for good or ill, will be more lasting than momentary.[4]

Admittedly, this approach neglects potentially important subconscious factors. Nor does it attempt to differentiate among the various categories of war—wars of conquest, maritime war, civil war, and so forth—although the study itself is concerned primarily with aggressive war and not civil war, which is dealt with tangentially largely in the last chapter. Nevertheless, this definitional framework does set forth some fairly basic and transparent preconditions, and in the process raises several issues that demand further explanation.

It should be understood that, although the characteristics set forth here are somewhat abstract, they nevertheless reflect an actual chain of events. They point to a certain level of social evolution that had not existed before, and they imply that warfare is a mechanism that performs certain functions, which logically will vary in range and intensity as communities develop in different directions. In short, it is an institution. Indeed, the very notion of origins implies more than just a beginning; it suggests a finite life span terminating if and when useful functioning should cease. This is important because it addresses two of the most salient questions about war: Is it innate in humankind? and, if not, Is it inherent in civilization? Much of the following narrative will be devoted to demonstrating that the answer to both of these questions is no—that we were free of war for most of our existence, and that its onset and continuation were dependent on levels of ecological adaptation that were inherently transitory.

Ironically, the most difficult theoretical questions to answer about war and civilization literally turn the original propositions upside down. In fact, a contemporary understanding of the relationship between cultural and genetic motivation is hard-pressed to explain why any young man would be foolish enough to go to war in the first place. Then there is the enigma of civilization itself. Prior to humans, the only creatures who have managed to evolve truly complex social structures are insects of the order Hymenoptera, whose haplodiploid reproductive patterns make it possible to live in this manner.[5] On the other hand, human beings, clearly lacking any such mechanism, provide no obvious and fully satisfactory answers as to how we managed to form and live in mass societies. It is clear, however, that going to war, because of the distinct possibility of personal annihilation, constitutes perhaps the most "altruistic" of social gestures. Hence, insights into why people are able to join and fight in armies are likely to help us understand how humanity deals with god-kings, warrior classes, pyramid construction, long lines in supermarkets, and traffic jams.

Given the questions raised earlier, real insight into the roots of war seems to call for two basic lines of inquiry. First, it is logical to hope that the life sciences can provide useful clues and potential analogues. In this context it will be important to look into those specific elements of human nature and evolution that could eventually have an impact on our careers as warriors and as members of complex societies. We emerge from this examination as profoundly ambivalent beings—creatures shaped by millions of years of hunting and gathering, and then thrust irreversibly into something entirely different. In the process, humans became agents primarily driven by their cultural adaptations, at once frustrated and empowered by what amounts to a new evolutionary track.

Having reached the threshold of this metamorphosis, it would seem appropriate to shift gears and begin thinking about war largely on the basis of cultural mechanisms. Almost immediately the question of origins takes on further complexity, for it becomes apparent that war's spread must be tracked and explained, and this in turn leads to, if not a multitude of origins, at least several discrete categories. In relatively short order (by approximately 2000 B.C., war in its primary Old World venue became a significant element in the operation of the four distinguishable types of societies that would emerge to dominate the post-Neolithic era—pastoral nomadism, city-states, imperial tyranny, and maritime enclaves. In each case, however, the role of warfare was different, being itself a primary manifestation of the basic technological and ecological adaptations of these societies.

But these casts of war hardly existed in isolation; rather, they were proverbially interactive, generating and regulating a set of political dynamics that would come to hold sway over Eurasia and remain essentially unchanged until the proliferation of firearms. In some respects the operation of this larger system can be analogized to the working of celestial bodies on the galactic level. At the core, containing the bulk of the matter (in this case people), are the huge agrarian imperial tyrannies. Only metastable, they deploy their massive armies in an eternal quest for balance through aggrandizement. Meanwhile, they are both beset and energized by a pair of divergent entities, one in close proximity, the other operating at the margins. The former, city-states burning like young blue-white stellar furnaces, at once thwart imperial ambitions by their very energy and fight among themselves in balance-of-power amalgams that are the stuff of future hegemonies. On the outskirts lurk the pastoral nomads, scattered across Inner Asia like a thin but highly charged dust cloud that periodically sends waves crashing over the imperium to at once terrorize and vitalize, to plunder and kill and form new dynasties. And, finally, almost akin to blobs of antimatter, are the maritime elements, also existing at the edges, using the buoyancy and reach of water to gather wealth and culture, and employing warfare so infrequently and abstractly as to remind us of its potentially transitory nature.

There were other permutations—for example, feudal remnants of imploded empires reforming into secondary balance-of-power systems based on medium-sized monarchies. But the basic quadripartite system sufficiently describes aggressive warfare's early evolution to serve as a basic frame of reference and focal point. Nevertheless, it would be a mistake to ignore the importance of what lay below, the system's foundations buttressed by the twin pillars of agriculture and animal husbandry. Perhaps more than any other factors, these two defined warfare's possibilities and limitations. Consequently, in what would become Latin America, when one of these elements was largely missing[6]--an assemblage of domesticated animals sufficient for true pastoral nomadism—warfare developed in a distinctive and significantly less virulent fashion. This too will be looked at in detail. But it immediately suggests something more pertinent to our own lives and futures.

Americans inhabit the most agriculturally productive nation on earth, yet fewer than one in fifty Americans live on farms raising crops and tending animals.[7] If war in its classic manifestation truly is representative of agrarian and pastoral life, an institution fundamentally useful in righting disequilibriums inherent to these modes of existence, then there is good reason to question its viability—not just today but beginning with the onset of the Industrial Revolution. In an earlier book, *Of Arms and Men*,[8] I suggested that the relentless growth in the destructiveness of weaponry would soon make the conduct of effective war not just difficult but virtually impossible. While weapons still appear to be war's most immediate enemy, further inquiry indicates that aggressive warfare has been steadily undermined along a number of other dimensions, particularly in the developed world; that its original purposes have disappeared with no appropriate replacements having emerged. If indeed the Second Horseman is slipping from his mount, his bloody ride without destination and nearing an end, the circumstances of his demise are certainly worth considering. And a comparison between the rider in the bloom of youth and the teetering cavalier of today promises to highlight the contrasts and provide valuable insights into the coming fall.

III

Easier said than done. The course outlined in the preceding section demands that we march down a path far more clearly defined in theory than in fact. Indeed, the road leading back to our prehistory rather quickly degenerates from a well-paved highway into a rutted wagon track filled with detours leading nowhere and strewn with the wreckage of theories long since abandoned by previous wayfarers. The further we go, the worse it will get, until a point is reached when about the only guideposts are a few fossil skulls and some suspiciously chipped rocks. Welcome to the deep past—proceed at your own risk.

This is barely exaggerating. Much useful reconstruction has been accomplished in the two or so centuries since it first began to be recognized that our ancestral line and the world in general were a great deal older than Archbishop Ussher's biblically mandated 4004 B.C. start-up date. Yet the true character of the enterprise is still subject to considerable misunderstanding. In part this has to do with the very nature of archaeology, which after all is concerned primarily with things[9]— tangible relics that can be held in the hand and that impart an aura of solidity and absoluteness to the conclusions drawn by those who have dug them up. This is deceptive. Actually, the edifice of modern archaeology is clad only occasionally with more than a tissue of evidence, and gaps exist stretching for interminable periods where there is nothing—no suggestive bones, no relevant rocks. Common sense tells us this is simply nature in action; after all, most things die or are discarded without leaving a trace. But probability argues the contrary. Something is almost always preserved, and time is a great collector. The truth is, much more pertinent evidence exists than ever will be, or ever can be, recovered. And this is less a function of nature than of how archaeologists must go about their business.

Real-world digs are, in the best of times, extremely arduous, costly, and time-consuming. Consequently, only a tiny fraction of the literally thousands of sites where humans are believed to have lived can ever receive more than the most cursory examination. Even locations subject to intensive study are seldom more than partially excavated.[10] What will be uncovered turns on a combination of logic, intuition, and luck, and is hemmed in by difficulty. Thus, no matter how promising an area along the Yellow River or on the Nile Delta, it will be shunned if it is believed that archaeological pay dirt is buried beneath hundreds of feet of silt. Sites lying below major areas of contemporary habitation—the Aztec capital of Tenochtitlán under Mexico City, for example—present analogous constraints. And barring unforeseen technological breakthroughs, it will always be so. Meanwhile, contemporary research trends have emphasized ever-more-intensive exploitation of a limited amount of evidence, with a legion of specialists studying everything from pollen samples to fossilized feces.[11] This has certainly added depth to the understanding of certain areas, but it has done little for others. Take the problem—especially pertinent for this study—posed by people with no fixed habitation, hunter-gatherers or, in particular, pastoral nomads whose exploitation of the horse allowed them to wander huge areas and whose material culture was strictly limited by the necessities of portability. Such people are all too likely to remain invisible to conventional strategies for gathering evidence.[12]

Archaeologist Mark Nathan Cohen suggests that his colleagues literally call a spade a spade—accept the limitations of the shovel and admit that archaeological data are essentially samples—and that the real strengths of the profession lie in making informed conclusions on the basis of limited evidence.[13] Actually, archaeologists have always sought to flesh out what they have found with generalizations drawn from other disciplines, and over the last thirty years there has been a concerted effort to address a broader range of issues.[14] Nonetheless, the roots of the profession are buried deep in the tangible, and this orientation persists, though in a more subtle fashion.

Consider, for example, the continuing impact of the three-age system—the nineteenth-century notion that prehistory was composed of three successive periods, a Stone Age, a Copper/Bronze Age, and an Iron Age. This scheme, based purely on the physical composition of a relatively narrow range of recovered artifacts, not only laid the foundations of archaeology as a discipline but imparted a certain vision of progress to the whole enterprise.[15] While the centrality of the three-age system has long since been pushed aside, the ceaseless process of gathering artifacts and the sheer momentum of the original categories (and the subcategories that followed) ensure a residual but often overlooked role for the scheme in setting the profession's terms of reference. And because so many of the artifacts, particularly the metal ones, happen to be weapons, the three-age system continues to impose what is probably an excessively material orientation on the study of war and its origins.

Meanwhile, its high-tech successor, carbon 14 dating, although constituting a monumental breakthrough in establishing absolute chronologies, has only served to further perpetuate the focus on tangibles. The physicists' gift to archaeology, it is as clinical as it is ingenious. For now artifacts, or at least organic artifacts, are valued in part as laboratory specimens whose radiocarbon content promises something like an exact numerical age (subject to recent recalibration)[16] a figure almost as skeletal as the bones and splinters on which it is based.

But breathing life into the long-dead is no easy matter, as witnessed by the still more complex and difficult set of problems posed by the use of ethnographic data. Shortly after the emergence of serious archaeological research, it came to be understood that the study of contemporary societies employing very simple subsistence strategies might provide a unique window on human prehistory. Surveys of so-called Stone Age cultures, gathered primarily by anthropologists, were enthusiastically embraced—quite probably because they were presumed to represent a tangible, factually based record. Unfortunately, this record proved to be shot through with cultural biases, not the least of which had to do with assumed levels of violence. Such notions were duly reflected in archaeological reconstructions that tended to speak in terms of "savages" and "barbarians" while further imposing the concept of progress on the transitions from one level of adaptation to another.

Although ethnographic information is presently gathered and used in a far more sophisticated fashion, there is no consensus as to its precise relevance beyond rather Delphic warnings such as "A hunter-gatherer today cannot simply be equated to his or her Paleolithic or Mesolithic forebear." But when searching for the origins of war, this is no admonition to be taken lightly.

For example, the relatively low levels of observed intergroup violence among recent hunter-gatherers can logically be attributed in part to selection pressures that relegated particularly shy, peaceful folk to areas unsuitable for other forms of economic exploitation—an unlikely process in the deep past when everybody was a hunter or a collector. Consequently, it is quite possible that generally higher levels of aggression between bands might have existed then than now.

Conversely, a number of anthropologists have recently concluded that the extremely violent profiles presented by certain Latin American slash-and-burn agricultural societies—a pattern not necessarily paralleled by early Neolithic communities with equivalent subsistence technologies—was largely a function of contact with colonizing whites and, more particularly, with their weapons and war-making practices, which spread even faster than they did.[17]

Where does this leave us? Speaking like a true archaeologist, James Mellaart warns against the application of information derived from any contemporary society in one locality to a similar culture that once occupied another region.[18] This seems too categorical. Certainly, caution is in order, but it also appears that the basic modes of existence do set some fairly universal boundary conditions that can be usefully projected across a range of past societies to flesh out the skeletal structure provided by artifacts alone.

But if the archaeological relevance of ethnography is subject to mixed reviews, less concrete but highly illuminating approaches drawn from the fields of biology and psychology are frequently ignored or rejected. In the first instance this has a good deal to do with past excesses. The study of human prehistory was dramatically influenced by the revolutionary climate surrounding the publication of Charles Darwin's *Origin of Species* in 1859. Yet the popularization of evolutionary thought and its implications for humanity would be largely the province of disciples such as T. H. Huxley and, particularly, Herbert Spencer, whose emphasis on struggle and "survival of the fittest" turned human development into a sort of perpetual battlefield and justified all manner of class and racial dominance.[19] Today, over a century later, the stigma of this unfortunate episode has yet to be fully removed from biologists interested in history and prehistory.

This borders on the ridiculous. Among Darwin's most important and lasting insights was that human beings are very much a part of the larger natural order, that we have no special claims of exemption from its processes.[20] Consequently, any true explanation of humankind's amazing transition from genetic to primarily cultural evolution and all this implies simply must make sense in terms of modern biology.

As noted earlier, warfare poses special problems in this regard. But biology also promises answers to certain key questions. For instance, it appears that combat performance is highly dependent on certain behavior patterns— small-group bonding to face danger, various manifestations of aggression, and so forth—which are best accounted for as adaptations that took place earlier in our evolutionary history. While certain of these adaptations are suspected of being genetically transmitted and therefore innate, they are also conceptualized as being simply predispositions and are in no way inevitable without appropriate environmental stimulation.[21] Nevertheless, the notion of genetically programmed behavior, particularly as it relates to violence, is typically viewed with suspicion—and not just by strict environmentalists. Thus the commentator at a recent and otherwise productive conference on the anthropology of war could dismiss the importance of innate aggressive patterns as "not so much wrong as irrelevant."[22] Behind this facade of objectivity is likely to lurk the specter of taboo biological thinking is dangerous and will lead us once again astray. Certainly, it is controversial and can easily be overemphasized. But it simply cannot be ignored in a study of an institution concerned with fighting and dying and dominance such as war unquestionably is.

It is also a principle of modern biological theory that, although populations and their genetic composition are important considerations, evolution still must be initiated at the level of the individual and his or her genes. There is a strong analogy here to the study of ancient societies and their institutions. One reason archaeological reconstructions appear so skeletal and lifeless is the reluctance of the re-creators to impute anything but the most basic motivations to the former inhabitants. If they emerge at all, they appear almost as automata. Not unexpectedly, this approach generally is defended on the grounds that the physical evidence simply will not support a more elaborate framework of behavior. Fair enough from a methodological standpoint, but it ignores the fact that the unparalleled sophistication of human thought and actions is critical in many respects to the societies we have created.

For instance, people are presumably unique in their awareness of the inevitability of death, and this in turn can be assumed to be a key factor in the rise of a uniquely human institution, religion. But archaeological theorists, as Geoffrey Conrad and Arthur Demarest point out, traditionally have had great difficulty integrating religion into dynamic models of civilization, considering it simply a stabilizing element.[23] This may be, but it does little to explain why cultures such as those of ancient Egypt and Peru devoted very substantial portions of their labor and resources to the support of what amounted to corpses.

War, like religion, was and is intimately related to the awareness and possibility of death, and is therefore bound to be deeply affected by the psychological status of the participants. Nonetheless, with the notable exception of Victor Davis Hanson's *Western Way of War*,[24] military histories dealing with ancient subject matter devote little attention to accounting for the motivation of anyone except perhaps the leadership. This simply cannot be justified in a search for the origins of war, when the question looms as to why a prospective soldier might join an army and go off to fight in the first place.

Gradually, it seems, scholars are coming to realize that the analysis of complex social systems must reach down to the level of the individual members. But the implications are probably more revolutionary than is now imagined. Thus, when comparisons of various economic and ecological strategies are made from the perspective of the individual, it becomes logical to ask not just whether they were better fed and housed but whether they were happier and more fulfilled. In this context the evolution of human society and institutions such as warfare begins to take on an entirely different cast.

Those familiar with the nature of historical research may find the previous discussion more than a bit ironic. For, as a group, historians are reluctant to make even the narrowest of generalizations unless supported by heaps of written sources. Consequently, they avoid questions of prehistory and the origins of basic social institutions, choosing to take up the chase only when societies become literate. In the case of war, written sources do provide some help, but largely as reflections of earlier events.

In particular, myth and comparative linguistics traditionally have been thought to hold great potential as mirrors of the deep past. Unfortunately, long years of intensive study have failed to bring either picture clearly into focus, producing instead information that is either vague and difficult to apply or incompatible with other mutually consistent sources of evidence.

Nevertheless, there is something there. The spread of the Indo-European language group does clearly parallel the diffusion of the horse and the imposition of

more male-dominated, warlike societies that might be expected to arise from nomadic raiding and conquest. Similarly, persistent myths of the Flood can be assumed to represent collective memories of the several-hundred-foot rise in sea
level precipitated by the melting of the last Ice Age's continent-spanning glaciers.[25]
So too do the frequent tales among agricultural peoples of an earlier, better time
seem to reflect an Upper Paleolithic landscape teeming with game and a lifestyle
in consonance with humankind's evolutionary past. More to the point, it has often
been said that civilization, particularly its Western manifestation, harbors at its
core an irrational fear of attack, the source of which is sometimes attributed to
the inner dynamics of the culture and the values it promotes. This may be true
in part, but also it could well be an artifact of a time when cultures that knew
nothing of war suddenly began suffering unprovoked attacks by terrifying strangers
whose hatred they could not fathom and whose military prowess they would never
forget. Nor will it be forgotten here.

The trail of the Second Horseman marks but a faint imprint upon the path of
prehistory. Following it will require not only care but resourcefulness and imagination. There will always be the danger of being misled, of pursuing a plausible
but false track. But ultimately there can be no turning away from useful but
unconventional sources of information, or war here will be reduced to little more
than a sterile sequence of events and rubble and discarded weaponry—an armored
cadaver. The alternative is to follow the trail in every reasonable direction it may
lead us. In doing so, we must first look downward and consider a marching
column made up of the tiniest of warriors.

{2}

THUNDER BENEATH OUR FEET

I

The scouts had returned but a short time earlier, causing the warriors to stream out of their quarters. As the warriors assemble near the entrance, alternately restrained and encouraged by their slaves, their meticulously polished copper-colored armor gleams in the sunlight. Suddenly they are off, moving out in a compact column following a path, carefully marked by the scouts, that will lead them directly to the territory of their designated victims.[1] Upon reaching the target colony they attack immediately, apparently completely surprising their prey, many of whom lapse into shocked passivity. A number of defenders fight ferociously, slashing relentlessly at the legs of the attackers. But the raiding warriors—bigger and better fighters—are soon piercing the armor of the defenders at will. Yet their primary aim is not to kill. Rather, they are intent on plunder, stealing the defenders' infants and bringing the living booty back to their own colony. Here the youngsters will be raised as slaves, their lives dedicated to serving the warrior elite—fetching them food and even feeding them. For their part, the warriors while at home live a life of languorous ease, alternately beckoning their slaves to serve them and burnishing their armor.[2] Indeed, they are veritable parasites, imperial paladins whose lives are dedicated to an endless quest for colonies to raid and slaves to capture.

Here in its most nascent form do we find the origins of true war.[3] Virtually all the prerequisites are present: preparation, a group orientation, governmental and military organization, palpable economic and political goals, and a focus on lasting rather than momentary results. Yet the practitioners are only a few centimeters long. They are *Polyergus rufescens*—Amazon ants. For perhaps as long as fifty million years,[4] members of this species, along with their distant cousins *Formica*

15

lugubris, *F. polyctena*, pavement ants, and the rapacious *Eciton burchelli* and *Dorylus*, have fought in ways analogous to human combatants.[5] That they do so does not appear to be coincidental. Rather, it flows logically from their way of life, much as it does with human warriors.

There are differences, of course, the key one being that these ants have no choice—their genes predestine them for a martial existence. As with virtually everything else in their social lives, genetic advantage determines their willingness to join armies and to fight. Ants are the result of haplodiploid reproduction, female offspring of a central queen, sharing three-fourths of their genes with their numerous sisters.[6] It is this affinity (one-fourth higher than the normal coefficient of relationship of one-half between parent and offspring) that binds the colony together, making advanced sociality possible.[7] Such creatures sacrifice for the group because it is their best chance to perpetuate their own genes. In this context death in battle is a trivial matter compared with the success of an army composed of genetic near replicates. For ants, the whole is truly more than the sum of the parts.

There are other disparities. One elemental behavioral driver of warfare among humans is revenge, itself a function of our extraordinary capacity for long-term memory. Ants, being ants, have no comparable facility, and are therefore without the urge to avenge depredations long past. And unlike humans, who are remarkable in this regard, ants have evolved into a great many species—approximately ninety-five hundred. Thus it can be argued that what appears to be warlike is in fact only one species preying upon another—at most well-organized hunting. To some degree this is a matter of perspective. Yet even conceding this point, it remains true that a number of species—pavement ants, African weaver ants, and several members of the *Formica* clan[8]—do in fact conduct organized combat against colonies made up of their own kind. Moreover, it is also the case that warlike interaction frequently occurs between species that are closely related—the red imported fire ant versus the native American fire ant for example.[9] It even seems possible to maintain that ants in general are unique enough that their overall patterns of group aggression are useful and suggestive in and of themselves. This does not mean that variations should be overlooked, just that underlying patterns which transcend species boundaries should not be discarded strictly for what amounts to definitional reasons. Only certain ants exhibit behavior that honestly could be called warlike. Nor is it possible to isolate one variety of ant that displays all the characteristics of human warfare. Nevertheless, variations in war-making behavior can be expected to reflect differences in ecological adaptation. This is as true of people as it is of ants. What will be presented here must necessarily be a composite. It will not be perfect, but this imperfection should not obscure one very crucial point. It is often said that only human beings wage war. This is wrong. We are not alone. And from our tiny sisters-in-arms there is much to learn.

II

Ants and termites have been called the superpowers of the insect world, probably constituting up to three-quarters of the insect biomass in a number of tropical and temperate environments.[10] For most of their ancient coexistence these two have been locked in a coevolutionary arms race that has resulted in the most complex weaponry and battle strategies known in the animal kingdom.[11] In particular, certain species of both ants and termites have developed anatomically deviant "major" or "soldier" castes specialized for fighting, frequently through larger size, more powerful mandibles, or the ability to manufacture and deliver lethal chemicals.[12] But, although termites have emerged from the struggle very well armed and organized, their role as proverbial prey of the ants has pushed their evolution primarily in the direction of defensive and not truly warlike behavior. Ants, on the other hand, have developed a huge variety of both offensive and defensive techniques, using them in almost every conceivable manner to obtain food, protect their nests, and conduct aggression—often in a manner that can be considered warlike.

If there is one central factor that can be correlated with warlike behavior among ants, it is numbers. Indeed, as two of the human world's leading authorities on ants, Bert Hölldobler and Edward O. Wilson, tell us, wars are "commonplace among species with large colonies."[13] This basic determinant, which extends across both sedentary and nomadic species, could be explained in several ways, depending largely on perspective; for the moment, however, it is sufficient and accurate to say simply that there is strength in numbers. Indeed, if there is a primary rule of battle among ants, it was probably best articulated by Confederate General Nathan Bedford Forrest, who attributed victory to arriving "first with the most soldiers." Thus battles tend to develop along lines based on success at "recruitment," largely through the laying down of pheromone-based secretion trails.[14] Not unexpectedly, martial encounters frequently devolve into matters of attrition with casts of thousands. Most easily observed are the contests waged by common pavement ants, featuring masses of biting, struggling brown workers, locked in melees for hours on end and reinforced steadily with recruits following the chemical path to battle.[15] Similar struggles, building in intensity as numbers increase and generally determined by sheer density of fighters, are waged by weaver ants and their close relatives *O. smaragdina*, along with several other species.

There are, however, a number of significant variations. Among the most interesting are the tournaments staged by multiple competing colonies of the honeypot ant. In these contests opposing forces summon their workers to combat areas, where hundreds of them perform highly stereotyped fight-dances. When one colony becomes stronger and is able to recruit more participants, the display tournaments cease, with the weaker colony being sacked, the queen killed or driven off, and the vanquished workers either defecting to the winners or being incorporated as slaves.[16] Usually, the process continues until one colony prevails over all the others. Bert Hölldobler believes that elastic territorial patterns cause honeypot ant colonies to have such frequent hostile interactions that actual violence would simply be too costly to the worker force. Consequently, ritualization has evolved as the most economical strategy of settling intercolonial rivalries.[17] This is suggestive, since human city-state warfare has tended to develop in the same direction, with combat that is often chronic becoming highly stylized and hemmed in by rules oriented toward mitigating violence.

Honeypot ants are also capable of deception. Termites are an important food source for these ants, and should a rich supply be unearthed near a rival colony, the discoverers are likely to stage what amounts to a diversionary attack. Several workers rush home, where they recruit a force of perhaps two hundred, which then swarms over the nest of the potential competitors, keeping them thoroughly occupied while the termites are retrieved.[18]

This tactic is carried still further by *Pheidole dentata*, to the point that it can be called preemption. In this case, *Pheidole*'s problem is the voracious red imported fire ant, which treats them as prey and can finish off a colony in an hour. A species made up of major and minor workers, *Pheidole*'s best chance is to spot a prospective attack early. Should minors discover a party of fire ant scouts, they immediately return to the nest and recruit a large force, which then rushes to the scene, where *Pheidole* majors chop the scouts to pieces. In this way is fire ant "intelligence gathering" thwarted, and a devastating attack forestalled.[19]

At the most basic level of interpretation, *Pheidole*'s motivation is quite transparent; these creatures are simply interested in survival. But the elaboration of tactics, the reality of mass participation, and the relative frequency of warlike behavior across species all bid us to consider the phenomenon more abstractly, to ask what role war among ants might play as a fundamental part of their adaptive strategies.

When viewed from this more general perspective, it becomes apparent that warlike activities are manifested by two essential types of ant societies: sedentary and nomadic. In the first category, which constitutes by far the greatest number of species and participants, territoriality looms large as a central factor in colony interaction and internal regulation. However, territorial behavior in general tends to be somewhat misunderstood, often being thought of as absolute rather than fluid and situation-specific.[20] Thus, natural selection has pushed ants to establish and protect territory only when its size and design are economically beneficial to the colony.[21] It follows that territorial strategies vary widely among species according to such factors as foraging techniques, distribution of food supply, and egg-laying capacity of the queen, with combat acting as a key determinant of colony configuration, spacing, and, ultimately, success.[22] In this context it is important to remember that haplodiploid reproduction ensures that the death of a worker represents only a potential labor and energy loss, and not the sacrifice of a reproductive unit. Indeed, frequent combat deaths can even act as a positive factor in a colony's adaptive strategy, not only through protection and the maintenance of resources but by helping to control the age and vigor of the population.[23] Thus ants are theoretically more likely to risk violence in conjunction with territorial behavior than are solitary animals—a proposition amply borne out in fact. Indeed, in the case of several species with massive colonies—*F. yessensis*, *Wassmannia auropunctata*, *Pheidole megacephala*, and the red imported fire ant— territorial expansion becomes so rapacious that it might accurately be termed imperial, resulting in the elimination of all other ants within a local area.[24]

That warlike behavior acts as an arbitrator among a number of key endogenous and exogenous ant variables helps to explain another remarkable manifestation of warfare among sedentary species, and that is slave taking. While the slaver *Polyergus*, described at the beginning of the chapter, is probably the most exaggerated example of this phenomenon, a number of other sedentary species have come to include slave taking as part of their territorial raids. *Dulosis*, as the practice is known among biologists, always involves species that are relatively closely related[25]—a logical enough affinity but also one that serves to illustrate the link to truly warlike behavior. For slave taking and its violent accomplishment are both parts of a dynamic process directed toward maintaining the internal vitality of the colony through the introduction of outside, or "bonus," energy.

In many cases enslavement is accomplished by the forcible appropriation of larvae and pupae, which in the somewhat more divergent species probably stretches the term and more accurately could be called coercive domestication. In ants of the same species or among very close relatives, however, enslavement involves both immature and mature workers.[26] In one particularly interesting example, *S. alpinus* workers are accompanied by enslaved pavement ants while they conduct raids on the latter species' nests. Not only do the pavement ants participate in the fighting, but they help to carry back the brood and surviving workers of their conspecifics to the home colony.[27] Certainly, enslaved ants are not always so cooperative, and there are even documented examples of slave revolts![28]

But it is important to avoid any notions of conscious intent here or in any other aspect of ant warlike or social behavior. Ants cooperate because their genes make them cooperative. But what is suggestive with regard to human societies is the forms that such behavior takes and their relation to the functioning of the social structure. Thus the phenomenon of a few species of sedentary ants having come to practice cannibalism in conjunction with warfare—honeypot ants and *F. polyctena*—clearly is best viewed as a matter of simple energy acquisition at this level[29] but nonetheless remains suggestive of the anthropophagist cast of Aztec military campaigns in a chronically animal protein–poor environment. Similarly, the ant equivalent of a coup d'état—an intruder queen entering an alien nest, assassinating the resident queen, and then securing adoption by her former workers who blindly raise a brood of a different species[30]—is plainly just another manifestation of the behavioral possibilities of haplodiploidism. Nonetheless, it is worth juxtaposing against the stolid acceptance of forcible dynastic change in large agricultural tyrannies and the general momentum of such societies in the face of military conquest.

It is somewhat ironic that the very aggressive behavior of a few species of nomadic ants—the best known being *Eciton burchelli* and *Dorylus*—should most often be equated with human military forces and their activities. In fact, the depredations of the so-called army ants are more akin to mass predation from a functional standpoint, since it is directed against any and all species encountered during their voracious marches. Nonetheless, the shape these campaigns take is so startlingly similar to the mass migrations of human pastoral nomads that even careful observers such as William Morton Wheeler are moved to flights of rhetoric:

> The driver and legionary ants are the Huns and Tartars of the insect world. Their vast armies of blind but exquisitely cooperating and highly polymorphic workers filled with an insatiable carnivorous appetite and a longing for perennial migrations . . . suggest to the observer who first comes upon these insects in some tropical thicket, the existence of a subtle, relentless and uncanny agency, directing and permeating all their activities.[31]

In actual fact that agency consists of a pattern of ecological and genetic adaptations based primarily upon group predation and nomadism—the two having reinforced each other and been served by queens able to lay huge quantities of eggs very rapidly. The net result is massive populations of predators so efficient and cooperative that they literally strip their paths of all resident animal life, thereby necessitating continual movement.[32] These wanderings generally take the form of major migrations to and from primary bivouac sites, followed by secondary forays radiating out from the base camps. If there is a pattern here, it appears to be directed by little else than available food supplies—so much so that a group of doryline or ecitonine army ants set down on a clean flat surface will take to marching in a circle until all die from hunger or exhaustion![33]

Here again, the factors that enable army ants to behave the way they do have little in common with any of the roots of human social behavior. Nevertheless, the manifestations of militant nomadism in ants do bear a number of similarities to the aggressive activities of humans following a pastoral existence. As with their human equivalents, the nomadism of army ants appears to be a secondary development, apparently having evolved out of what were essentially sedentary beginnings.[34] But the very predatory character of raiding in both cases is largely a function of the requirements of portability dictated by life on the move. Territory, slaves, and long-term dominance are essentially meaningless—victims are subject to immediate exploitation, and there is little incentive in keeping them alive. The aggressors are driven to raiding in part because they are magnificently equipped to do so but also because their way of life demands it—with the ants it is their terrible hunger, with the humans it its the unstable population dynamics of the animals they herd.[35] In either case it imparts a compulsive and unpredictable quality to their depredations. What we are dealing with, respectively, are wandering masses of mouths, eating their way across the landscape. The larger the group, the faster pasturage or available insect populations are depleted, and the more compelling the logic of movement and aggression. So it is that these voracious ant armies number in the millions, just as major outbreaks of nomadic aggression were characteristically preceded by intertribal congregations.[36] If there is strength in numbers among the sedentary, there is only hunger among the nomadic.

III

The supposition that these manifestations of warlike behavior among ants—nomadic rapacity, slaving, territoriality, tactical elaborations such as diversionary attacks—all have human correlates should not be misconstrued. These phenomena probably say far less about human nature than they do about war as an adaptive mechanism at a certain level of ecological exploitation. No one can seriously maintain that humans and ants are, at base, very much alike, but parallels in the objective conditions of their lives could force behavioral similarities. Thus the very facts that both creatures wage war and that war is instrumental to the functioning of certain ant and human societies raise a fundamental issue: are there viable analogies to be drawn between the social structures of these entirely disparate life forms, or is this notion misleading and anthropomorphized?

Certain leading ant men demur. Edward O. Wilson, Bert Hölldobler, and a number of other entomologists prefer to view ant colonies and the societies of other social insects primarily as superorganisms.[37] Guided by a growing understanding of caste determination, communications, and feedback, these scientists tend to equate the operation of a colony to that of a single large animal, while analogizing the behavior of its individual constituents to the cellular components of that animal. In terms of cybernetics and systems dynamics this is certainly a useful approach, supported, in fact, by the maintenance of a specific nest temperature by some ants, honeybees, and certain termites and by the very nature of genetically mandated cooperation.[38] Nonetheless, there are also clear differences, the key one being that mature workers are physiologically independent entities that frequently act as external agents of the colony. They are specialized but not nearly so specialized and diverse as the cells of a complex organism. Thus even Hölldobler and Wilson call this approach "a heuristic device," implying that there are other ways of looking at ant colonies.[39]

In the end it is probably a matter of perspective. Just as human cultures could be viewed as exhibiting at least some of the features of macroorganisms, the activities of ant colonies can also be compared profitably to certain key manifestations of historical societies. Thus William Morton Wheeler, the dean of modern myrmecology and an early supporter of the superorganism concept, also noted some undeniable "resemblances between human and ant societies," maintaining that they constituted "a very striking example of convergent development."[40]

In fact, a comparison of the colonies of certain sedentary ants and early human imperial civilizations reveals some remarkable similarities. Both are organized in rigid hierarchies capable of managing very large populations, at times numbering hundreds of thousands and even millions. Such large-scale societies naturally demand stable, intensively exploited resource bases. While colonial ants are generally foragers, several species have demonstrated the capacity to form symbiotic relationships with certain fungi and animals sufficiently reliable and profitable to approximate the human dependence on agriculture and husbandry.[41] Indeed, biologist David Rindos maintains that harvester ants cultivate in virtually the same manner as people—preparing special beds, fertilizing and weeding the crops, and then harvesting them as the colony's principal food.[42]

But probably even more basic in the larger scheme of things, very large populations presuppose continual toil. Both types of societies are supremely labor-intensive, lavishing enormous amounts of worker energy not only on the necessities of support but also on construction programs, both functional and seemingly nonfunctional. The latter category is intriguing, since the somewhat mysterious human penchant for monumental architecture, particularly pyramids, is paralleled in the ant world by certain species that construct substantial symmetrically shaped mounds above their nests. While the mounds are believed to be related to microclimatic temperature regulation, researchers emphasize that the structures remain in a continual state of flux, with workers constantly moving material and repairing even the slightest damage.[43] It has been suggested that human-built pyramids acted as "energy sumps," helping to disburse excess intensity generated by such large populations.[44] An analogous dynamic could partially account for the mounds of the ants, who have been described by a number of observers as compulsively busy.

In any case, the importance of homeostasis in such large societies and the sheer difficulty of maintaining such an equilibrium should not be overlooked, nor should we underestimate the role of war as an ultimate balance wheel. Hölldobler and Wilson consider territorial aggression as an "important and possibly premier mode of population regulation among ants," but they also believe it to be part of a system of controls that must be considered as a whole.[45] Much the same could be said about human warfare, particularly in very populous agriculture-based societies but also in pastoral cultures where people were not nearly so numerous but the numbers of domesticated animals were frequently very large. At this level societies based on massive populations were only metastable, steady over the long haul, perhaps, but characterized by dramatic up-and-down shifts—proverbial demographic roller coasters. For ants as well as people living off symbiotic relationships with a few food sources, the effect of a number of population-influencing factors such as disease, predation, climate, and unfettered reproduction tended to be dramatic and uncontrollable in terms of both impact and time of occurrence. Of course, very large die-offs are not uncommon in nature, but only the social species had the organizational potential to combat them. But a mechanism was required, and mass aggression, being subject to some modulation and the possibility of initiation on demand, qualified in this regard. This does not mean that war was a precise equilibrator, only that it was inherently more manageable than the other Sturm und Drang variables that periodically broke over such societies. And it follows that war would have assumed a key role as a stabilizer. In times of hunger others could be preyed upon and stolen from. When populations rose, new territory could be appropriated. When numbers were driven down, supplemental labor could be captured. This was the logic of raiding, imperialism, and enslavement, whether the perpetrators were ant or human, and this is why the origins of war must begin here. For ants were the first to exploit these relationships, and in doing so became the original warriors.

There still must be something deeply troubling about comparing ants to humans. In part this is a matter of sheer pride: behavioral equations between ourselves and what amount to automata are bound to seem not just implausible but at some fundamental level insulting. Offended *Homo sapiens* are welcome to take refuge in the argument that only some ants live in complex societies and that the composite presented here actually warps the total myrmecological picture. Nonetheless, it remains true that those ants existing in complex social amalgams *do* exhibit patterns of behaviors which we recognize as familiar and which residents of the Shang dynasty or Tawantinsuyu (what we call the Inca Empire) would have found even more familiar. So the question remains: Why?

To be honest, it is not fully understood why creatures so dissimilar seemingly act so much alike. But it does appear to be based on the feedback relationship between population dynamics and a pattern of activities initiated at a certain level of ecological exploitation. It would be stretching things to say that humans weighing tens of kilograms and ants weighing fractions of a milligram occupied the same ecological niche. Nonetheless, it seems that a self-reinforcing chain of events begins to unfold once creatures settle down and start to exploit the environment in a certain way at a sufficient level of intensity. Whether ant or human, tightly interlinked factors of organization and cooperation make larger populations possible, which in turn serves to further intensify the former elements until diminishing returns establish a new behavioral/demographic plateau. War is simply one of several systemic mechanisms involved, and its development at this level is probably quite predictable. But it is not inevitable. The fact that other social species do not practice organized warfare makes this apparent. This, too, is not precisely understood, but it is probably significant that both ants and humans are terrestrial, while bees and wasps fly—a factor that logically would have important territorial implications. Mass aggression, on the other hand, appears well suited to enforcing the territorial demands of ground-bound creatures like ants and people.

Before leaving the microcosmic world of ants, the reader should once again be warned that, at best, the examples used here are simply analogous to human warfare, parallel behavior patterns based on somewhat similar problems. Ants do not choose to wage war; they are essentially automata. They evolved to this stage through a process of mutation, and like their sisters marching endlessly around M. C. Escher's Möbius strip, they are marooned there by the architecture of their genes. We, on the other hand, learned to do these things, swept along by the process of our own cultural evolution. And what was learned could be unlearned. But before any of this happened, before we came to ride with the Second Horseman, we had a long and very different path to walk.

{3}

THE SOUL
OF A HUNTER

I

The camp was buzzing. At last the hunt had gone well, or that, at least, was what the men were claiming. A mammoth had been cornered in a ravine and killed. But the youngest of the party, a boy, really, had been trampled. His broken body now lay hidden from view by the skins of the hunters' lodge, but everyone knew he would soon die. The women were furious. Their mood did not improve that night when the men of the party had insisted on performing the Spear Dance at the time of sharing. "Fools!" the boy's mother cried out. "For barely a moon's supply of meat you boast and posture. But your silly hunt and your vanity have cost us our son." Her words precipitated an argument that lasted until dawn. It was not the first or even the worst in recent memory. It was a bitter time for the band, a time of decision.

Five summers had passed since the reindeer had stopped coming. No one could remember anything like it. The hunting party had wandered farther and farther from the Home Path in search of big game, but it had done little good and had caused trouble with other bands who seemed somehow more numerous. No one had gone hungry. The group had simply moved more often, and the women spent extra time gathering edibles. A few men had even quit spear hunting and took to stalking smaller prey near the camp with bows. This had particularly enraged the spear hunters and led to quarrels that resulted in two men leaving the band. Meanwhile, the women wanted everyone to move to the river during the salmon run. They claimed that the men should stop their wandering expeditions and instead catch the plentiful fish. This, too, outraged the men. "You don't understand what that would mean. We take our strength from the big animals. They are what bind us together. If we cannot hunt them we cannot be ourselves." This had seemed like a vain excuse to the women. But time would show that the men did have a point.

This vignette focuses on a critical fork in humankind's long path of development. The circumstances outlined here would have taken place over hundreds and even thousands of years, but during that time span countless little dramas such as this one would have transpired. In some cases the changes augured here would prove only temporary. The reindeer and the big animals might return, and things could continue as before. But in the long run something fundamental was happening. A way of life millions of years old was drawing to a close. Our days as hunters and gatherers were numbered, and a future of pyramids and shopping centers awaited us. But our earlier existence would leave its mark. For just as the child is the father of the man, we retained, it seems, the souls of hunters and of collectors. For this was the context in which we evolved, the life and the environment that stimulated us to speak and to grow the big brains that made so much else possible. To assume that we would or could have cast this heritage entirely aside is to deny a very important dimension of our own humanity. At the core we are what we once were.

But retaining some measure of objectivity in describing this way of life and its influence is a particularly tricky proposition. Traditionally, it is a topic that has alternately attracted and repelled, a profoundly polarizing intellectual venue. Thus, views on the subject do not simply reflect contemporary social theories[1] but tend to be cast in broadly positive or negative terms. Speaking for the detractors almost 350 years ago, Thomas Hobbes probably crafted the classic description of prehistoric squalor: "No arts; no letters; no society; and which is worst of all, continual fear, and danger of violent death; and the life of man, solitary, poor, nasty, brutish and short."[2] But at approximately hundred-year intervals, Hobbes's Paleolithic thug would be reborn in the guise of Rousseau's noble savage, only to be cast again into the primeval muck of evolutionary retardation by the social Darwinians and finally plucked up and showered off by today's environmentally aware generation of scholars. Thus, he has become a decision-making and cost-benefit-analyzing hunter, as well as a member of "a practiced and ingenious team of lay botanists."[3]

But if the Paleolithic truth probably fell somewhat short of this ecologically correct rhetoric, the contemporary view does appear to better reflect actual conditions and the totality of existence. Life may have been short, but it was not necessarily nasty or brutish.

II

The Pleistocene epoch—the geological period that encompasses Paleolithic culture—was a time of genetic ingenuity and profusion among mammals, especially big ones. The overwhelming reality, the back-and-forth march of mile-high glacial ice sheets stimulated the evolution of a lavish array of megafauna—animals whose very size was their central advantage in a climate where heat retention paid big dividends. So, among those joining the more orthodox mammalian assemblage were moose nine feet high at the shoulder, beavers as big as bears, giant ground sloths, mastodons, and mammoths[4]—a movable feast for human predators with the wits and courage to help themselves at nature's groaning board. This, in essence, was the Eden remembered in the myths of so many grain-based cultures—not a garden of earthly delights or even Club Med, but for creatures who evolved from foragers and hunters, close enough to paradise.

But life two hundred centuries ago was hardly idyllic. It was dangerous, filled with lethal pratfalls and virtually devoid of effective medical techniques. Hunting, especially for big game, undoubtedly took its toll among the men, and as for the women, saddled with childbirth, mortality was so high that one sample of seventy-six Upper Paleolithic skeletons did not yield even a single female judged to have lived beyond age thirty.[5] On the other hand, if injury and infection could be avoided, our Stone Age ancestors were probably quite hardy. Recent analysis of their bones and teeth has revealed an episodic pattern of feast and fast, with food shortages apparently being short-lived.[6] Meanwhile, their high-protein diets made for big bodies, the men averaging close to five feet eleven inches and the women around five feet six inches. Lots of meat and fat also promoted disease resistance, as did living in a cool climate and in small isolated bands, effectively thwarting the Fourth Horseman, epidemic.[7] Yet keeping warm would have been a constant preoccupation in an environment raked by glacially chilled north winds. There would have been plenty of fur, and a demonstrated ability to build permanent shelter which stretches into the deep past, perhaps as far back as four hundred thousand years.[8] However, as Peter J. Wilson notes, domesticity was avoided until fifteen or twenty thousand years ago.[9] There were clearly practical considerations, but this avoidance also points to something important.

The essence of this way of life, the part that would have made it satisfying, was mostly nonmaterial. For survival demanded that existence be meticulously adjusted to the opportunities presented by continual movement, and this in turn required that tangibles be kept to an absolute utilitarian minimum. The quality of the human capital, on the other hand, was likely to have been high. Ethnographic data drawn from a wide variety of recent hunting-and-gathering societies indicate that individual freedom and self-sufficiency are not abstractions; they are expectations reflected in every aspect of daily existence, and there is no reason to believe this was any less true in the Upper Paleolithic. However, in such societies the resultant individual was in no way an island; he or she was very much bonded to an intensely social group, which in almost all instances can best be analogized to an extended family of perhaps twenty-five to fifty members. In this context independence meant avoiding becoming a burden to the rest, and preserving the openness and fluidity of social relationships.[10]

Judging by today's hunter-gatherers, personal property would have been extremely circumscribed, and sharing the rule of thumb. Decision making was likely to have been a matter of consensus, with all the adults given the opportunity to contribute their opinions, for human relations in such societies are inevitably founded on a substantial measure of equality.[11] Similarly, interaction between the sexes, despite the male's superior strength and control of weaponry, probably took place on a basically level plain.[12] Most likely there was considerable sex-based role differentiation—men may have been the hunters, but women as gatherers were probably the steadiest providers. In the end equality would have been founded on group membership—that and personal attraction. For ethnographic reporting and the relative profusion of figurines of nude females among Paleolithic artifacts indicate that sexuality was a major preoccupation.[13]

The problem with such descriptions, however, is that they render abstract what was precisely the opposite—a sort of endless encounter session immersed in a whirlwind of chatter. For speech, the one behavioral aspect that truly sets humans apart, is likely to have been highly developed by this time.[14] More to the point, it was undoubtedly vivid, filled with natural imagery, and at times poetic. For this was a way of life that honed the senses and demanded an extensive knowledge of nature, which would have been reflected in every aspect of self-expression. The verbal equivalent would have been endless stories and even epics, tales that would have surprised and delighted us with the richness and sophistication of their language.

But before waxing too eloquent on the charms of Paleolithic humanity, it is important to realize that to a great degree these attractive societal features functioned to avoid and resolve conflict among what were unquestionably active, aggressive, crafty, and long-memoried individuals living in an atmosphere of extreme and unending intimacy. This is why sharing was likely to have been scrupulously enforced, all opinions considered, equality practiced, and talk endless. But still, judging by contemporary hunter-gatherers, tensions would have built and a certain amount of what today is known as domestic violence probably occurred. Yet mayhem is greatly feared in such societies and beyond a certain point cannot be tolerated, so there is an ultimate safety valve vested in the ability to simply leave the band and join another.[15] Because outbreeding is a genetic necessity in such small human populations, hunter-gatherer groupings are relatively open with respect to the comings and goings of members. But it can also be assumed that this fluidity of memberships has a good deal to do with keeping this "exit option" open.[16] But, however well this may have worked in the Paleolithic context, there were still two underlying sources of potential conflict, which, if they didn't threaten the immediate equilibrium, did spell major change in the future.

Among the females it was infanticide. For even living in the midst of the Pleistocene bestiary, hunting and gathering could support only very thin population densities, probably no more than one or two people per square mile[17]—the sole exceptions being a few cases such as in southwestern France and the central Russian plain, where very unusual conditions allowed for larger local concentrations. In spite of this, however, in all of France the human population probably never exceeded twenty thousand during the Upper Paleolithic. This is a very low figure, especially in light of statistics indicating that an average woman is fully capable of carrying up to eight pregnancies to term.[18] Birth-related mortality for both mother and child was undoubtedly high, but the absence of the great epidemic killers of the young would have raised survival prospects significantly. The need to find mates outside the immediate band and the combination of prolonged nursing and a high-protein diet could have widened the gap between pregnancies, but at this point other means of birth control were likely nonexistent.[19] Thus, when all the factors are considered, it has been estimated that there were as many as 50 percent more live births than were needed to sustain populations at Upper Paleolithic levels.[20]

The only available counterpoise would have been infanticide, ranging from outright murder to simple neglect, focused on females—for they are the engines of reproduction, the levers of the population curve. Mothers killing daughters, or at least acquiescing to their deaths—this was not a demographic strategy calculated to produce happy campers.[21] The physical costs of nine months of pregnancy, added to the terrible psychological burden of destroying one's own issue and then multiplied across the women in a population, must have generated a heavy sum of grief and self-loathing, a veil of tears hanging over the meat eaters' paradise.

So long as there was no alternative, the burden of female infanticide was probably borne with sullen resignation. But when the possibility of change appeared, the opportunity to settle down and exploit a food source steady enough to support larger populations and help stem the tide of baby killing, it is a safe bet that women were in the forefront of those ready to make the transition.

There were also contradictions in the lives of the males. There is little reason to assume that anything approaching true warfare was part of the Upper Paleolithic social landscape.[22] Territory, property, and labor—the real drivers of the Second Horseman—were simply not sufficiently relevant to the world of the nomadic hunter-gatherer for war to take hold at this point and reach the status of a useful and self-perpetuating endeavor. Nonetheless, it is likely that the conditions encountered in pursuit of this way of life, particularly in the Pleistocene, had encouraged the evolution of what would become the behavioral foundations of war. So if hunting did not make men warriors, it probably provided them with the necessary equipment.

They were already professional killers—and not just run-of-the-mill exterminators. The men killed on an epic scale, taking on the largest and most dangerous animals. That was where the meat was. Indeed, so big and lethal were their prey that against many of them it was mandatory to hunt in groups. This was critical, for in these groups were the seeds of armies—or at least platoons.

Anthropologists point to what is apparently an inherent proclivity among human males to bond in pods that scrupulously exclude women.[23] There were certainly practical reasons why big-game hunting should have become a male-only pursuit. And over the millennia the continuing experience of confronting big, lethally aroused animals must have forged hunting parties into teams specialized to face danger, bonded at the individual level to risk everything in pursuit of the mutual objective and in protecting members in peril—the latter being further reinforced by the likelihood of kinship affinity. But what really separated this kind of group from an equivalent band of male chimps—intensified its special characteristics and amplified its lethality—were two uniquely human assets: speech, which allowed for unparalleled planning and coordination, and weapons.

Other animals were certainly well armed, but humans were the first with the skills and imagination to create their own killing instruments in a methodical and self-conscious fashion.[24] By this time our arsenal would have become fairly well developed and, perhaps of equal importance, functionally divergent. On one hand, there were a series of standoff weapons—the bow, the alatl (javelin thrower), the bola, and the sling—safe, ingenious, but limited in effect and good only for smaller prey. Killing really big, thick-skinned animals required close confrontation and eventually either penetration or heavy blows to the head. This was the province of the heavy spear and the club, the natural armaments of the big-game hunting teams. Nor would this conceptual dichotomy of arms be a transitory phenomenon. It would remain an important characteristic of future military forces and the way in which they would and could fight. Meanwhile, the sex-typical nature of the hunt had ensured that the use and possession of both categories of arms had become almost exclusively the province of males.[25]

What would have emerged was a sort of brotherhood of killers—the raw material for the cohesive small units upon which all armies one day would be built. But meanwhile this fraternity of death dealing was absolutely without precedent in terms of hunting efficiency, as the excavated bones of around one hundred thousand horses driven over the cliffs at Solutré in the Rhone valley and similar finds at Torralba and Ambrona in Spain all testify.[26]

Indeed, the very success of this male-only, weapons-bearing killing mechanism raises some fundamental questions about its ultimate impact. First there is the issue of human violence. On the face of it, the basic conditions of Upper Paleolithic life do not appear to presuppose a high level of bloodshed. The demands of outbreeding likely would have biased relations between local bands in a friendly direction. Some violence could be expected over sexual issues and mate transfer, but overall peaceful reciprocation would have been more efficacious. On the other hand, the innate human wariness of strangers, which has been established experimentally,[27] would have worked to limit contact between unfamiliar bands and those moving into new territories. So long as populations remained thinly distributed and there was plenty of room for retreat, it is logical to assume that withdrawal naturally would have been preferred over hostility and violence. Nevertheless, the potential certainly was there, and some scholars point to Upper Paleolithic archaeological evidence of slaughter such as the smashed skeletal remains of a group of twenty-nine people at Sandalja II in the former Yugoslavia.[28] While subject to differing interpretations, such evidence should be carefully considered, particularly from a psychological perspective. It has been fairly well established that there is a strong human inhibition against killing other people.[29] One day this would be overcome—overwhelmed is a better word—by what psychologist Eric Erikson calls "pseudospeciation," the notion that one's enemies are not really human and therefore are appropriate objects of lethal violence.[30] Now it makes sense not only that the hunter's experience with slaughter would have preconditioned humans for this ultimate transition but that outbreaks of multiple murder might have taken place in the Paleolithic context. Nevertheless, there is a fundamental distinction between sporadic acts of mass violence and true war based on political and economic motives that are basically inappropriate at this level of ecological adaptation. The proclivity to kill people in large numbers was certainly a precondition for war, but it was not necessarily the real thing.

But if the very efficiency of the male hunting machine did not yet mean war, it would prove fundamentally disruptive in another way. There is a view abroad, supported by a substantial body of evidence, that portrays the hunter-collector as a careful killer, a true game manager,[31] which is absolutely at odds with the prodigal acts of slaughter noted here. Further inquiry, however, reveals no essential contradiction, only the necessities imposed by the hunting of herd animals. For the act of congregation makes such animals by definition scarce.[32] It is the same logic that causes fish to form schools—contacts will be fewer, and when they come the predator will be overloaded, having only have a limited time and ability to kill.[33] This is why hunting animals will appear at times wantonly destructive. They have been conditioned by one of the iron laws of the jungle: kill all you can while you can. But for humans blessed with big brains and the ability to share and remember their ideas, there was no law that couldn't be bent. They learned to anticipate migration patterns, stampede herds over cliffs or into bogs, cut off their escape, improve tactics and weaponry—a growing bag of tricks that must have steadily raised the kill ratio.

Meanwhile, climatic factors were also working against the megafauna, for as the earth warmed and the glaciers headed north, the conditions that favored really big animals steadily disappeared. Forests of birches and evergreens invaded the tundra and grasslands that had nourished the great herds, and as their ranges vanished so did they. Survivors like reindeer, musk oxen, cattle, and horses cut their numbers and migrated to colder areas, but numerous species—mammoth, woolly rhinoceros, giant elk, and many others—simply became extinct.

While causation remains a subject of considerable scientific controversy, it stands to reason that humans played some role in pushing many species over the edge.[34] Mark Nathan Cohen notes that the die-off of large mammalian herbivores was particularly sudden and dramatic in North America, where over two hundred genera of Pleistocene animals disappeared. This event, which roughly correlates with the abrupt arrival (c. 25,000 B.C.) and subsequent infiltration of humans, can logically be explained by an inability to adjust quickly to advanced hunting tactics. On the other hand, in the Old World, where animals would have had much more time to accommodate to our evolving skill as predators, the die-offs were slower and less inclusive.[35] Nevertheless, the results were the same. The big animals were going away.

This was happening at a time when there were likely to have been noticeably more mouths to feed. For the evidence indicates that human numbers had been growing. Estimates of rates and totals diverge widely,[36] but it does appear that something was happening. To a limited degree this may simply reflect relative plenty in localized areas, which had allowed a few populations to rise rather dramatically, but in most groups it was probably more a matter of mechanisms. Although infanticide could control expansion within the group, outbreeding and the "exit option" encouraged fractionation—breaking off to form new elements that would have quickly expanded to band size. Gradually—very gradually—the space between bands would have narrowed, and as this happened the inherently flexible conception of territory would have begun subtly sliding from a very loose and permissive "we belong to this territory" toward an exclusive "this territory belongs to us."[37] And with less room to maneuver, the innate wariness of strangers would have been more likely to prompt a defense of territory rather than simple retreat. Once again, this did not mean war, but it was moving slowly in that direction.

Everything was shifting, and as the tectonics of ecology ground toward a new stasis, humanity was left clinging to a pinnacle in the process of collapse. But unlike so many other animals, this was not a question extinction. We were equipped to deal with change like no other. When this world disappeared, we would quickly find another and mold our behavior to its demands so well that after a time it would hardly seem possible that we had been or done anything else. But in other ways we would never escape our past.

In retrospect, the Upper Paleolithic can be seen as a sort of culmination of everything we had become, a climax of five million years of development. The human line had hunted for a long time, but Pleistocene ecology allowed this aspect to become exaggerated at the very time that evolution was putting the finishing touches upon its masterpiece. There is no controversy here; all agree that the Upper Paleolithic ended with modern humans, *Homo sapiens sapiens*, at center stage.[38] And if the chasm separating us from our former life looms wide, it is still just a short distance away—perhaps one-five-hundredth of our total evolutionary path. Our present conditions may influence us greatly, but we are still products of that earlier world. Everything we think of as most human—our faces and bodies, our thoughts and emotions, our speech, our very brains—all bear the stamp of this heritage. We remain, as Ortega y Gasset called us, "municipal Paleolithic man."[39]

This is not often considered, but it helps to explain our dissatisfaction with what came to be known as civilization as well as our dreams and aspirations. More to the point, it helps to explain war or at least the behavior necessary to wage it. Human warfare and the types of societies it serves are institutions—conscious elaborations on basic means of subsistence—without which they are likely to become irrelevant. But the factors that allow us to participate in organized aggression are more deeply rooted and far older. While introducing them in the Paleolithic context serves to describe them as they existed at our species' point of departure, it does not do justice to their functional role or dynamic interaction. To do that we will have to look further back, examine briefly our species' evolution and the development of the characteristics that made us unique.

III

Recent evidence indicates that the long transition from ape to human was less of a logical progression than had previously been imagined. Sequential and autocatalytic developmental models—for example, the tightly linked notion that upright posture freed the hands, which in turn evolved to use tools, thereby enriching experience to the point of stimulating the necessary brain growth to make speech and abstract thought possible[40]—are now being de-emphasized in favor of a more random accumulation of traits cast together over time by evolution. Rather than an inevitable or at least coherent chain of developments, human nature is emerging as a sort of mélange dictated by expedience.[41] Although this is less orderly and explicable, it is perhaps a more fitting family history for a creature whose most pronounced feature is, after all, improvisation. But it is also apparent that not everything has been driven out of sequence.

Certain key bodily changes still appear to have taken place early, though their role as a catalyst is more clouded. At any rate, shortly after australopithecine hominids began the human journey by venturing out of the forest and onto the savannas, they were walking on two legs. All the physical evidence—the placement of the spinal chord at the base of the skull, the breadth of the pelvis, and, most startlingly, a clear set of 3.75-million-year-old footprints etched forever in the fresh volcanic ash of Tanzania—indicates upright posture and a stride much like our own, though perhaps less refined.[42] Not only feet but hands had already assumed an essentially modern form, implying considerable manual dexterity in a creature that remained apelike in many other ways. Since the fossil record from between four and eight million years ago is basically blank, it is difficult to know whether these were sudden or gradual developments or what exactly prompted them.[43] But they do appear appropriate for a creature who initially may have wandered about the grasslands, eking out a living by gathering seeds, an activity that would have placed a premium on digital manipulation.[44] Otherwise, *Australopithecus* and his revolutionary successor, *Homo habilis*, were not particularly impressive physical specimens, slightly built, between four and five feet tall, and largely bereft of natural defenses—not even a decent set of canines like those of baboons. And the savannas were dangerous, filled with big cats and other carnivores that made a living preying on the teeming herds of herbivores which were the mainstays of the food chain. Very likely the little hominids were also subject to predation, and their long arms indicate frequent retreats to the trees.

But hunger and opportunity beckoned them toward change. For at some time these protohumans developed a taste for meat. Most probably they began modestly, munching on insects, something many primates do. On the savanna, however, much bigger protein bonanzas were available in the form of carcasses temporarily abandoned by their sated killers. Of course, these pungent prizes would also be the object of some considerable competition from other species, and our hominids would have been strongly motivated to improve their defenses.[45] Besides increased cooperation, this entailed a recourse to sticks and stones. Since chimpanzees are known to hurl rocks and wield branches,[46] this was probably not unprecedented, but the level of skill likely was. For the nimbleness and balance of bipedal posture combined with a hand preshaped for grasping and throwing gave hominids a natural advantage in this sphere. In a sense we were built to be armed—and also to use tools. For experience employing rocks as hammers eventually would have revealed that the razor-sharp flakes that sometimes broke off could be used to dismember carcasses far faster than the teeth of other scavengers, allowing large quantities of meat to be butchered and quickly removed to safer locations.[47] This technological breakthrough would have handsomely reinforced the hominids' efforts to improve their humble panoply of sticks and stones, creating, sometime between 2.5 and 3 million years ago, the world's first armaments industry.[48]

Now they were ready to take the offense, to become hunters and not merely exploiters of carrion. Exactly when this occurred, or how fast it progressed to big game, remains a matter of some controversy, with contemporary scholars tending to dispute Raymond Dart's original assessment that early hominids quickly began killing large animals.[49] But this remains a question of degree, and most concede that the key step which launched us on our relentless march to the status of top predator was likely taken early, albeit with small or injured prey.

It was a critical transition; armed and combative, our remote ancestors became creatures to be reckoned with. And while the image of "killer ape" popularized in the sixties and seventies, most notably by Robert Ardrey, has been toned down considerably, the weight of contemporary scholarship probably stands behind Grahame Clark's assessment "that man found himself and emerged as a dominant species first and foremost as a hunter."[50] As we shall see, there were certainly other vectors driving human evolution. But hunting opened a galaxy of possibility, giving our line access to animal spare parts that would eventually allow the exploration and exploitation of virtually the entire earth. If we lacked fur, we might clothe ourselves against the cold with the pelts of more hirsute species. Bones and tusks and antlers might be recycled into better weapons; carrying containers could be fashioned from skins, making gathering far more efficient; even artificial shelter became a possibility. More subtly, but perhaps of equal importance, the curtain of fear that is the prime differentiator between predator and prey had been lifted. The blind instinct of flight was no longer an absolute. Our ancestors were granted the luxury of calculating the odds, of considering the possibilities of attack and retreat. Just how stimulating hunting actually was to the development of prehuman intelligence remains debatable, especially considering that a number of very efficient and resourceful predator species are not particularly bright. Nevertheless, it does seem clear that becoming a hunter would have marked a fundamental transformation in our self-conception. And this is no trivial matter.

For at this point it is safe to say that our ancestors were aggressive in the full sense of the word. This statement is easily misunderstood, however. It is not meant to indicate that they were highly violent, a judgment that remains open to question. It simply makes note of the likelihood that they manifested what are the two basic forms of aggression—that which is directed against other species and that which takes place within the species.

Aggression and its impact on human evolution are not topics looked upon with much favor at the moment—indeed, in certain quarters they are anathemas.[51] To a considerable degree this stems from two sources: the drive-discharge model of Konrad Lorenz and the very complexity of the phenomenon, particularly when viewed from a causal perspective. In the first instance, Lorenz's theory that aggression is basically an instinct that builds up like steam pressure to be released in cathartic bursts, along with his willingness to apply it to human affairs, has come under a great deal of criticism—much of it well founded.[52] Although Lorenz's defenders argue that he never meant to imply that aggression was a blind imperative or that it wasn't deeply conditioned by learning,[53] the concept was further undermined by the inability of subsequent research to define an alternate pattern of causation that was clear and comprehensible.[54] The roots of aggression, when investigated in terms of motivation and initiation, are so complex and intertwined with other behavior that they have emerged as a sort of intellectual blob so fraught with qualifiers as to be virtually useless for analytic purposes.

Nevertheless, when aggressive behavior is simply described, certain patterns do emerge which give every appearance of transcending species boundaries. More to the point, human aggression (and presumably that of our forerunners) is not only recognizably consistent with that of other primates but bears some clear functional resemblances to these more general patterns.[55] At the heart of this schema is the dichotomy between predatory and intraspecific aggression.[56]

In the case of the former the proposition is simple—to kill and eat the prey, which does everything possible to avoid such a fate. Beyond that there are basically no rules or mitigating factors. Sheer pragmatism prevails—silence, stealth, and surprise are the favored modes of operation, and with few exceptions the predator will choose as its victim the youngest and most helpless.[57] Restraint will come only through fear or satiation, and even so there remains, as noted earlier, pressure to kill wantonly. Yet the instruments applied, whether tooth or claw or spear, are notable for their simple generalized nature and their basic stability over time.[58] Predators will frequently hunt in groups and at times will address their prey en masse. Sex is no consistent determinant; in some species females will join the males or in certain instances—lionesses, for example—will hunt among themselves. Predation then emerges as a utilitarian, basically free-form activity, primal and utterly pragmatic.

Aggression among animals of the same species is more complex, typically arising over the issues of reproduction, territory, and dominance—the three being basically interrelated.[59] Killing sometimes occurs, but it is of no particular genetic advantage and therefore not central to the objectives of the opponents.[60] Instead, there is a decided tendency toward ritualization, with combatants normally following a clear pattern of rules and employing their defense mechanisms in a symmetrical and nonlethal fashion—antler versus antler in deer and moose, for example.[61] Ganging up is rare. Rather, these are formal duels between single combatants, their essential role being to settle individual disputes. Not surprisingly, gender is an important consideration. And while conspecific aggression among females does occur, it is much more the province of males. It follows that the distribution of defensive implements is heavily skewed in the male direction. Moreover, this sort of fighting does not emerge in earnest until maturity, with the requisite weaponry often being classed as secondary sex characteristics.

This sort of combat is confrontational by nature, and because the evolution of intraspecific aggression is generally in the direction of increased ritualization,[62] bluff often ends up playing an important role. In certain instances violence is completely eliminated, with the duel devolving into a pattern of threat gestures and competitive appraisals. Noise and visual impressions, particularly those related to size, are important here, with animals responding in ways that make them appear louder, bigger, and more frightening—even developing defensive implements so large and intimidating as to be useless for anything other than comparison.[63] Ritualization also has a spatial dimension, with combat sometimes being staged on a habitual venue either segregated from the females or located where they can observe without becoming involved—the sage grouse being a particularly notable performer in this regard.[64] While these characteristics certainly do not extend across all species, they do represent recurring themes—particularly among birds and mammals—and are clearly differentiated from predatory behavior. For if predation can best be related to matters of pure survival, intraspecific aggression is much closer to a game—a dangerous game, perhaps, but still a game.

Not only is it apparent that our ancestors engaged in both forms of aggression, but it seems likely that our subsequent behavior, particularly as it relates to violence and killing, bears the stamp of this heritage. This is not meant to imply that we are locked into compulsively repeating specific patterns of aggression, only that empirical evidence seems to indicate certain predispositions that, if stimulated, will influence development. In almost every respect human behavior is a matter of learning and culture, but in certain areas it has become increasingly hard to dismiss the presence of an evolutionary residuum. Aggression is one of those areas. Presumably, if the subsequent environment did not call them into play or continued as before, these influencing factors would have remained latent or, alternately, focused on the specific areas that originally caused them to evolve. But in creatures as flexible as humans, they were potentially transferable, and one day, when our level of ecological adaptations made war a logical development, these two nexus of aggression would be available as the behavioral raw material. Indeed, the ensuing ride of the Second Horseman can be charted as a series of oscillations between the two basic poles, with participants growing more predatory as war grew voracious and acting more as players when conflicts stabilized.

But just as these patterns were no behavioral straitjacket for individuals, neither did they prevent human aggression as a whole from taking on some very distinctive features. For example, there is no reason to suppose that, prior to the point at which humans began specializing in big game, females would not have been involved with scavenging and then hunting smaller animals. This would help to explain why modern women, though clearly less aggressive than men on average, do not seem inhibited by rules in this area and act in a predatory fashion when finally aroused to violence.[65]

Following the same logic, it may be that weapons specialization among males came relatively late and partially as the result of problems posed by their lethality. For it seems that the introduction of arms must have raised a considerable challenge for the hominid social structure. It has even been suggested that certain early hominids met extinction because they proved incapable of controlling weapons use among their own kind.[66] The survival of our strain, on the other hand, is prima facie evidence that our ancestors learned to keep the violence under control. One way this end might have been served would have been the transference of weapons into the hands of males, who were inclined to ritualize and mitigate their use against conspecifics. Since females were likely less aggressive in the first place, the net effect might have been somewhat marginal but, when combined with the gradual segregation of the hunt to males alone, could help to explain why only one sex ended up as weapons bearers.

All of this was fluid, however, for the key and unique aspect of our ancestors' aggressive traits, and indeed all of their behavior, was that it was much more subject to change than that of other animals. For we were beginning to develop in an entirely new way—at first slowly but then gradually accelerating toward a point at which we would become not just top predator but a creature of unprecedented potential. To see how this change occurred, however, we must focus on the organ that changed the most.

Sometime before 2,000,000 B.C. the brains of our ancestors began growing very rapidly, gaining nearly a cubic centimeter on average every two millennia until they had nearly tripled in volume.[67] Nobody knows for certain why this happened, but paleoanthropologist Dean Falk has come up with a plausible explanation for why it had not happened earlier. The savannas were hot, and the brain is very intolerant of temperature increases. Until an arterial structure capable of cooling larger masses of brain tissue evolved, an effective volumetric limit was placed on further growth. With *Homo habilis* such a vascular radiator seems to have appeared, clearing the path for a dramatic expansion.[68]

But if the road was now open to Emily Dickinson and Albert Einstein, the identity of the driver still remains in question—though not for lack of suspects. Traditionally the candidates have clustered around the stimulation provided by physical and ecological adaptation—bipedalism, toolmaking, and hunting—once again reflecting the emphasis characteristically placed on activities that left tangible evidence. However, the development of language, being more ethereal and far harder to pinpoint, usually has been dismissed, relegated to a very late stage of human evolution.[69] This is curious since the appearance of true language, like the brain's expansion, was a spectacular and largely unprecedented event that might logically have been tied to the growth of the organ responsible for making us such chatterboxes.

This may be changing. Although many anthropologists and brain specialists continue to support the late evolution of human speech, an alternative view seems to be emerging. For recent endocasts of *Homo habilis* (latex impressions of the interior of the skull, revealing signs in the bone of blood vessels and nerve structures) indicate that—unlike australopithecines, whose brains were essentially apelike—this hominid's cerebral cortices were more like those of humans.[70] More critically, *Homo habilis* seemed to have had a Broca's speech area, making a rudimentary human-like language a decided possibility.[71]

This is exceedingly important, for language is the basic prerequisite for culture, and culture provides the best explanation of how one day we were able to buck all genetic precedent and begin living like ants. For unlike communication among animals, human language permits the formation of symbols without immediate reference to sensory stimulation. Thus people can think and talk about virtually anything, anytime, independent of physical and environmental causes. It is this autonomy or "openness," political theorist Roger D. Masters explains, that has frequently been cited as the reason why cultural practices are able to become independent of genetic causation.[72] In this regard it may be significant that both language and the genetic code are amenable to alphabetic representation, and at a certain level can be seen simply as parallel information systems.[73] Meanwhile, if the brain's evolution was profoundly and continuously influenced by language, then this serves to demonstrate what a powerful and deeply rooted force culture must be—a force strong enough to break rules billions of years old, to metamorphose evolution and speed it up by an order of magnitude—to make us what we are. Yet explaining how culture enabled us, in the space of something over a hundred centuries, to undergo a transformation that would leave most of us living sedentary lives in mass agricultural societies, but still remaining essentially the same creatures shaped by millions of years of wandering, and foraging, and hunting, is no trivial matter.

It is no accident that early humans lived in pods with a maximum membership of around fifty. Besides the numerical limitations imposed by "prime movers" such as foraging, there are clear genetic constraints on group size.[74] Humans, like all mammals save one—the astonishing naked mole rat—are fully sexually reproductive, causing genetic affinity in outbreeding populations to decrease on average by a factor of two each generation.[75] Put simply, this means that not only degrees of relatedness but also the genetic motivation to perform altruistic and life-threatening acts for the good of the whole must fall off rapidly in large populations. In effect, this imposes some rather inelastic parameters around the degree of social cooperation that can be expected and is observed in fully sexual species.

Herding might seem to contradict this rule, but it is in fact a phenomenon of self-defense, not sociality, with members exhibiting no loyalty to the group, only a desire for inconspicuousness and the strength afforded by numbers.[76]

Much more interesting and suggestive is the phenomenon of the naked mole rat, nature's only mammalian experiment with reproduction approximating that of the social insects. These three-inch-long animals live in colonies of as many as three hundred members, burrowed in the dry soil of East Africa. The ruling queen is the sole female who reproduces, mating with one to three dominant males, generating a population that on average shares 80 percent of its genes. The result is a group exhibiting a remarkable degree of cooperation (except during interregnums), working together to dig tunnels in the rock-hard soil and performing heroic and self-sacrificing acts of colony defense.[77] These tiny creatures appear to be literally the exception that proves the rule—as long as genetics dominates behavior, the sole route to advanced sociality is through the narrowest of breeding channels.

Yet humans, without compromising their sexuality, did ultimately find a way around the genetics-driven numbers crunch, learning to form loyalty-based collectives—true mass societies and their most naked manifestation of self-sacrifice, armies. But how exactly this was managed remains unclear.

Some have tried to extend purely genetic altruism (known technically as inclusive fitness) to unrelated individuals through forays into game theory models such

as the "Prisoner's Dilemma."[78] These, however, require tit-for-tat personal reciprocation and fail to adequately account for the rise of what would amount to
faceless bureaucracies demanding sacrifices in exchange for delayed and often
abstract and unequally distributed benefits.

In an attempt to resolve these shortcomings, a number of theorists have invoked
the notion that societies compete and evolve at the level of the group, and that
individuals within them will sacrifice because they and their kin will be better off
if the collective entity succeeds.[79] This, however, ignores the natural advantage of
"selfish" individuals in such situations. Consider, for instance, the leverage of the
draft dodger left amid the female population, while the army-bound "altruist"
risks not only life and limb but his chances for reproduction—a situation that, if
carried to its logical conclusion, could lead literally to the extinction of self-sacrifice. Although these examples greatly simplify some complex and sophisticated
arguments, they do serve to illustrate the necessity of explaining the existence of
mass society, and particularly warfare, at the level of the individual and in terms
that account adequately for his or her participation.

One key reason why genetics-based accounts of human behavior raised such a
furor has been the critics' intuitive understanding that we are truly thinking beings
whose intelligence has emancipated us from the constraints placed on animals.
Their objections, however, were continually confounded by an inability to meet
the ethologists and sociobiologists on their own ground and provide an alternative
model that explained how rational processes could transcend and not just rationalize genetic imperatives. This is now beginning to change as it comes to be
understood that, like computer hardware, the brain's evolution was accompanied
by a parallel software revolution.[80] For, as already noted, language made it possible
for culture to break free of genetic motivation in a way that was impossible for
other animals, and once this happened human behavior could begin to evolve in
truly novel ways.

To account for this, Robert Boyd and Peter Richerson recently proposed a
"dual-inheritance" model of human evolution that, while not casting genetic motives aside, provides alternate cultural routes and explanations for behavior that
might seem implausible from a purely genetic perspective.[81] Specifically, it highlights certain mechanisms of cultural evolution that serve to explain why development is, relatively, so rapid and builds such momentum. It is notable that,
unlike genetic inheritance, the recipient is not limited to parental acquisition but
can receive cultural indoctrination from a variety of sources and then pass it
horizontally to contemporaries. Moreover, cultural evolution is "Lamarkian" in
that the information accumulated by an individual may be affected by life experiences and then passed on in this altered form.[82] This means that the "mutation
rate" of culture is inherently much higher than in genetic evolution, where change
is basically uninfluenced by acquired characteristics and can occur only through
the differential reproduction of variant individuals. While there are other variables,

these factors alone illustrate how culture could build so rapidly, become so elaborate, and emerge as a force strong enough to lead individuals in directions not dictated by genetic advantage.

This would be critical to the eventual appearance of warfare. However much preadaptations such as weapons bearing, small-group and kinship loyalties, and a proclivity for slaughter prepared us for combat, unless individuals were willing to risk their lives and heredity for a corporate entity fundamentally without genetic affinity, then the whole notion of warfare among humans becomes not just impossible but absurd.

There is another question of some significance. Why have we not changed more? If cultural evolution eventually became so rapid and powerful, why didn't the large agricultural tyrannies in which so many of us lived for what amounts to thousands of years turn us into far more compliant and obedient creatures? Instead, although necessity may have demanded that we cooperate with such regimes, it is still possible to detect a sullen sense of discomfort, a longing for a way of life that permitted basic equality, the freedom to say what we thought and go where we wanted, and even an absence of war—yearnings that help explain why despotisms were so prone to sudden collapse and why in the modern era humanity would move decisively toward a life that better reflected these values. Indeed, by implication it can be argued that through it all we retained the souls of hunter-gatherers.

Yet again the question arises: How was this possible in mass societies whose organizational logic and ideology would have continuously operated to eradicate our ties to this earlier existence?

Logically, at least, the simplest solution would be genetic, that these values or at least some component of them were transmitted through the operation of heredity. The problem, however, is that there is presently no evidence that behavioral predispositions this complex can be transmitted genetically, nor is there even any detailed understanding of how such mechanisms might work. Thus we are led to a cultural explanation, in this case focusing on the primary levels of social interaction. For virtually regardless of the constraints placed upon them, humans always seemed to find ways in their interactions with friends and loved ones to maintain and reinforce the intense sociality represented by their earlier existence. Rulers and their minions might set up grand social hierarchies to tax and coerce them—but in their private lives people give every evidence of continuing to treat each other much as they always had. And the attendant cultural maintenance of values would have continued from generation to generation, subtly perpetuating human nature much as it had originally evolved, even when subjected to what amounted to hostile surroundings.

But we are getting well ahead of our story. For the initial development of human characteristics and our capacity for culture still took a long time and would also be paced in part by the more gradual evolution of the brain, which was tied to genetics. The evidence indicates that like many nonlinear phenomena, culture progressed roughly according to a sine curve—gaining momentum very slowly over a long period,[83] reaching a takeoff point, and then accelerating very rapidly. Hominid vocalization, for example, probably remained for a very protracted period much more a simple means of communication and coordination than a repository for accumulated knowledge, ritual, and values, the essence of culture. Religion and intricate patterns of reciprocation—the two other key mechanisms that one day would tie individuals firmly to their cultures—should have been similarly slow in their development. For although the brain's evolution, particularly in the area of memory, made an awareness of individual mortality and patterns of long-term obligation factors of potentially great significance, the constraints of a hunter-gatherer lifestyle mitigated against their elaboration. It follows that a slow maturation of culture also would have postponed differentiation among cultures, a premise that seems in consonance with the notable sameness of not only the artifact base but also the symbolic residuum up to the Mesolithic and even beyond. If cultural differentiation is at the heart of ethnocentrism, which in its most naked "we-they" form is often considered a prime instigator to combat,[84] then it is logical to assume that this behavioral impetus to war must have remained at least somewhat muted for a very long time.

There is a problem, however. Recently geneticists Allan C. Wilson and Rebecca Cann have suggested, on the basis of mitochondrial DNA (DNA outside the cell nucleus that accumulates mutations at a steady and measurable rate), that all modern people are directly descended from *Homo sapiens* who lived in Africa only two hundred thousand years ago and subsequently supplanted earlier humans everywhere else.[85] While Cann has suggested several alternate means of displacement, including disease and an unwillingness to interbreed, the implications for protracted and genocidal violence are hard to ignore. Although the Wilson-Cann thesis has received considerable support, alternatives have been proposed by physical anthropologists and other researchers using mitochondrial DNA evidence to account for a less abrupt transition to modern humans.[86] Still, in its purest form, the displacement thesis does not square well with the arguments presented here. As to the possible means of displacement, population concentrations were not great enough to make epidemic a good candidate, nor does breeding bias against hominids lacking "some modern trait, such as advanced language skills"[87] seem convincing. This leaves violence. Yet the motivation and organization to wipe out what amounted to the rest of humanity simply is not apparent at this point. It is true that the material evidence reflects something of a creative explosion beginning in the Upper Paleolithic, but this can be explained by cultural evolution finally having reached the takeoff stage just as well as by displacement of earlier hominids by modern people. Categorical statements are risky, but the weight of evidence and logic indicates that at this stage war and genocide were things of the future, not the past.

One day we would become true "organization men," more loyal to the vectors of culture—governments, armies, and religions—than to the dictates of our genes. This did not necessarily presuppose wild enthusiasm. For the heritage of an earlier, freer way of life would render humans forever ambivalent. But we would go along—in part due to compulsion and the logic of events, but also because we could go along. Evolution had done its work. The hardware was ready; only the software awaited completion. But further development would demand a more settled lifestyle where material culture could build upon itself. This day was coming. The extinction of the megafauna and human population increase made it inevitable. And with it would also come increased violence, more vicious and frequent than it had been. But the question remains: Was it war?

{4}

FALSE ALARM

The five men had been waiting since long before dawn, concealed along the path leading from the river. They tensed at the sound of someone approaching, drawing their bows soundlessly. Then as the figure passed they let fly. There was a scream. The young mother had been hit twice—the first arrow piercing her neck and partially severing her spinal chord, the second driving through the infant she carried and then into her chest, locking them together in death's embrace.

It was unclear how or why the trouble had started, only that it had begun many seasons before. But there was nothing vague about the hatred and fear all felt for the two settlements that flanked them at the mouth of the river by the great salty sea. The feud hung like a dark cloud above them, threatening death when they least expected it. And when it struck there was only vengeance and the blood of the enemy to wash away their shame and even the score.

There had been attempts to end it, many of them. But always a stolen woman or even an insult had rekindled the fire. Three summers before, all of the groups, after days of talk, had agreed to become brothers again. Then violence had erupted during the feast of celebration and four people, one of them a child, had been killed. All sides remembered it as an act of unprecedented treachery, and since that time both enemy settlements had preyed upon them with increased ferocity.

"We must leave this place before we are all killed," the brother of the latest victim argued the night of her disappearance.

"This cannot be. There is plenty of salmon in the river. The shells of our ancestors are here," one of the elders retorted, referring to the great refuse pile of discarded mollusks built up over the generations.[1]

"We must make them fear us," another chimed in. "Then they will stop."

"They will never stop," the young man shot back. "Besides, there are deer in the woods and no floods. There is plenty for us if we will only look for it."

"Why should we look, when everything we need is right here?" the elder replied.

"Because we will die."

After two more murderous incidents the group would split apart, with the larger body moving inland and the remainder abjectly joining one of the settlements where they had kin. Within months the cycle of violence would begin anew, the protagonists being the two former allies.

The refugees fared better. After some travails they established a range in the oak forest sixty miles inland, where they adopted a more mobile lifestyle based largely on acorns and red deer supplemented by a wide variety of smaller game and edible plants.[2] Their numbers remained low, but so did the incidence of violence.

The passage of three centuries would find them still in the general vicinity, subsisting in much the same manner. The two settlements at the mouth of the river, on the other hand, were now fifteen feet below the surface of the sea,[3] abandoned long before in the face of a breakdown in the salmon run and oyster beds induced by the great glacial meltdown.

The scenario outlined here serves to illustrate the elusive nature of our topic—the difficulty of isolating one chain of events as the definitive origin of war and the criticality of definitions to the process. On the face of it, the events at the mouth of the river give every impression of real war—a campaign ruthlessly pursued, clear winners and losers, and withdrawal by the latter from an economically attractive area. The picture is deceptive, however. The conflict was probably episodic, based more on emotion than on systemic causes, for the antagonists were likely to have been driven primarily by a desire for revenge, not for more shellfish and salmon—though this could have been an underlying factor. Nor were the attacks aimed at much beyond inflicting bodily injury and death. But, perhaps most important, there was a way out, an acceptable alternative to staying and fighting. For the inescapable nature of war's snare would eventually be predicated on nowhere else to go and its corollary, a population too high to go elsewhere. Neither was true in this case; nor were they likely to have been true in most inhabitable areas until considerably later. Populations were growing during the period we call the Mesolithic,[4] probably quite dramatically in coastal environments capable of supporting a settled existence, but across the entire spectrum of occupation, populations remained thin enough at this point to limit intense territorial behavior and to provide options for those in need of a change of venue.

II

The Mesolithic, as the name implies, was a transitional phase and therefore potentially of great significance in understanding the factors behind humankind's great metamorphosis. Traditionally, however, it has not been the subject of much interest and was usually dismissed as a sort of nadir between the full flowering of the Upper Paleolithic and the coming of agriculture, the so-called Neolithic revolution—humanity stuck in a holding pattern while it learned to exist on a diet emphasizing lower trophic levels along the food chain.[5] As usual in archaeology, this paucity of research can be traced to an equivalent paucity of tangibles. For the most interesting sites would have been along riverbanks, lakeshores, and coastal locations when the water levels were much lower, and subsequent inundation rendered them almost impossible to find.[6] So the Mesolithic remained basically a hiatus until very recently, when a combination of new techniques and ethnographic insights led to a reexamination of the period.

In essence, what has emerged is a growing consensus that complex hunter-gatherer societies—those emphasizing the intensive exploitation of a narrow band of food sources in resource-rich areas—had begun to manifest most of the social characteristics generally attributed to the coming of agriculture: sedentism, technological innovation resulting in higher productivity, population growth, incipient social stratification and formal leadership, increased regulation of human relations through ritual, the expansion of trade networks, resource competition, increased territoriality, and, most pertinent here, the appearance of what is frequently referred to in this literature as warfare.[7] The latter judgment is backed up by undeniable evidence of increased interpersonal conflict, including both pictorial representations of combat[8] and clear indications of violent death among skeletal remains.[9] Finally, it is important to note that ethnographic data support not only this basic developmental syndrome among complex hunter-gatherers but also similar conclusions as to the appearance of war.[10] Yet what seems quite obvious from one point of view becomes less so when the focus is shifted.

There is little question that recent work done on the Mesolithic is highly interesting and suggestive,[11] for it provides a very important perspective on the causes of social complexity, a subject of endless debate among those interested in prehistory. And, as Glenn Sheehan, a student of Eskimo societies, points out, it tends to illustrate that complexity is more a function of the interaction of a number of self-amplifying variables than a matter of any one prime mover.[12] If such a thing did exist, it probably was a localized and abundant food supply—nearly any food supply, though some are clearly better than others. For once humans had reached the point that culture was the central evolutionary trajectory, large and complex societies awaited only a support mechanism, and this was a very flexible requirement. As archaeologists Olga Soffer and Paul Mellars have shown, a few societies living off superabundant big-game supplies in southwestern France and central

Russia apparently began to undergo this transition during the height of the Upper Paleolithic, settling down and adopting characteristics wildly out of sync with the general patterns of life during this period.[13] The Mesolithic brought increasingly numerous examples of the phenomenon, this time based on more prevalent but still localized marine resources. But the pattern of development was fundamentally similar and would continue to be so.

This is important, for it serves to illustrate the link between humans and the social insects. Ironically, it appears that in the most fundamental sense culture and mental sophistication really only made it possible for people to live in large societies; the actual form these societies took was determined primarily by their economic underpinnings, which until the Industrial Revolution were usually keyed to the exploitation of a stable food source. Of course, this is a sweeping generalization that discounts the very important ways in which human intelligence conditioned complex societies. But still at the most elemental level it serves to explain why human social structures resembled those of ants for so long—in essence we both were what we ate, or at least products of how we got it.

Nevertheless, in the Mesolithic context—and the Paleolithic, for that matter—there was a catch. For the food supply upon which social complexity was based also acted as a limiting factor, being narrowly available, difficult to store, basically impossible to intensify, and prone to run out.[14] The net effect was arrested development. For one reason or another—the initiating factors were probably somewhat interchangeable—people would settle down around a good fish run or a large mollusk bed and the various drivers of social complexity would begin working on one another. But before the process could go very far and build a really substantial social superstructure, the food supply would begin to collapse beneath its weight. In little pockets of territory societies would rise, pass through incipiency, and then fall, utterly unable to transcend this intractable barrier. This is why there were no shellfish cities or mammoth megalopolises. Without a basic food source capable of massive expansion and intensification, the self-reinforcing feedback loops of social complexity must always break down. This is also why there was probably no such thing as warfare in the sense that we understand the term. For like the other elements in the system, there was neither the means nor the necessity for full institutional maturation.

But increased violence under such circumstances was entirely plausible. For one thing, even moderately larger, more densely packed concentrations of people almost certainly must have raised the level of interpersonal tension within the group and among individuals confronted by the effects of incipient congestion and sensory overload.[15] And just as this discord might have been directed outward as one mechanism of control, the monopolization of a relatively attractive and stationary resource base was bound to draw external competitors, some of whom might have been inclined toward forceful usurpation. The net effect, even in an immature system, would have been a substantial increase in the levels of suspicion, hostility, and, ultimately, interpersonal violence. Indeed, there is pretty clear skeletal evidence for instances of massive mortality during the Mesolithic in Scandinavia and on the upper Nile.

Nonetheless, both logic and the available evidence seem to indicate that this violence would have remained for the most part inchoate and spontaneous, perhaps reflective of underlying causes for conflict but not directly and consciously motivated by them. Rather, physical aggression would have focused on revenge and usually would have taken the form of sporadic though brutal attacks on individuals—the classic profile of blood feuds, not warfare.

Yet this conclusion is clearly at odds with a number of authorities who, frequently relying on ethnographic data gathered from advanced hunter-gatherer and horticultural societies, attribute the birth of warfare to the kinds of societies that first appeared in the Mesolithic.[16] On the face of it, this perspective seems reasonable enough. Certain contemporary horticultural societies, the Yanomami, for example, along with recent salmon-and whaling-based Amerindian cultures in British Columbia and Alaska, do appear highly violent and, in the latter instance, waged planned and purposeful military campaigns.[17]

But these ethnographic assumptions have recently been challenged on the grounds that sudden contact with European colonizers, even indirect contact, dramatically raised the levels of violence exhibited by many native societies. Anthropologists R. Brian Ferguson and Neil Whitehead argue that the introduction of a variety of instrumentalities, ranging from infectious diseases to firearms to divide-and-rule politics, amounted to a sort of "cultural Heisenberg effect" that greatly intensified intersocietal violence at the very time it was being observed and recorded.[18] It follows that, in their pristine state, such societies may have been considerably less warlike, and in this respect a better approximation of actual Mesolithic conditions.

The appearance of true war at this point also can be questioned on the basis of relative population densities and distribution. While human numbers do appear to have been climbing, it stands to reason that most of these gains would have occurred among more settled communities dependent on marine resources. On the other hand, Mark Nathan Cohen's assumption that the carrying capacity of the hinterlands reached a ceiling for foragers during the Mesolithic sorely lacks corroboration in the archaeological record and now appears questionable.[19] There is support for some limited population growth here, and the smaller, more closely distributed groupings necessitated by increasingly intensive foraging patterns likely paved the way for more rapid expansion later. But in the meantime, there is insufficient evidence to believe that the Mesolithic "countryside" was as yet filled up, or that territorial boundaries had grown significantly less flexible.[20]

This is important because free space is the enemy of true warfare. In a highly influential 1970 article Robert Carniero introduced the concept of "environmental circumscription"—attractive agricultural areas surrounded by mountains, deserts, or otherwise unproductive and inhospitable territory—as critical to the irreversible emergence of war and cultural complexity.[21] But this does not appear to have been the case in the Mesolithic context. Moreover, the marine resources supporting localized sedentism were simply not robust enough to allow for really massive population growth. Thus, settled communities remained relatively small and below the scale that would prevent a return to a more mobile lifestyle. Indeed, it appears that there was always considerable movement back and forth between these settlements and the hinterlands.[22] The net effect, as indicated by the vignette at the beginning of the chapter, was a safety valve that remained open to whole communities, or at least significant elements, beset by unacceptably high levels of violence. By taking the "exit option" they might escape the cycle of mayhem with a reasonable expectation of finding a new life—perhaps somewhat more rigorous and not without hostile interaction, but decidedly less dangerous. But the day was coming when this would no longer be true, when agriculture would all but close the escape hatch. For farmers would become too numerous and their skills too specialized to be able to look elsewhere for a better life. In Eurasia a few might slip through the cracks to follow herds of domesticated animals. Otherwise, humankind would be stuck down on the farm, forced to defend their territory and accept all that this implied. But in the Mesolithic this lay, for the most part, ahead of us.

III

Nevertheless, there remains a very important, indeed potentially critical, source of evidence to be considered—cave paintings that have been used repeatedly to support the notion that war really did emerge in the Mesolithic. The best known of these representations come from one source, the hills above the eastern Spanish coastal district, or Levant. And while opinions remain divided as to the exact age of the Iberian representations, a Mesolithic time frame is generally accepted.[23] One thing is clear, however: they do constitute the first indisputable evidence of human combat, and therefore merit close attention.

Several items of interest emerge. While scenes of fighting are usually displayed together by proponents of warfare's Mesolithic origins,[24] they actually constitute very much a minority of the total, most of which have to do with hunting.[25] Of the Spanish paintings, several that have frequently been termed "troops of warriors" could just as well be hunters dancing or in pursuit of game. Besides these, only four deal unambiguously with human violence. The first two are notable in that they depict a single victim hit with multiple arrows, in one case by ten archers. The final two Iberian examples—a small scene of only seven warriors from Morela louisiana Vella and a much larger grouping of at least twenty-nine figures from "Les Dogues" shelter in Castellón—are obvious representations of combat. But in both cases the action is confused and without unambiguous signs of organization.

The participants are clearly on the run, perhaps hoping to rip off a few quick shots before retreating. A similarly chaotic tableau seems to emerge in one of the few non-Iberian examples of what is said to be warfare—a rock engraving from Kobystan just off the western shore of the Caspian Sea. But there are problems here, not simply with dating the scene firmly in the Mesolithic but also with the inclusion, just above the supposed combat, of a line of dancers or spectators, raising the question of whether this is not simply a representation of some sort of ceremony.[26]

Questions of interpretation have also been raised in the analysis of rock paintings apparently depicting combat among the ancient pre-agricultural peoples of Arnhem Land along the north coast of Australia. In a very recent analysis Paul Taçon and Christopher Chippindale argue that these scenes change markedly from one-on-one fighting (possibly as old as 8000 B.C.) to tableaux they believe depict group combat beginning around 4000 B.C. Significantly, the authors note that this date coincides with an inundation of the habitat which could have led to increased competition for territory, but also might have resulted in the kind of localized coastal resources capable of supporting increased population and growing social complexity.[27] At any rate, the massed "battles" depicted do seem characteristically confused, and some seem to have simply resulted from an accumulation of figures painted over time.

Corporately, the images are revealing, but less so than some would have them be. For the subject matter will bear only so much speculation, and for the most part this falls well short of true warfare, better approximating the activities surrounding blood feuds. For example, the Spanish scene showing ten warriors exulting over the single victim shot full of arrows most plausibly depicts the aftermath of an ambush. Nonetheless, the image has repeatedly been termed an "execution," and in one case the perpetrators are characterized as "organized into a firing line . . . one of the most significant features of early, organized warfare."[28] Military historian Arthur Ferrill also sees evidence of disciplined fighting forces attempting a double envelopment (a tactic that uses a weak center to lure an opponent into a trap between two strong wings) in the image of the seven warriors from Morela louisiana Vella.[29] Common sense argues that such an interpretation stretches even liberal rules of evidence beyond the breaking point. In this regard it is worth noting that in every single image of combat or violence the warriors are shown fighting solely with the bow, or, in the case of the Australians, the boomerang and javelin. This is significant because the bow is a long-range weapon particularly suitable for hit-and-run attacks and fighters lacking in motivation. On the other hand, had the cave paintings pictured tightly packed warriors locked in close combat employing spears, there would be entirely more reason to conclude that these were determined soldiers willing to fight to the bitter end.

But this is not to say that the images from the Spanish Levant represent nothing new or significant. For one thing, it should be noted that the violence depicted here almost certainly took place among foragers living on the outskirts of the more settled marine-based communities. And while theory indicates that these groups were probably less violence-prone than their more sedentary equivalents, the pictures demonstrate that intergroup hostility in the hinterlands was a very real and possibly growing part of life. Of particular interest is the battle scene from "Les Dogues," which is the single example of mass combat in the entire collection. Here is a clear indication that cultural imperatives had become sufficiently strong to support true group fighting, for individuals to risk their lives and genetic futures for something beyond their own inclusive fitness. On the basis of this evidence it does appear that combat among humans was changing; it is difficult to say how rapidly, but the direction is clear. And if true warfare had not yet come, there is every reason to believe that it was now humanly possible. What was missing was a really robust economic foundation for sedentism, a food source capable of expanding widely and dominating the countryside. But this was not far off. For if warfare, like nature, abhors a vacuum, something was out there waiting to fill the void.

{5}

PLANT TRAP

I

As he dug, beneath the blazing sun, he thought of the weeds and how much he hated them. He had no doubt that each kind--and there were too many to name—had a will of its own, a demon's soul determined to ruin his life and the life of his village, to choke the emmer wheat, the barley, and the lentils, the good plants that brought them sustenance. So he dug all the harder, rooting out weeds until exhaustion overtook him and the time came to return to the village. He would be back, but the weeds would be waiting.

That night he lay thinking, kept awake by the aching of his teeth and the mild fever that came to him sometimes. There had been no rain for twenty days and twenty nights, and in the fields only the weeds seemed green. All had prayed faithfully to their Mother, but the days only seemed to grow hotter.

At times like this he wished he could simply walk away from the village, a fantasy that inevitably led to thoughts of his brother, a shepherd whom he seldom saw during this season. Most villagers seemed to see little difference between those tending the flocks on the outskirts of the village and the wild ones in the hills. But he knew this to be wrong. He still spent a good deal of time with his brother during hunting season, and he found himself increasingly attracted to his life out on the edge of the settlement. For his brother still stood straight, unstooped by the endless toil of planting and weeding and cutting and threshing. His brother's children had milk, and there was more meat, the very thought of which made him salivate. But when he raised the subject his brother always laughed: "You don't know what we have to do. We must find them pasture and water to drink. We have to follow them and watch them always. We must keep the wolves and wild ones away. They are like children. And we are as much their servants as you are the servants of your fields. It is our fate; there is nothing else."

But he was not so sure, lying there deep in the night prodded by a vague sense of unease and the ceaseless throbbing of his teeth. Something was wrong—something so basic and colossal that it was beyond words or even conscious thought. But like the weeds it was always waiting, haunting his dreams when he finally slept.

The man would never join his brother and the flock. Instead, he would die three months later of an acute fever brought on by an abscessed tooth. Less than twenty-nine years old, he was by today's standards undersized and malnourished, his spine misshapen by constant bending and his gums a mass of infection. This composite human was also the product of a transformation in subsistence patterns that until recently had been considered an unqualified blessing—an agricultural or "Neolithic revolution."[1] But if revisionists now see it as neither revolutionary nor much of a blessing, agriculture remains the foundation for civilization and ultimately warfare. For it would be on the backs of men and women like the one described that pyramids would be built and empires would rise and fall. And although we might question the human and ecological consequences, it was this partnership between plant and animal that made it possible for our species and the few domesticates that we nurture and cultivate to dominate the face of the earth. For as much as anything we rule by sheer weight of numbers. This is why the origins of war must be explained in terms of how we became first farmers, then pastoralists, and finally warriors. While it will become clear in the analysis of the New World that this sequence is not mandatory, the evidence indicates it to be the path along which true war emerged in its earliest and most pervasive guise.

II

Consider a wheat field in the middle of Nebraska, a very large wheat field with a single crop stretching as far as the eye can see. Each morning a farmer or a few farmers rise early and work all day using hundreds of thousands of dollars worth of sophisticated machinery to ensure the field's well-being—adding nutrients to the soil, eliminating competitors, even providing water during dry spells. True enough, once the field ripens all the plants will be cut down and eventually eaten. But they also will be carefully replanted. Who, then, reaps the benefits—just the farmer, or is it also the wheat?

This example encapsulates a new way of looking at the origins of agriculture and domestication in general pioneered recently by a number of researchers, most notably plant biologist David Rindos.[2] He argues that what was once viewed as a brilliant human invention was actually a relationship accidentally conceived and then cooperatively reinforced through a process not that different from the mechanisms governing the symbiosis between ants and certain fungi.[3]

The key to such relationships is mutualism, with both plant and animal evolving in ways that intensify the partnership. Both benefit and are able to radically increase their numbers, but each becomes utterly dependent upon the other. In human terms this spelled a long stretch down on the farm and an incentive to defend territory like none that had come before.

This line of reasoning is supported by several striking features of our transition to agriculture. The first is its universality. In the period between 8500 B.C. and A.D. 1 the great majority of humans made the transition from wild food to planting and harvesting domesticated crops—a span of only eight and a half millennia in the more than four-million-year history of our line. Moreover, this took place all around the world, and in several quarters transpired independently with no apparent outside stimulus. If this was an invention in the true sense of the word, then it occurred to a great many people at around the same time.[4] But even more remarkable is the fact that no major crop (or food animal, for that matter) has been domesticated since the early days of agriculture, and even our best modern attempts to improve existing crop species have been less than totally successful.[5] Here again, either our distant forebears had fantastically green thumbs or there were relatively few plants inherently suited for domestication—ready themselves to play an important role in the process.

Most sources are inclined to believe that the transition to agriculture first took place in the Middle East,[6] with the initial steps usually attributed to a group subsequently known as the Natufians. Originally conventional hunter-gatherers specializing in gazelle, these Natufians supposedly happened upon several rich stands of wild cereal (emmer wheat and wild barley) growing in the hills above the Mediterranean in what is today southern Lebanon sometime around 10,500 B.C.[7] Quick to recognize a good thing, they are assumed to have begun harvesting the fields and eventually settled down around them. "After all," one authority asks rhetorically, "where could they go with an estimated metric ton of clean wheat?"[8]

Reasonable enough, except that it was probably not that easy or simple. The fortuitous wheat field—thick with shoots, all obligingly germinated and ripening at the same time, each rachis heavy with grain—was an inherently implausible phenomenon, Rindos would argue.[9] Plants in the wild simply don't work that way; they are conservative in their life cycles, with only a fraction being active during any one growing season as insurance that some will survive in times of environmental stress. Also, they don't waste energy on large edible portions unless there is a good reason for them being there.[10]

A more likely scenario entails our Natufians or some other group in the region coming upon a fairly extensive but not particularly rich stand of grain—in fact, just nutritious enough to make it worth harvesting and processing but not carrying very far.[11] Thus a base camp would have been a logical thing to establish right at the source for as long as the grain held out. When it was exhausted the humans would move on, not to return until the grain had a chance to reaccumulate.

Over time, however, something remarkable would begin to happen; grain yields would start to climb. This was no miracle, just natural selection at work. For the very act of being harvested would have greatly increased the chances of a cereal plant becoming reseeded. Thus, even though this may have taken place accidentally, humans constituted, from a plant perspective, a high-quality dispersal agent. So much so that they began competing for our attention with carbohydrate bait. For those shoots with the largest edible portions were naturally the most likely to be harvested and, therefore, reseeded.[12] To further amplify their chances at replanting through wastage, enterprising plants might also give up seed dormancy and take to ripening annually during narrow bands of time—effectively glutting the landscape with grain.[13] In this way they would come to dominate the fields. The only requirements were opportunistic plants with relatively plastic genetic structures and, of course, cooperative and hungry humans.

These humans, meanwhile, would have acted appropriately, gradually settling down on a more permanent basis in the midst of the ever-richer fields and becoming increasingly specialized in their foraging. While there is evidence of an earlier but prematurely interrupted example of this sequence having taken place in Egypt, the pattern is clearly and continuously reflected in the plant-oriented artifacts drawn from Natufian sites, such as grinding stones, mortars, pestles, and small sharp microliths with a peculiar sheen indicating that they were used as blades on sickles.[14] But the very large number of animal bones (mostly gazelle) found at Natufian sites also indicates that hunting remained an important preoccupation.[15] Yet trend lines in a very different direction were already evident. Semipermanent Natufian dwellings have been excavated, usually with grain storage pits inside or nearby.[16] Also, occupation sites averaged over three times larger in area than those of previous cultures,[17] an obvious sign of population growth. Meanwhile, Natufians continued to expand, until after fifteen hundred years they began pushing into ecologically marginal tracts, quite possibly as a result of increasing numbers.

Internally, there were also intriguing signs that the process leading to social complexity had taken hold—suggestions of nascent social ranking, more personal possessions, and evidence of long-distance trade. And while there are no archaeological indications of warfare, group sizes appear to have been growing toward the point of sustaining endogamy (breeding within a local population), a threshold that would make it less necessary to get along with neighboring populations.[18]

Sometime before 8500 B.C., however, Natufian culture appears to have entered a crisis stage when significantly drier conditions forced cereal resources to retreat from water sources used by humans, leading to a dramatic decline in the number of settlements. At this point survival could have prompted the remaining Natufians to lend Mother Nature a hand by sowing seed grain in what formerly had been self-germinating fields. For, after nearly two thousand years of intensive involvement with wild cereals, Natufians were probably very much aware of propagation, growing cycles, and the conditions necessary for successful cultivation.[19] And if it didn't happen here, then it could well have taken place slightly later at Mureybet II, a settlement of over two hundred houses, which archaeologists James Mellaart and Hans Helbaek believe was just too large to be supported solely by wild grain collection.[20]

At any rate, it was unlikely to have been much of an innovation. Indeed, there are those who maintain that, as sophisticated foragers for millions of years, the members of our line long understood that plants could be grown for food—it was just too much work when there were other more attractive food sources available.[21] At this point the cereals being cultivated remained relatively close morphologically to wild strains, differing primarily in the moderately larger size of their edible portions and in germination frequency.

However, the stage was set for further modification. For the plants, in the process of becoming even more attractive to humans, would evolve in ways that would dramatically increase their dependency. Many would cast aside their protective mechanisms against predators such as natural pathogens, devoting all their energy to producing more and tastier food product.[22] This, in turn, would lead to true gigantism in propagules (the seed and edible tissue package), along with the development of a tough rachis that suppressed scattering and made reaping easier.[23] But these changes also reduced or eliminated the chances of natural dissemination, thereby placing the affected plants well on the way to "betting the farm" on their two-legged benefactors.[24] One by one at various points around the globe, enterprising carbohydrate manufacturers—emmer, einkorn and barley, millet and rice, and maize, along with a cadre of fruits, legumes, and tubers—took roughly the same evolutionary path, becoming, at least from their own genetic point of view, man's best friends.

Not so with a pack of other opportunistic species, which, by mimicking crop plants in their early stages of growth, managed to invade cultivated areas as pure parasites.[25] The resulting legion of weeds would introduce an unprecedented level of drudgery into humankind's subsequent economic activities and symbolize the darker side of our ancestors' ever-more-entangled relationship with colonizing plants.

Something fundamental was happening, and it was not necessarily to our benefit. For it is a stark commentary on the system that the modifications in cereal plants brought about by domestication (a smaller number of larger reproductive units, long and uniform maturation, high energy investment in offspring) would be balanced by changes in the opposite direction among humans subject to an agroeconomy—decrease in stature, earlier reproduction, greater number of offspring, and less parental investment.[26] For skeletal evidence ranging from the Natufians through the Neolithic (the period encompassing early agricultural cultures) indicates that plant-dependent human beings were less robust than their Paleolithic ancestors—a head shorter, plagued by dental problems, subject to malnutrition and dietary deficiencies, and prey to a variety of infectious diseases that were largely the product of living in closer quarters.[27] The sole advantage, at least from a species perspective, was that there would be more of us; Marvin Harris estimates a forty-fold increase in the Middle East from 8000 to 4000 B.C.[28] But here again, burgeoning populations and the resulting transformation of the landscape would see the chances of reassuming the life of a hunter-gatherer diminish, quite literally, to the point of no return.

It was as if we had been lured aboard an ecological teeter-totter, the dynamics of which increased the well-being of a select group of plants at the expense of our own, at least on an individual basis—a veritable plant trap. And it would only grow larger and more enveloping. For unlike earlier means of subsistence, agriculture was eminently suitable for both massive expansion and intensification. But it is also important to realize that although agriculture was globally persistent and acted as a stable base for long-term population growth, it was on a local level subject to wild swings in productivity, to blights and climatic perturbations, which could and frequently did have disastrous consequences for dependent populations. This source of instability, when combined with the problem of sedentism-induced epidemic disease, imparted a volatility to agricultural cultures that, while not intuitively obvious, was nonetheless fundamental to their makeup and behavior. However, even in bad times the system was destined to spread. For in the face of famine, those portions of Neolithic populations forced to seek new residences brought with them and reestablished exactly the same crop plants that caused the initial collapse, a process that continued basically until all suitable land was brought under cultivation.[29] Meanwhile, the sole alternative for all but a relative few would be further intensification through irrigation in those areas where it was possible, and with it would come still greater populations and dependency. And so the tendrils of the plant trap coiled ever tighter around those who had once roamed free.

It is probably appropriate that among the very first examples of truly domesticated barley and emmer wheat, plants with a tough, nonscatterable rachis, were those found in the Jordan valley amid the storied ruins of Jericho at a strata datable to around 8000 B.C.[30] For Jericho was a place very much ahead of its time, a portent of what humanity would become. It was large, far bigger than any known contemporary settlement, a ten-acre site with a population estimated at between two thousand and thirty-five hundred.[31] Hunting was practiced, since only the bones of wild animals have been found here; it is also likely that trade was conducted in the asphalt, sulfur, and salt products of the Dead Sea. But neither could have supported such unprecedented numbers; only agriculture could do that.[32]

But with numbers and wealth came something even more startling at this early date: Jericho was fortified, surrounded by freestanding walls the mythic echoes of which can be heard in the story of Joshua's famous horn. Twelve feet high and six feet thick, the wall is estimated to have exceeded eighteen hundred feet in circumference, consisting of more than twelve thousand metric tons of limestone.[33] Add to this a massive twenty-five-foot stone tower weighing in excess of one thousand tons,[34] and what emerges is an extraordinary and unprecedented public work—a monument, apparently, to fear.

Yet one recent skeptic, archaeologist Ofer Bar-Yosef, has argued that the walls of Jericho are no such thing but instead simply constitute a huge attempt at flood control.[35] While such an explanation would certainly fit better with the central thesis proposed here as to the true origins of war, it remains unconvincing. For one thing, there is little sign of any deeply incised channels, which would indicate frequent and dangerous flash floods.[36] Secondly, while walls might be useful in deflecting swiftly flowing water away from a populated area, the massive tower would be essentially useless in this regard and still, therefore, requires explanation.

Indeed, the weight of evidence and common sense both argue that the structures at Jericho were meant to keep out people, not water, and must be accounted for on these grounds. What, then, made Jericho so attractive and the lives of its inhabitants sufficiently dangerous to have induced them to invest an estimated fifteen hundred man-days of labor in what amounted to a stone insurance policy?[37]

A number of authorities, including McNeill and Anati, believe it was the salt-sulfur-asphalt trade and the wealth it brought to Jericho that acted as a magnet to marauders.[38] But if this was truly the case, the town logically would have been located closer to the source; even then, the question would still remain: Wealth in what? For Jericho was absolutely unique, and at this point what constituted wealth to a commercial entity made up of sedentary inhabitants might not necessarily have been that attractive to surrounding hunter-gatherers.

The major exception would have been food. For in a time of desiccation, James Mellaart argues, surplus crops, seed, and instructions on how to grow it "would have been more valuable than gold."[39] The uniqueness of Jericho apparently lay in its fortunate location, virtually atop an extraordinarily reliable aquifer that debouched down a series of foothill fans, distributing life-giving water across a broad area—a sort of natural irrigation system. Not only was the resulting oasis an obvious candidate for cultivation, but it was likely to have drawn a host of thirsty game animals.[40] It seems, therefore, that the walls of Jericho were built to protect a horn of plenty, surrounded by a dry and hungry world.

If this was indeed the case, then these fortifications are a clear indication of agriculture's potential to create conditions leading to warfare of a most unrestrained kind. As we shall see, people would build walls around towns when they believed their homes and families were in jeopardy. Nor would all war necessarily involve such a possibility. Historically, warfare has assumed a predatory, genocidal cast only under certain circumstances: in cases of extreme political turmoil, religious animosity, or conflicts between very different economic systems; in anticipation of privation;[41] and in instances when new weapons or methods of fighting were introduced. Admittedly, this does encompass a broad range of possibilities, but it is nonetheless true that combat frequently would stabilize for long periods, with the fighting remaining or becoming increasingly ritualized and limited in both nature and extent. Typically, under such circumstances the violence would follow patterns analogous to intraspecific aggression in other species and would not involve the abuse of females, children, or other noncombatants. This does not appear to have been true at Jericho. Its walls stand in mute testimony to the probability that everyone living there was a potential victim of outside aggression; that the danger was aimed at the society and not just its fighting men. So it seems, at its very inception war manifested what were to be among its worst possibilities.

Yet what was probably most significant about the walls of Jericho is that they apparently constituted a false start, a local aberration followed by a hiatus of twenty-five hundred years. For there is no evidence of any other fortified sites in the Near East, at this time or thereafter, until around 5500 B.C.[42] Nor would there be other clear signs of organized violence much before this date. Apparently, the hostile conditions that led the inhabitants of Jericho to build walls around their lives would not quickly be repeated. Jericho, then, constituted more of a harbinger than a true beginning. And while it is reasonable to assume that warfare eventually would have evolved independently out of the cultural and institutional matrix made possible by agriculture, historically the process in the Old World would be hastened by an outside catalyst. This, however, would demand domestication of an entirely different sort.

III

Consider now the beasts of the field—the cattle, the sheep, the goats, and the horses. Like the plants that we eat, they are daily sacrificed to our hunger. In times past and in much of the world even today they are compelled to join us in our toil, suffer our whims and our abuse. Unlike the plants, however, they have central nervous systems and are mobile. Why, then, the question looms, don't more of them simply run away? There are numerous answers, ranging from lack of foresight to fences, but one that is frequently overlooked in this day of animal rights activism is that they want to be with us, have chosen to live among us.

As with plant biology, the nature and process of animal domestication are being reconsidered in the most fundamental way. Like agriculture, animal husbandry is coming to be viewed not as an invention but as an example of coevolution—in this case a true biological partnership. For unlike crops, whose drift toward domestication would have been driven on their side purely by the random forces of natural selection, the mammals involved here were self-directed entities capable of making clear choices—one of which, it is argued, was the decision to join us.

There are certainly skeptics. But, once again, they must explain why it was that, with one exception, the domestication of all the key work and food species took place almost simultaneously between 8000 and 5500B.C. and that subsequent efforts to bring animals into the fold have all basically failed.[43] Either our Neolithic forebears knew something we don't about selective breeding and behavior modification or virtually all the animals willing and able to throw in their lot with us did so at this point.

Dogs were the first, the exceptions that joined forces with us well ahead of the rest—around 10,000 B.C., during the Mesolithic.[44] The timing makes sense, since, unlike subsequent domesticates, they came from a line of carnivores and predators at a time when most humans were still preoccupied with hunting but when the lumbering, easy-to-stalk megafauna were largely gone and the future pointed toward smaller, more agile prey. Our new partners' speed, endurance, and keen noses, combined with human memory, cunning, and weaponry, gave access to game neither could hunt alone.[45] In one sense the process was quite natural, based on the same logic that leads to mixed-species flocks on the African savannas, mutual advantage through sharing strengths and compensating for weaknesses. Nor were humans necessarily devoid of prior skill at controlling animals. There is intriguing evidence of herd manipulation as far back as the Upper Paleolithic.[46] But there was a key difference here: while in human company, wolves had become dogs. They looked and behaved differently—the face and muzzle were characteristically shortened; they became proverbially subservient and playful. Indeed, they had been transformed into something else, as would all the animals who followed them into our fold.

How did this happen? What were the mechanisms involved? Controlled or at least differential breeding immediately jumps to mind, yet natural selection raises problems in this context. Whereas food crops were basically annuals, the generational cycles of the animals involved were a number of times longer. This put an inherent speed limit on change, sufficient to make it difficult to account for the physical and behavioral transformation among the various species in the time allotted—basically around twenty-five hundred years.

The answer to this Darwinian puzzle may well lie with one very significant reservoir of variation that was potentially available—the change that all birds and mammals experience in the course of developing into adults.[47] Because the range of differences among a fully grown population is small in comparison with the variance between any of those adults and an average juvenile, there is the possibility of tapping and locking in a variety of "youthful" characteristics should adulthood be reached prior to full maturation. Moreover, this might be accomplished through relatively small changes in the genes that control the rates of individual development.[48] This phenomenon, which is well documented in the evolution of many domesticated species including humans, is known as *neoteny* (literally "holding youth"). While neoteny actually operates in a complex fashion, impacting separate organs and elements of behavior variably and in some cases not at all,[49] the net effect is to inject a youthful aspect and demeanor into the affected animal. This is significant because, as Stephen Budiansky, author of *The Covenant of the Wild*, a fascinating book on animal domestication, explains, young birds and mammals show, in addition to a basic dependence, a notably greater curiosity, a capacity to learn new things, a characteristic lack of fear, and a willingness to associate and play with other species—exactly the kind of traits that might lead to and allow for domestication.[50] As with the food plants that became domesticates, the animals involved were opportunistic species adept at adapting to changing environments. In this case their reserves of variability were likely a product of the wild climate swings of the Pleistocene, but they also would prove crucial to domestication. In short, these were species preadapted to life among humans.

Herein may lie the key to how the process actually took place. It is probably more than coincidence that the second and by far the greatest wave of animals to join us was made up almost exclusively of herbivores. Whereas humans had once been hunters, creatures to be feared and avoided, suddenly we had become the masters of edible vegetation, still dangerous, perhaps, but now the possessors of nutritious crop stands and food stores that must have drawn these curious, opportunistic plant eaters like a magnet.

Gradually, almost accidentally, the relationship would have been intensified and formalized—a process that could have progressed from hunting and eating what were viewed as nuisances to the capture and control of significant numbers for later consumption. Quite naturally, the most aggressive animals would have been killed first, the tamer ones kept and eventually allowed to reproduce.[51] Budiansky cites experiments by Russian biologist D. K. Belyaev which indicate that selective breeding for docility has the effect of unlocking and intensifying a range of neotenous traits.[52] At any rate, skeletal evidence clearly indicates that continued association with humans did coincide with significant physical change, and the resulting breeds were demonstrably more dependent, docile, useful, and apparently more attractive to humans—the latter phenomenon seemingly reflected in the Disney Studio's progressive juvenilization of the beloved cartoon character Mickey Mouse, chronicled recently by puckish biologist Stephen Jay Gould.[53] Meanwhile, in return for their cooperation, domesticated animals would receive better and more regular nutrition, protection from other predators, and a chance to reproduce at a rate far greater than in the wild. The net result would be a fantastically successful partnership, when measured purely in terms of fecundity, with humans and the relatively few species who have chosen to live with us now estimated to constitute a full 20 percent by weight of everything presently alive.[54]

An archaeological example of how the process might have taken place has been uncovered at Tel Abu Hureyra, the largest known Neolithic site in Syria and one of the first permanent settlements in the world.[55] Subsequent to the appearance of domesticated grain around 7500 B.C., gazelle hunting continued together with agriculture for at least another millennium. There is clear evidence, however, that both sheep and goats had been domesticated by about 6500 B.C.[56] Nonetheless, for a considerable period the numbers of domesticates remained low in comparison to gazelle bones. Thus, if Tel Abu Hureyra was indeed typical, the change from hunting animals to keeping them was likely to have been a gradual, perhaps even unconscious, transition.

As species were absorbed one by one, their subsequent roles would have progressively emerged. Pigs, an excellent source of meat and ready to eat almost anything, became, in essence, garbage collectors—consuming crop waste, rotting vegetables, table scraps, as well as human and animal excrement—thereby providing a means of controlling village filth and the attendant hazards of infectious disease.[57] Also, the true herd animals would have only gradually revealed their full usefulness—sheep developing their woolly fleece progressively during the Neolithic; the full nutritional value of goats and cattle awaiting the spread of lactose tolerance among adults and particularly the evolution of more digestible milk products; and the traction capacity of bullocks going unused until the invention of the plow.[58] Even so, once established, these species were destined to multiply rapidly, their reproductive capacity reinforced by protection from predators and the neotenous reduction in the age of sexual maturity.

Yet increasing numbers of ruminants were bound to lead to problems of crop damage, either through consumption or simply by trampling.[59] There was an obvious solution, however: the removal of the herd animals to the margins of the settlement.[60] There they would be free to graze and browse in the care of those who increasingly became specialists in animal management.[61] And just as agriculture modified its environment, there is evidence that as early as the seventh millennium B.C. steady grazing was exerting selective pressure on the distribution of plants in a manner that prevented the regeneration of arboreal resources and literally created pasturage. Meanwhile, the larger the herds, the greater the incentive to capitalize upon the animals' mobility to move still further afield.[62]

It follows that, by degrees, the Neolithic community entered a process of social and economic bifurcation, as two separate ways of life began to form around domesticated plants and then animals. Nonetheless, it is important, nomad archaeologist Roger Cribb warns, to avoid drawings lines of exclusivity too completely around these two developing cultures, especially at this point.[63] For although nascent pastoralists were increasingly mobile, they were not yet really nomadic and likely would have continued to interact regularly with what was probably still viewed as a home base.

This situation appears to be reflected at Tepe Tula'i, a seasonal campsite discovered in Khuzestan and dated tentatively at around 6200 B.C.[64] Faunal remains show a predominance of ova-caprines (sheep and goats) and point, in Cribb's words, "to a degree of separation between pastoral and agricultural subsistence strategies."[65] Quite probably the shepherds at Tepe Tula'i temporarily left the village environment and led their flocks to what were being turned through grazing into steadily greener pastures. In all likelihood, they would have returned home at regular intervals, but the degree of their independence at this early date should not be overlooked. Moreover, it is reasonable to assume that much the same thing was taking place elsewhere. But it is also important to realize that archaeological evidence of this sort of existence is extremely difficult to locate—Tepe Tula'i was found largely by accident—and that is one reason why the significance and even the presence of what would become pastoral nomads have been consistently underestimated by a profession still focused on tangibles. Nevertheless, something important was happening here. Independence had not yet been declared. But arguably it was coming.

IV

Subsistence strategies robust enough to support large populations (either human or ruminant, depending on the context) were a necessary prerequisite for the emergence and perpetuation of true warfare in our species. Yet the ultimate political implications of these economic adaptations—agriculture and pastoralism—along with the concomitant psychological and sociological adjustments among the respective populations would take a long time to reveal themselves. For it is important to reiterate that, with the notable exception of Jericho, the evidence indicates that Neolithic societies in the Near East managed to remain relatively free of endemic violence for upwards of two millennia after the domestication of food crops and the establishment of true sedentism. This long stretch is highly suggestive and bids us to reflect further on the nature and possibilities of the cultures involved.

Recently, there has been a tendency, particularly among female scholars, to portray early agricultural communities in a favorable light—simple, innocent, and basically virtuous.[66] While the picture can be overdrawn, considering what emerged later the penchant is understandable. And although timing and nature remain hotly disputed, it is generally conceded that the evolution of agricultural communities in the Old World was violently, even catastrophically, impacted by outside forces, which significantly altered their subsequent development. Nevertheless, there were certainly elements of continuity—dynamics inherent to agricultural societies that, if left to mature in isolation, would have produced results similar to, though perhaps less exaggerated than, those which actually occurred in Europe and Asia. For, as noted earlier, cultural complexity based on direct agroeconomic exploitation consists institutionally of a series of tightly linked factors that not only appear to call each other into being but operate in a manner that is mutually reinforcing until they are intensified well beyond the point of diminishing returns.

The archaeological evidence indicates that this process had already begun within the settlements of the Neolithic Near East. In general, the dramatic increase in the practice of agriculture and the attendant growth in population throughout the region would have led to the beginnings of societal organization through scheduling and specialization.[67] Housing, which began as loosely clustered circular huts and progressed steadily toward ever more tightly grouped rectangular mud brick structures, also points to task specialization as well as at least some growing requirement for defense.[68] By around 6200 B.C. places like Oal'at Jarmo and Alikosh were showing signs of a much enriched technological culture, with a considerably wider inventory of tools and artifacts. At Tepe Guran we see the introduction of that all-purpose hallmark of sedentary existence, pottery.[69] This was reflected not only in utilitarian vessels but also in the proliferation of cult figurines, frequently depicting nude females with exaggerated breasts and buttocks. Also feeding growth in the stock of material possessions was an ever-widening trade network, which by 6000 B.C. was transporting valuables such as turquoise and cowries across several hundreds of miles.[70] Finally, graves began to reflect not only increased personal possessions but also apparent differentials of wealth.

All of these trends seem to have reached a kind of initial fruition at Çatal Hüyük, a brilliant Neolithic community that thrived in what is now southern Turkey between around 6250 and 5400 B.C. Consisting of perhaps a thousand houses and a population on the order of five to six thousand, Çatal Hüyük is one of the largest and clearly the wealthiest Neolithic site ever found. While it had no defensive wall, the standardized rectangular dwellings were packed together in pueblo fashion so that the outer edges of the town would have presented a single face without a break.[71] James Mellaart, Çatal Hüyük's excavator, does not believe it was specifically designed to be fortified, but the plan certainly would have been useful in defending the place and could well reflect growing danger from without.[72] There were other signs of intensification and growing competition. Mixed among the remains of weapons usually associated with the hunt was an assortment of baked clay sling balls, ceremonial flint daggers, and mace heads—the latter being of particular significance since they are basically useful only in combat against humans, a judgment apparently supported by the relative frequency of skull fractures at Çatal Hüyük.[73]

The skeletal evidence is also revealing in another regard. In spite of the town's obvious material wealth and capacity to grow crops—something like fourteen food plants were under cultivation—there are clear indications of anemia and other problems that could be attributed to widespread dietary deficiencies.[74] While the clarity and nature of social stratification are difficult to judge on the basis of such limited evidence, prestige burials do point to something approximating a leadership class.[75] Also emblematic of governance was the presence of shrines and cult

objects in sufficient numbers to assume the existence of formalized religious practices at Çatal Hüyük. Yet it should not be overlooked that the themes of the
subject matter—animals, particularly bulls, and corpulent naked women—stretched
back to the Upper Paleolithic, reflecting a tradition very different from that of the
male sky gods who would eventually thrust their way into the Eurasian pantheon.

In this regard and others, it is important to realize that for all the signs that the
consequences of social complexity were taking hold, Çatal Hüyük still basically
exemplified continuity and an essential kinship with Neolithic village life. Iconography indicates considerable equality between the sexes. Despite evidence of certain economic disparities, the standardization of dwelling size and quality points
to parity and sameness of lifestyle in other respects.[76] There was perhaps more
violence, but there is no sign that the town's wealth was based on warfare and
freebooting rather than hard work and trade.

Indeed, if there were revolutionary forces at work at Çatal Hüyük, they were
the same forces that had been at work since shortly after people began staking
their future on plant domestication—dynamics not just destined to bring them
class, property, religion, and government but also changing the way individuals
thought about themselves and their lives. Human beings, designed to wander and
forage and interact freely, suddenly found themselves pinned down to a life of
work and routine and structure. Certainly, there would have been some perceived
advantages—an escape from or at least an easing of the terrible burden of infanticide; a growing sense, if not the reality, of security evoked by permanent dwellings and large stores of food; the pleasures and reinforcement of reciprocity that
come with material acquisition.

On an existential level, however, it is quite possible to argue that these would
have proved small compensation for the penalties involved in this sort of life. The
future that loomed for the great bulk of humanity would be that of a drudge—an
existence based on unending toil, the necessity to defer to small groups designated
as inherently superior, and fear. Peter J. Wilson argues that the primary conditioning force of sedentism, the act of living in a house, imposed upon our consciousness the general structure of public and private, and provides "a foundation
for ambivalence and suspicion among neighbors."[77] We would become creatures
hedged in by walls, separated from one another and, eventually, from the outside
world. Meanwhile, the specialization and dependency encouraged by agriculture
would make it difficult to leave the community, thereby obstructing the "escape
option" that had always operated to relieve interpersonal tensions. Psychologically,
this is the stuff out of which armies and wars arise, especially when combined
with the need for territory. For agriculture demanded that we protect not only
our homes and possessions but the land necessary to grow food. And, as the
world suitable for agriculture filled up with people and farms, the proclivity to
wander would come to be increasingly frustrated by a patchwork of exclusivity.
Meanwhile, within such societies the need for stability amid growing disparities

of wealth would lead to the formalization of social stratification and its enforcement through coercion. The problem here was that it was basically an open-ended proposition, with no inherent limits on elevation or degradation along the path that lies between god incarnate and slave. The economics of things demanded that most sink rather than rise, so that the self-image of the masses would necessarily be that of inferiors.

So it was that the human spirit would be progressively enmeshed in a web of institutions, obligations, and objective conditions that would create a profound discordance between daily life and our intuitive expectations of what life should be. Only our species' profound flexibility and adaptability, along with a primary loyalty to cultural evolution as manifested in the elements of speech, reciprocity, and religion, made it possible to live such a life. Yet the deep memories harbored in our primary social interactions—subtle transmitters of our basic moral structure—would have continuously demurred, providing an unobtrusive but indestructible counterpoise. Yet on the level of events the logic of agriculture would prove overwhelming, sweeping the bulk of humanity along in a tidal wave of expansion.

There was, however, intrinsic to the first tranche of domestication, the possibility of escape, but only for a relative few humans—accompanied by a great many animals. For the diversity and suitability of the domesticated fauna in the Old World were sufficiently great to pose the possibility of pastoral independence from the agricultural hegemony. The basic combination of sheep, goats, and cattle provided a potentially flexible economic mechanism, with the relative numbers of each species being readily adjustable to take into account varieties of terrain, vegetation, and availability of water. Respective life cycles and mortality were also subject to some balancing in terms of risk, with sheep and goats acting as a speculative element, quick to reproduce but prone to die-off, and the less prolific but heartier and longer-lived cattle offering an element of stability.[78] The logic of mixed-species herds was further reinforced in terms of diversity of product. Indeed, with the development of more easily digestible milk products such as yogurt and koumiss, such herds could potentially fulfill nearly all the basic human needs, including food, clothing, and shelter, the latter in the form of fabric or felt tents. Meanwhile, the difficulties of animal control would have been mightily alleviated by the ministrations of the ever-present dog, whose predatory instincts were curbed, apparently through neoteny, to the point of only chasing and maneuvering flock animals.

The process of leaving, as noted, was sure to have been a halting one. But the basic necessities were accounted for, and the escape hatch should have opened fairly early. Consequently, it seems entirely possible that shortly after 6000 B.C. there were groups of pastoralists in the general region of the Middle East who had basically divorced themselves from their agricultural incubators and were living as true nomads. Certainly, there are scholars who would label this premature. While conceding that independence even without riding animals has been ethnographically demonstrated, they argue instead that the symbiotic forces holding the two subsistence strategies together (essentially the exchange of animal products for grain and produce) would have been strong enough to prevent a complete break at this point.[79] Almost without exception, however, these skeptics ignore the psychological dimension, the attractions that autonomy and a life on the move would have offered to a species with a long heritage of wandering.

Life out here, however, would have been a far different proposition from that of the hunter-gatherer. For pastoralists, like agriculturalists, remained fundamentally producers, not consumers.[80] Indeed, herds have been described as "fields on the hoof," and in terms of care and concern this is probably an appropriate characterization.[81] The life of the herdsman was and would remain difficult and problematic, with pastoralism emerging as an even less stable means of subsistence than agriculture and only capable of supporting relatively small populations of humans in any case.[82]

Nevertheless, key aspects of our former lives would have been truly resurrected. These nomads were able to move around as they chose—subject to the needs of their animals but still basically free. Territory and its defense would remain essentially meaningless. They would live in loose communities that were continually fractionating, in part to relieve internal tensions. Disparities of wealth likely existed, but they would have been largely restricted to differences in animal holdings. This factor, combined with mobility, would, by necessity, restrict the possibilities for social stratification. Theirs would be an open and relatively tension-free lifestyle.

But in one critical respect this profile is deceptive. They were not destined to be peaceful. Quite the contrary. Since their basic possessions, animals, were highly removable, they probably stole from and raided each other virtually from the beginning. But the brunt of pastoralist aggression likely would have been directed elsewhere. For the dynamics of their existence argue that they would come to view the agriculturalists as prey and targets of opportunity. In the eyes of nomads, the tillers of the soil were lesser beings, and would be treated accordingly.

{6}

RIDE OF THE
SECOND HORSEMAN

I

It started as a game. Boys, showing off, had taken to vaulting on the backs of captured mares corralled in the box canyon by the river. Initially, most flew off the terrified beasts, virtually the instant they landed. But a few of the more agile lads managed to steady themselves by grabbing a mane and were taken on a wild but short ride that always ended in the dirt. At times it seemed to the elders that every boy in the settlement was nursing a horse-induced injury and a mouthful of excuses why it prevented him from tending the flocks. "Horses were meant to be hunted and eaten, not ridden," parents complained. More than once the horses had all been killed and the meat dried just to get rid of the nuisance.

But always, some young ones would lure several more of the curious creatures away from the river where they came to drink and into the canyon so they could be trapped. Word spread fast, and by nightfall there would be boys moaning and limping about the village. What the adults failed to notice was that the rides were growing longer, if no more controlled. One day a father was amazed to observe his son riding round and round, his long black hair streaming in the wind. It was still a useless and dangerous pastime, but that night the father had not failed to brag of his son's exploits. He would have been a great deal more enthusiastic, had he any idea what was on the boy's mind.

For he and several others had found that sometimes they could induce changes in a horse's direction by jerking the mane to either side. Now the son had fashioned a noose with two long ropes attached, which, after a great deal of exertion, five of them succeeded in slipping over a mare's nose. When he leaped aboard and began experimenting with the rope reins, he was amazed to find he could not only maneuver the horse laterally but, by pulling back on both, could cause the horse to stop, basically in its tracks. Within a week the boys had ridden several mares across the river and out onto the steppe. From this day they would seldom be seen on foot.

Centuries later the boy would be revered as a messenger from the gods. For he would have changed entirely the lives of those who cherished his memory. No longer would they grow crops, along with the cattle, sheep, and goats. Now they would raise only stock, and not at the forest's edge but out on the great plain, the once-forbidding steppe, which was now their home. Here they had become a people on the move. For herding could be accomplished on horse-back, and water and fresh pasturage located by scouts riding far ahead of the main body.

Meanwhile, lives spent on the backs of horses opened possibilities that once would have been considered inconceivable--like shooting a bow accurately at full gallop. And the bows themselves had changed, growing shorter but dramatically more powerful as wood was gradually supplemented with glued layers of sinew. The resulting horse-man-bow combination produced a fighting instrument of un-precedented capabilities—a sort of ancient ultimate weapon, able to outshoot and outrun any opponent with ease.

Soon enough, the young men had taken to raiding the scruffy settlements at the edge of the steppe for grain and cattle and women. Just as frequently, they stole from each other's flocks. But as they wandered farther afield, they brought back stories of much larger and richer communities far to the south. As the flocks settled into winter pastures, more and more the men's attention turned to these fabled targets of opportunity. The irresistible song of plunder—talk of wealth and the beauty of far-off ladies—fed upon itself, until one day the fighting men set out to find their destiny. Opportunity beckoned, and in the meantime the sheep would be safe with the women.

The riders would have had no suspicion, but they were about to be joined by the Second Horseman. For recent discoveries point to such raids by pastoralists—initially on foot but ultimately on horseback—as plausible catalysts for a chain of events that led to the emergence of warfare in the form we find most familiar, the strain that came to dominate the civilizations of Europe and Asia. This was, as noted earlier, something of a historical and ecological accident. For as we shall see later, warfare in the New World, lacking a pastoral component, would evolve in a significantly different and decidedly more ritualized fashion. But in the end it would be overwhelmed by its older and more virulent counterpart, to be re-membered as little more than an anthropological curiosity. Yet it is in the differ-ences among the various forms of warfare that a critical lesson is to be learned, for they show us that war is but an institution, dependent on extant conditions. And when those conditions cease to exist, the Second Horseman's ride must grind to a halt.

II

For pastoralists the process of separation must have proceeded unevenly across the Middle East and the regions to the north. In most areas of human habitation, herds and those who tended them probably continued to hang around the outer edges of farming communities. But as numbers grew and pastures matured, a certain amount of geographic concentration must have taken place, accelerating the evolution of a distinctive shepherd culture and a growing sense of true independence. But there remained several basic impediments to its full flowering, particularly if truly unfettered nomadism was to be achieved. Without effective pack animals, possessions would have to be kept to such an absolute minimum that it must have inhibited the development of even a highly austere lifestyle. To some degree this problem may have been mitigated by the early use of rams and other herd animals for pack transport, an expedient for which there is some evidence.[1] But this would have been far from ideal. Still more fundamental was the fact that these people remained pedestrians living lives based on mobility. To effectively exploit the vast land mass that stretched to the east, they would require a much faster means of getting from place to place. They needed mounts.

To put it mildly, our understanding of the origins of horse domestication and riding is in a stage of rapid transition. But it is already apparent that the general trend is in the direction of a much earlier date for horse riding. Until very recently, opinions as to the origins of equine domestication were split between those who believed that asses, onagers, and then horses were first used as draft animals among agriculturalists in the Middle East[2] and those who stood by the hypothesis that they were first ridden in central Asia and later hitched to vehicles in Mesopotamia and Egypt.[3] But whether the cart was placed before the horseman or vice versa, it was generally supposed that these innovations did not take place much before 1500 B.C.[4] This has now changed dramatically. In what amounts to a brilliant piece of high-tech archaeological detective work, David Anthony, Dimitri Telegin, and Dorcas Brown have recently succeeded in catapulting the date of horse riding back more than twenty-five hundred years, perhaps to as early as 4200 B.C.

Their story centers on Dereivka, a hamlet around 155 miles south of Kiev in the Ukraine. The site, which is located on the banks of the Dnieper River in a transitional zone between forest and steppe, was excavated repeatedly by Telegin. He found it to be one of hundreds identified with the Sredni Stog culture, and radiocarbon dating placed it at around 4000 B.C., give or take a few centuries.[5] Life there very much resembled that of the group represented in the previous re-creation; the people practiced both agriculture and animal husbandry, and they ate a lot of horse. Indeed, as a percentage of food refuse, the remains of horses were about twice as prevalent as they had been in earlier regional cultures.

But there were some other unusual features of the horse bones recovered at

Dereivka. For one thing, there was a considerably larger representation of males than might have been expected in a herd kept solely for food. Odder still was a ritual burial site containing, among other things, the skull and foreleg of a seven- or eight-year-old stallion and two perforated pieces of antler that seemed to be the cheekpieces of a bit.[6] Potentially, this was of major significance, for there could have been only two conceivable uses for such a bit: either the animal was employed in pulling a vehicle or it was ridden. The first possibility could be dismissed on anachronistic grounds, since the wheel would not be invented for another thousand years, until about 3300 B.C.[7] So this left only riding.

But there were serious problems with the evidence. The antler cheekpieces, though plainly similar to others used as circumstantial indicators of riding in the Middle Bronze Age, still did not constitute a bit—only potential parts of a bit. Without a complete example of such a device, the possibility would remain open that these pieces served some other purpose.

After considering their options, the team of Anthony, Brown, and Telegin concluded that if they were ever to prove their point, the evidence would have to come, quite literally, straight from the horse's mouth. For the actual manner in which a bit operates suggested a promising line of investigation. When properly adjusted, a bit should remain in the space between a horse's incisors and premolars, enabling the rider to exert control by applying pressure against this very sensitive part of the mouth. Horses, however, quickly learn to fight this by using their tongues to lift the bit back onto the premolars—"taking the bit between the teeth."[8] Yet the heavy pressure exerted by the horse's enormously strong mandibular muscles combined with the action of the bit slipping back and forth over the prow of the tooth causes a characteristic spalling of the enamel as well as a distinctive beveling, which together are generally referred to as "bit wear." After studying a wide range of contemporary horses for the nature and extent of bit wear using an electron microscope,[9] the investigators turned again to the Dereivka stallion from the ritual burial site. Not only was premolar beveling visible to the naked eye, but when the tooth was placed under the electron microscope it revealed exactly the same pattern of occlusional spalling as that of a modern bit-worn horse.[10] The conclusion was unavoidable: the Dereivka stallion had been bitted and therefore ridden at least six thousand years ago.

Such an innovation, David Anthony argues in another important article, almost certainly would have had a critical impact on a stock-raising people living near the edge of the steppe.[11] For out on these great plains life-supporting resources could be quite rich locally but were typically separated by long stretches of poor, very inhospitable territory. Plainly, this was no place to lead one's herd without having a good idea of what lay ahead. Yet horse riding, because it increased the range of movement by a factor of at least three,[12] made possible the kind of long-range scouting necessary for safe and effective exploitation of such a vast and uneven environment. While the archaeological evidence (such as it is) appears to indicate that penetration of the very deep steppe took a fairly long time and was contingent on the invention of wheeled vehicles,[13] subsistence along a broad swath of rimlands could have come almost immediately. Once out here, the lives of the participants would have been transformed nearly as quickly. Using the analogy of Native Americans, whose exploitation of horses took place largely independent of the Europeans who actually introduced the animals, Anthony is able to make a convincing case that the combination of vastly improved transport and a new environment would have been sufficient to very rapidly produce a fundamentally different lifestyle perhaps within a century or two.[14] In the Old World context this new existence would have included a great deal more mobility, to the point of true nomadism, and adoption of stock breeding as the prime means of subsistence, and an increase in the appearance of weapons, along with other signs of individual and group aggression. Even given a very limited amount of evidence, the archaeological record appears to support these suppositions.

While chronologies are subject to some dispute,[15] it seems that by the late fourth millennium B.C. the apparent successor of the Sredni Stog culture, the Yamna, had become sufficiently mobile that, in Anthony's words, "it could be assigned to no particular place but only to a very broad steppe region."[16] Similarly, the archaeological evidence indicates that both Sredni Stog and Yamna cultures placed heavy and increasing emphasis on horse breeding and animal management in general. Finally, the graves of each people not only grew steadily richer but yielded more and more weapons. For example, projectile points were eight times more frequent at Sredni Stog sites than they had been in the predecessor culture.[17] Even more suggestive are a series of mace heads carved to represent horses, recovered from Yamna sites.[18] But probably the most intriguing evidence that something fundamental was happening is indirect. A sedentary river valley culture known as Tripolye C-I began to form huge and aberrant settlements of up to a thousand dwellings at the end of the fourth millennium B.C., only to be replaced by the Yamna. The best explanation for these macrovillages is that they constituted extreme defensive concentrations of people pushed together by outside pressure.[19] It seems reasonable to assume that the pressure we are talking about arrived on horseback.

The problem is that it would have come much too late and too far north to help account for the proliferation of walled towns and other signs of warfare in the general region of the Neolithic Middle East beginning after 5500 B.C. and gaining momentum over the next several thousand years. At best, this line of evidence only pushes the date of mounted nomadic pastoralism and attendant raiding by mounted warriors back to sometime in the fourth millennium B.C.

Far to the south, however, Neolithic towns had already begun to find the need to protect themselves with walls. There are several potential explanations. In the past, scholars have tended to attribute such indicators of war exclusively to indigenous factors—an intensification of social complexity and a growing competition among the sedentary agricultural communities themselves, the classic statement being Carneiro's.[20] While not denying the role of such internal motivators, it can still be said that this paradigm falls short in one critical respect. It fails to address why war in the New World, where independent pastoralism was missing,[21] should have developed in an entirely more ceremonial fashion, which did not normally require the construction of walls around population centers.

On the other hand, a more satisfactory sequence of events might be constructed if raiding by local pastoralists is thought to have had a major impact on the early spread of Middle Eastern Neolithic walled towns. Such a process would have begun with forays conducted on foot by groups inhabiting the highlands and other fringe areas not suitable for agriculture, gradually gaining momentum as concentrations of more independent shepherds congealed and proliferated. But is this plausible?

Pastoralism is a complex phenomenon. Even at this early date it probably encompassed a range of dependencies and degrees of mobility, with some practicing what amounted to sedentary animal husbandry and remaining very much a part of Neolithic communities, while others engaged in only seasonal transhumance. Nonetheless, it also appears that the central variables involved in the shepherd's life were self-reinforcing, or, as archaeologist Roger Cribb explains, "The greater the degree of pastoralism, the stronger the tendency toward nomadism."[22] Logically, under the right conditions this dynamic would have produced independence—true pastoral nomads, even though still on foot. It also seems plausible that the conditions encountered by such groups eventually would have resulted in a sustained pattern of fairly intensive raiding.

This hypothesis, however, can be criticized on several grounds. First, the recent scholarship surrounding pastoralism has tended to employ ethnographic data to stress the synergy between pastoralism and agriculture. Under this paradigm the two elements are seen as having a primary interest in cooperation—taking advantage of the logic of specialization to establish a fundamental trade based on the exchange of animals for grain.[23] In the context of this market-based explanation, it is further emphasized that the greater instability of pastoral life and the difficulty of obtaining alternate sources of vegetable foodstuffs would have made the pastoral dependencies on this trade all the greater.[24]

There are problems, however. Not only was the market mechanism somewhat problematic and presumably underdeveloped at this point in history,[25] but, more to the point, pastoralists would have been most in need of agricultural products just when their herds had been driven down by bad fortune and they had the fewest animals to exchange. Disease, predation, and periodic overgrazing made such episodes practically inevitable, thereby necessitating some other means of obtaining not just grain but animals to restock their herds. Later horse nomads, as Bar-Yosef and Khazanov point out, would solve the problem by exerting their substantial military advantage to obtain what they needed, by force if necessary.[26]

But equivalent foot nomads, the two scholars maintain, had no such military advantage, arguing for this reason they could not have survived and, therefore, probably did not exist.[27]

Their reasoning is flawed in one critical respect, however. They miss the rather obvious point that the foot nomads' military advantage was vested in their very mobility.[28] They would have lived nowhere in particular and had no essential territory to defend. Their targets were stationary, while their own assets were movable. They could plan an attack on a known location, while the farmers, in turn, could not know exactly where to seek revenge—provided, of course, that the pastoralist perpetrators could get away cleanly.

This was not a given. For while pedestrian attacks—granted the element of surprise—are quite plausible, escape, as Anthony points out, is another matter entirely, especially if the attackers were laden with foodstuffs and captured animals.[29] Safety, therefore, would seem to have directed that once the attacks were under way, everything possible be done to discourage pursuit. And so it would have made sense to compound the victims' bewilderment with terror—to kill wantonly all who might follow them, to render the community helpless, and to destroy in a way that might induce shock.

Most probably necessity would have been reinforced by inclination. Even those who emphasize the symbiotic dimension, such as Khazanov, are ready to concede the "negative attitude" that contemporary pastoral nomads hold for the sedentary—the "universal opposition between us and them."[30] This is unlikely to have been any less pronounced in the Late Neolithic. For each represented a very different way of life in a social equation that, although it may have accommodated cooperation under certain circumstances, also would have promoted very violent interaction in others.

The net effect would have been a sporadic pattern of aggression, sufficiently terrifying to result in the gradual abandonment of open village sites and the concentration of populations in fortified townships.[31] Archaeologically this can be shown to have taken place irregularly from the Late Neolithic through the Early Bronze Age in eastern Anatolia, the Levant, and the areas east of the Zagros Mountains extending into the northern reaches of Mesopotamia—a hit-and-miss process that likely reflected the mobility of the aggressors. Yet behind this geographic irregularity would have been clear economic and even ideological motivation—a syndrome destined to impose itself on the subsequent development of armed hostility in Eurasia and ultimately across the globe. This is why it makes sense to locate the origins of true human warfare in this initial confrontation between the sedentary and the nomadic—even if the latter arrived on foot.

And it would not always be the case. Previously, it had been assumed that there had been no horses in the Middle East until, at the earliest, the second half of the third millennium B.C.[32] The one recorded example of earlier horse bones—found at the walled fourth-millennium site of Norsun Tepe in Anatolia—was assumed to have been wild.[33] Subsequently, however, the Hungarian archaeologist Sándor Bökönyi documented horse bones dated around 3200 B.C. at five sites in a narrow band on both sides of the Euphrates River in southeastern Turkey and argued convincingly that they were domesticated.[34] In the last few years horse bones have been recovered far to the south in the Negev region of Israel from the Chalcolithic sites of Shiqmim and Grar, also dating to the fourth millennium B.C.[35] Since horses cannot be moved in any number without mounted herdsmen, there is good reason to believe they were brought into the area by true equestrian nomads.[36] While the presence of these animals can certainly be explained as the result of trading, the military advantage of those who introduced them would have been rather quickly made evident. Under the circumstances, it also becomes possible to hypothesize a pattern of raiding—this time by a far more formidable and terrifying adversary.

This is suggestive in a number of respects. For in the same time frame the acquisition of wheeled vehicles enabled the horse nomads of the north to penetrate the deep steppe, and in general to vastly extend the scope of their wandering. But perhaps of more significance, it also coincided with the rise of the state and the consolidation of much larger urban concentrations in Mesopotamia during the late Uruk period. While these sites were earlier thought to have been initially unfortified, there is now evidence that they were in fact surrounded by city walls.[37] Indeed, what was going on here could be considered to some degree analogous to the formation of the macrosettlements of the Tripolye C-I culture, especially if the influence of nomadic horse raiders is thought to be a possibility.

Admittedly, this constitutes a big "if". For there is no direct evidence of equestrian nomadic depredations in the area at this point—inscriptions describing them, or even relics such as those recovered at Dereivka—nor would there be until much later. Yet it is generally accepted that the more nomadic a group became, the paler its archaeological footprint.[38] Whether mounted or on foot, nomads wandered over large areas, pursuing a livelihood requiring virtually no specialized tools, and almost without exception their possessions were made from organic materials subject to rapid decay.[39] Pedestrian pastoralists at least lived in the region, and there are some signs of their presence, but horse nomads would have been alien sojourners, present for short periods and in very small numbers, their attacks scattered and sporadic. So the archaeological "smoking gun" may continue to elude us.

While the case for pastoral aggression—either pedestrian or equestrian—cannot ignore the conventional rules of evidence, it is also true that modern particle physics is based not on the capture of the subatomic specks under study but on inferences derived from the effects they produce in high-speed collisions. In this case we are dealing with what amounts to the impact traces of armed aggression. And the walls that farming communities built around themselves were massive and very tangible, far too labor-consumptive to have been constructed for anything but a vital reason.

Yet it also may be that they were the result of indirect as well as direct pressure. For war tended to spread in a manner analogous to contagion, its essential vector being fear. Once under way, hostilities often established themselves at higher rather than lower levels of violence. For among those lacking a ready avenue of retreat, aggression could be met only by submission or equivalent countermeasures. So it is quite possible that once the precedent of attacking whole communities had been set by pastoralists, it would have been continued by the original victims—jumping from agricultural community to agricultural community until circumvallation and eventually siege warfare became the norm.

In the absence of better information, the relative influence of the pastoral and sedentary in this process must remain a matter of speculation. But the substantial body of archaeologists who reject entirely the role of independent pastoralists in the intial spread of warfare is left to explain why human combat developed in an entirely more ceremonial fashion largely without walls in areas where autonomous shepherd cultures were indisputably missing. And if the early influence of horse nomads remains still more clouded, it is apparent nonetheless that their penetration of the deep steppe would rather quickly put the finishing touches on one of history's most fearsome military instruments.

III

For there is reason to believe that when left to their own devices, pastoral peoples would have developed in short order both the capacity and the compulsion to periodically break out of their customary nomadic orbit and descend upon sedentary folk like waves of energy. As with many of nature's greatest experiments, this culture emerged suddenly and then changed very little—a social organism so elegantly simple yet perfectly adapted to its venue and the needs of those it served that the pressures to evolve would dwindle with each of its few subsequent innovations until all further development ceased. For theirs was a society so deeply and irresistibly shaped by factors of environment and psychology that, as the centuries passed and ethnic group replaced ethnic group, each would end up living in exactly the same manner as its predecessors.

If there could be such a thing as the Zen of change, these nomads mastered it. But even this took time. And although the introduction of the horse appears to have been pivotal in extending pastoralist mobility, it was the acquisition of wagons and the domestication of camels after 3000 B.C. that probably allowed for the later infiltration of virtually all of Inner Asia, including vast stretches of very arid territory.[40] Material culture, even with wheeled transport, remained Spartan but in its own way highly sophisticated. Portable dwellings, particularly in the form of beehive-shaped felt yurts, provided shelter even under the harshest conditions of wind and cold.[41] Containers and utensils were fashioned of organic materials sufficiently light and rugged that they would not be replaced until the advent of modern plastics.[42] Indeed, practically the sole concession to aesthetics was weaving—the almost compulsive generation of textiles and carpets of unrivaled quality, but still portable and storable.

Yet perhaps the pastoral nomad's greatest technological achievement was his composite bow, a weapon unsurpassed in lethality until the development of relatively sophisticated firearms. Here again, it is a testimony to nomadic inconspicuousness that credit for its invention generally has been assigned to settled societies—this in spite of the fact that in both China and the Middle East composite bows appeared in the second half of the third millennium B.C., shortly after these cultures were first subjected to sustained contact with pastoral nomads bent now on conquest, not just raiding. Logic and the weight of analogy argue that the process of diffusion took place not from littoral to steppe but in the opposite direction, and that the composite bow is a good deal older than is now believed.

A. H. Pitt-Rivers, the father of modern archaeological excavation technique, argued in the late nineteenth century that the development of the composite bow would have been powerfully stimulated by a steppe environment where no suitable wood existed for bow manufacture.[43] This line of reasoning has been subsequently rejected on the grounds that societies that did have good wood for bows switched to composites nonetheless. Yet bows here were already highly effective, and there would have been little obvious motivation for changing them. On the other hand, such weapons might still have been introduced from outside and then adopted simply because they were better. More to the point, it would have been the absence of good wood for bows, not its presence, that was likely to have been the telling factor, particularly if the user rode a horse. For of all trees, only the yew, with its combination of elastic sapwood and very strong heartwood, is suitable for making really powerful one-piece bows, but they are by nature too long to shoot comfortably from horseback.[44] So in an environment of few trees and many horses, there would have been ample reason to experiment.

Once again, Native Americans provide an example of just how this might have taken place. Almost immediately after they began riding horses and moved out onto the Great Plains, tribes took to shortening their bows and then reinforcing them with sinew, using a glue derived from hides. Sinew was an ideal choice, since it was extremely resilient, having a tensile strength roughly four times that of bow wood. Soon many removed the wood entirely, substituting elk antler, which bore compressive loads better, to produce a weapon that was only a step short of the final and most sophisticated construction technique—all this accomplished in a century or two without external stimulus.[45] This process can plausibly be projected to have continued among Old World nomads until they had perfected a weapon fashioned from carefully tapered and glued layers of wood, horn, and sinew, which was likely to have been a good deal shorter than a typical one-piece bow but still possessing a gaping advantage in draw and power. Even more importantly, lives spent on horseback would have automatically generated the skill and balance to shoot such a weapon accurately and efficiently while on the move. Traditional opinion on Inner Asian nomads, typified by Stuart Legg, has generally rested on the assumption that the acrobatic horsemanship and specialized weaponry which made them so lethal must have required many centuries to develop.[46] But a case can be made that this seriously underestimates the power of the forces working on these horsemen, the overwhelming nature of the steppe and the irrefutable logic of pastoral nomadism itself. This was a life of manifest imperatives and without margin for error. Those who did not adapt immediately must have either perished or retreated from the prairies. And as the case of the Native Americans illustrates, the traits necessary for survival could appear very quickly. So it is that we can hypothesize that these herdsmen, in what amounts to the blink of an eye, found themselves in possession of an instrument of frightening military potential—frightening not only in the man-bow-horse synergy but also because of who these people were and how they lived.

Their stage would be a vast and pitiless one, marked by temperature variations as great as any on the planet and a fundamental aridity relieved only intermittently and capriciously by storms of awesome power. This was a world of extremes in which life and death were dictated from the skies, a place where storm gods must have been plausible arbiters of fate. Here human populations always must remain thinly distributed and small absolutely, capped by a resource base both meager and spasmodic. For pastoral prosperity could be transformed into utter catastrophe with but a single visitation of the dreaded *jud*, a freeze of such intensity that it could turn miles of rich pasture into a wasteland virtually overnight.[47] And ruminants, along with horses, died far more quickly than people when deprived of nourishment for any substantial period, so herds caught in the midst of such a disaster must inevitably suffer rapid and massive mortality. Add to this the threats of predation and disease, and it becomes apparent why the demographically speculative nature of shepherding would have yielded a predisposition for organized theft and a notable insensitivity toward death. For, as David Anthony points out, the quickest and most efficient means of building and rebuilding herds was frequently raiding the four-legged assets of other nomads—an activity guaranteed to stimulate further raiding along with the military skills necessary for success.

A hard and unpredictable environment bred people similarly disposed.[48] This was a race environmentally conditioned to oscillate dramatically between relative benignity and monumental aggression.[49] Over and over their victims, or at least the survivors, would express horror and incomprehension at the pastoral nomads' capacity for violence, their disregard for human life, their refusal to operate according to any accepted rules of military conduct. Many observers had difficulty in comprehending them as people at all.

The antipathy of the steppe nomad for sedentary agriculturalists has been the subject of frequent commentary by those familiar with the two cultures. Available anecdotal evidence, along with contemporary understanding of human evolution, points to this dislike as extending beneath the spheres of economics and ecological contrariety down to the roots of the psyche. For although pastoral nomadism was based on a network of symbiotic adaptations, it still was initiated through a basic rejection of agriculture. True enough, the pastoralist was tied to his herds much as the farmer was chained to his fields, but his tether was longer and less obvious. For if nothing else, life on the steppe was spacious, and aboard a mount with the speed of the wind a nomad might truly feel himself to be a free man, living a life comparable in the emotions it provoked to that which had shaped humanity for millions of years.[50] More than in agricultural communities, hunting and weapons training retained a central role in nomadic existence.[51] There would be some governance, but it would remain flexible, transitory, and largely built upon personal relationships and the extended family.[52] And ironically, although the coming of the nomad can plausibly be imputed as having had the effect of furthering the subordination of women in the agricultural sphere, out on the steppe the difficulties of life and the shared burden of survival ensured females a substantial measure of respect and even equality. These were proud people in large part because of the way they lived. Frequently pastoralists, particularly those living at the geographic margins, would resort to part-time farming, but almost inevitably, when given the opportunity, they would break the link and defect back to the nomadic life.[53] For it was, in their minds at least, a higher calling.

Agriculturalists occupied a lower moral rung in the eyes of the nomads; being plant eaters, like their own stock, they were viewed at once as a resource to be exploited and also with a measure of jealousy. For if it is possible to apply modern parlance to such a situation, they had "sold out"—traded material possessions for freedom.

So the stage was set. For the reasons outlined earlier, it is possible that, sometime in the late fourth millennium B.C. those living on the fringes of the Inner Asian plateau would have developed both the means and the motivation to have conducted devastating raids on the communities far to the south. Yet it is important to note that it would not have been necessary for such forays to have been either frequent or massive for them to have made a profound and catalyzing impression, even coming after a sustained period of pedestrian pastoral raiding. Populations of horse-riding nomads must have been very small and scattered at this point. But fear is not incremental—a single horrific event is much more likely to elicit an exaggerated response than a string of merely unpleasant ones. By its nature equestrian nomadic culture was an engine best suited to impact and traumatize—so much so that, even after thousands of years of subsequent exposure, victims would still have difficulty perceiving clearly even who these intruders were, how many of them there were (inevitably a horde), and where they might have come from. For those first attacked it must have been worse. But a single experience with these terrifying and vindictive horsemen could have been sufficient to have changed significantly the indigenous social dynamic—supercharging it, accelerating its spread to neighboring communities, and encouraging the development of significantly larger walled population centers. And while agricultural societies would learn only haltingly to ride horses and fabricate composite bows, it is the assumption here that the impact of these spectral nomads would have been sufficient to propagate war of a particularly virulent sort in the Old World's cradles of civilization.

IV

Yet the historical role of pastoral nomads would hardly be limited to that of instigators. They would remain instead at the right hand of the Second Horseman—a protean force threatening for at least four millennia to descend without warning upon the hubs of agricultural life to exact a price in blood, and ensuring in the process that war throughout the Eurasian sphere continued to be a matter of whole populations and not just rulers and their minions.

But a considerable span of time would be required before these virtuosos of mayhem fully assumed their larger identity. Circumstantial evidence indicates that they spread gradually across the Inner Asian plateau, infiltrating the deep steppe and finally emerging at the outer bounds of what was becoming China in the first half of the third millennium B.C.[54] Here they would reside, strewn across the entire region in a perpetual state of flux, a thin but contiguous mass.

This, in turn, would transform the entire dynamic of their aggression. So long as pastoral nomadic populations remained tiny disjointed entities, we can presume that their raiding was random and self-initiated. But later, when Inner Asia had been populated, albeit sparsely, their assaults would be propelled by a giant ripple effect, as one hard-pressed group displaced its neighbor until those at the edge coalesced and crashed against societies of the littoral.[55] Because these outer elements now had no place to which to retreat, the emphasis would shift from raiding to actual conquest, beginning in the early second millennium when groups like the Kassites and the Hittites descended upon various Middle Eastern polities to become their rulers.[56]

But while the continuing dynamic of conquest from the steppe would have the effect of reinvigorating, or at least replacing, agricultural dynasties, later pastoral nomads were in most ways ill suited to perpetuate themselves in sedentary environments. Not only were they fundamentally unsympathetic toward agricultural life and often brutal toward its practitioners, but the very act of conquest and domination robbed them in short order of their central advantage—the military skills automatically imparted by a life on the steppe and bound to deteriorate elsewhere. Meanwhile, the new rulers, illiterate and unacquainted with the skills necessary to administer such societies, were inevitably forced to turn to the same class of bureaucrats they had recently sought to overturn. And then, in a matter of a few generations, these usurper nomads would either be transformed into virtual duplicates of their predecessors or discarded, after a decent interval, as irrelevant. For whatever its psychological advantages, the culture of the steppe was simply too austere to encompass and direct the sophisticated agricultural tyrannies of the littoral in a way that marked a significant departure.

Nonetheless, the symbiosis would persist, based on the twin pillars of necessity and opportunity. For in spite of their Spartan self-sufficiency, pastoral nomads were proverbially poor in material things—luxury goods, weapons, and storable food in bulk—the enticements that would draw them irresistibly to the world of the agriculturalist. For their part, they would bring their textiles, and pelts, and stock—especially horses, in numbers and quality unattainable elsewhere. But should trading prove unsatisfactory, they might resort to simply taking what they wanted, for their military advantage was of such a magnitude that this always remained an option.

This was not simply a matter of tactical superiority; pastoral nomads constituted military power of a different order, an instrument of violence that transcended (or simply ignored) the most basic presumptions of armies representing sedentary societies. For agriculture would dictate that war among the settled would be essentially about territory, both on the battlefield and in a larger political sense. This in turn, stimulated a kind of aggression heavy with instraspecific overtones—confrontation, ritualized behavior, and symmetry in opposing armaments. Nomads would have none of this, or at least very little. War for them was basically an exercise in predation, its object to fool and then to kill. Their only code was a pragmatism almost inconceivable to the paladins of agriculture. Thus, in a famous passage from Herodotus, when Darius, the invading Persian king asks Idanthyrsus, his Scythian adversary, why he kept running away, the latter replies: "If you want to know why I will not fight, I will tell you: in our country there are no towns and no cultivated land; fear of losing which, or seeing it ravaged, might indeed provoke us to hasty battle."[57]

Most nomads would not have bothered addressing what to them was so obvious. Instead, they would continue to exist as a kind of military incubus, unapproachable but ever-threatening—an apparition both terrifying and outlandish to the eyes of those steeped in the military conventions of the settled. For such a force while on campaign would not have even resembled an army but rather a great, unruly animal mass, a kind of cattle drive with malicious intent, indistinguishable from a peaceful migration except for the abnormally high percentage of men and horses as compared to ruminants. And if in battle they revealed themselves as disciplined and tactically resourceful, the confusion and terror of their victims nearly always distorted the accuracy of their reporting.

For as much as anything, it was this very quality of phantasmal strangeness that conditioned the responses of the sedentary to these ghost riders. Periodically, there would be a rational assessment of this threat, such as Thucydides observation, over sixteen centuries before pastoral nomadism's climatic Mongol outburst, that if the dwellers of the steppe ever once united, then no nation in Europe or Asia could stand against them.[58]

But for the most part, agriculture's response is best understood on the level of inchoate fear. In the West, where their depredations remained more sporadic, nomads were known and remembered in myths such as that of the centaurs, half men half horses, a wild and unpredictable race. But in the East, where the danger was continuous and obvious, the responses were the more tangible.

The case of China was truly poignant but instructive. For the danger the Chinese faced was of such a magnitude that to have truly met it would have required the militarization and, eventually, the degeneration of their entire civilization. So they temporized and experimented—vacillating between appeasement and activism, creating at several points their own versions of armies on horseback only to see them desert in droves and drain the national treasury.[59]

The alternative was to build a barrier. And if monumental architecture was indeed imperial agriculture's way of working off excess energy, then the series of boundary fortifications constructed repeatedly across China's northern and western frontiers were singular manifestations of that urge, conditioned by the fears of a culture continuously bedeviled by the nomad. Those wondering at the huge outpouring of energy invested in a project of such dubious utility might well conclude, as did Owen Lattimore, that it was the most tangible evidence of China's will to resist and to draw a line between the world of the plant growers and those who roamed outside.[60] But in reality it was of little help as wave after wave of horse-borne invaders poured in. China and the settled world could only persevere, and wait for the flood to recede. This, in essence, was the dilemma posed by the nomad.

So it seems that the shepherds, having escaped the clutches of domesticated plants, returned vindictively to bring us true warfare. But it also appears that this was largely an accident of ecology and circumstance, and that agriculture also harbored the seeds of war, kernels certain to have sprouted naturally, though more slowly and in a less fulsome fashion, had there been no such thing as pastoral nomads. The dwellers of the steppe would continue to haunt the world of the settled. Yet they could never constitute more than a tiny fraction of humanity, and their brand of war was destined to remain an eccentric mutation of an institution that would spread among practically all of us. But to understand how this would occur and the functioning of what amounts to a central node of warfare, its beginnings must be explored elsewhere.

{7}

URBAN IGNITION

I

Uruk[1] was in shambles. The events of the past year had overtaken its
citizens like the waters of a flood. First had come the envoys of Agga,
son of Enmebaraggesi and king of Kish, demanding that the men of Uruk
cease digging wells and irrigation ditches on the territory in dispute between the
two cities. And, of course, Gilgamesh, the hothead lord of Kullab, had seized the
opportunity to loudly demand war before the council of elders, knowing full well
that he had supported the project precisely because it would bring such a response. But the elders saw through him and decided to negotiate with Agga.

But Gilgamesh, ever the schemer, went instead to the assembly, composed of
all the city's fighting men. And as usual here, things heated up quickly, fanned
by loose talk of how small and scattered were Agga's forces, and how the men of
Kish always hung their heads at the first sign of trouble. So war it would be, and
Gilgamesh, son of Ninsun, they chose to lead them to glory. Or that's what the
fools thought!

What really happened was anything but glorious. For rather than leading them
out in a phalanx to the plain, where they could fight Kish like men, Gilgamesh
hid behind the city's walls, turning over its defense to Enkidu, that nomad beast
from the steppe,[2] and the band of hatchetmen he kept entertained with local
whores. And they did nothing—absolutely nothing! Not even ten days had passed
before Uruk found itself besieged by Agga and the men of Kish, none of whom
looked to be hanging their heads now. It was Uruk's boys whose knees turned
to water, too weak even to carry them atop the walls to look down upon such a
host. And none was more craven than Gilgamesh, lord of Kullab, whining: "Who
has heart, let him stand up, to Agga I would have *him* go."[3]

85

And so one of Gilgamesh's men, Burhurturri, took up the cudgel. Out he went through the city gates, not to fight in single combat, as we all thought, but to parley! Well, according to Zabar, who was watching from the wall, whatever Burhurturri had to say couldn't have been very convincing, since Agga had him beaten to a pulp. And still the men of Uruk did nothing, just stood around terrified. It was only when Enkidu made a halfhearted move toward the gates that Gilgamesh, realizing that things were getting out of hand, ascended the city wall, where he could negotiate without fear of being pummeled.

How his tune had changed! It was Agga this and Agga that. "O Agga, my overseer, O Agga, my steward, O Agga my army leader. . . . Thou art king hero. . . . *Thou* prince beloved of Anu."[4] And since it was hot and the men of Kish stood little chance of breaching Uruk's defenses, they left.

This noble episode Gilgamesh proclaimed a victory and a vindication of Uruk's honor. And how Enkidu and his hatchetmen lorded it over the council of elders, and how they manipulated the assembly. For those same fools who were afraid to fight Kish delivered up to Gilgamesh the powers he claimed were the prerogative of the lord of Kullab in former days. What lies!

But now people are afraid to speak the truth for fear of offending the thugs, and every day Gilgamesh and his henchmen act more and more as if they own the city. Woe unto those with long memories, for Gilgamesh will be sure to have the scribes record his own version of these matters.

Doubtless our embittered observer and probably Gilgamesh himself would be disillusioned to discover that the events that transpired at Uruk were merely incidents in a much larger drama, the culmination of forces that directed the whims and ambitions of mortals along deep-seated channels toward a destination that was largely unavoidable. Being human, historical participants would always generate some unique permutations, elements absolutely beyond prediction. But viewed from afar, all but the most distinctive would be reduced to the variability of factors in a very large and powerful equation. For these were people caught in a plant trap, packed in by outsiders, and left to operate on each other. And the chemistry they produced would largely result from the reaction of these elements, not human individuality or creativity. Within limits variables might fluctuate, producing differing social profiles. But at this level of ecological adaptation, options were circumscribed by the hard necessities of survival and human possibility thwarted by famine, disease, and fear. This is why most of us came to live like ants.

II

It amounts to a litany of disaster. Halicar was burned repeatedly. Mersin was destroyed, leaving a mass of human remains in the wreckage. Arpachiya was smashed and looted by an unknown enemy. In Azerbaijan at the site of burned Hajji Firuz II, twenty-eight massacred corpses were found. In Thrace the settlement at Tell Azmak was also burned. With growing frequency between the Late Neolithic and Early Bronze Ages, town after town in the Middle East either was destroyed, built walled fortifications, or both.[5] Whole cultures, such as the Halaf, simply disappeared.[6] Meanwhile, the archaeological representation of weapons specialized for fighting among humans, such as sling balls and especially mace heads, increased dramatically. Yet these manifest signs of violence do not seem to reveal any discernible geographic progression, occurring instead on a more or less random basis. As we have seen, most contemporary archaeological authorities attribute these developments to the maturation of conditions intrinsic to the Neolithic communities themselves.[7] It has been argued here that the kind and magnitude of the sedentary response would have been more plausibly initiated by a pattern of pedestrian pastoralist attacks culminating in long-range horse-borne raiding. But whether this ultimately proves to be the case, the nature of the response is difficult to debate.

For the appearance of walls around population centers is a mute but eloquent testimony to the need for protection and the existence of outside pressure. And circumvallation, by nucleating communities and concentrating populations, very likely accelerated and intensified the factors characteristic of social complexity.[8] As this occurred it would have encouraged the rise of so-called central places— networks of relatively evenly spaced focal points for communications, organization, and distribution.[9] Although the theory of "central places" does not demand it, local conditions—the simple fact of safety in numbers—and energy conservation would have favored the consolidation of peoples in larger settlements. Working together, these factors would have begun to produce something that began to look less like a village and more like a city.[10]

While urban centers are not an absolute prerequisite for social complexity, few would dispute that they help. Cities are the vessels of civilization, acting like reactors in a chemical factory, heating, compressing, and intensifying the process. This much is clear. But the nature and sequence of the reactions going on inside remain subjects of endless debate. For like chemists arguing over the mechanics of catalysis, archaeologists have become obsessed with the causal chain that transformed simple farmers into members of vast social hierarchies.[11] So they argue over whether it was war, which led to consolidation, which then generated huge populations, bureaucracies, and social stratification; or whether it was irrigation, which necessitated massive waterworks, which demanded large populations and literate functionaries to order them about; or perhaps the engine of populations growth propelled it all, or simple ambition, or class conflict, or religion. So it goes, an ever-growing tangle of brilliant argumentation and equally ingenious demolition.

While the debate probably serves a heuristic purpose by sharpening arguments and encouraging the reevaluation of evidence, it could ultimately fail to reveal a prime mover. For a growing understanding of systems dynamics has sensitized many researchers to the likelihood that the factors involved in social complexity are mutually reinforcing and operate in concert through a series of feedback loops.[12] Some factors are bound to have been more influential than others, but any one of the major variables still might have started the larger cycle. Thus, pristine societies around the world could have been nudged toward social complexity by one of several initiators and still ended up functioning along similar lines, as they appear to have done.

This is not meant to imply that the evolution of social complexity was entirely a closed, autonomic process, or that the role of outside forces was strictly limited to that of a pressurizing agent. Walls may have kept out marauders, but they were clearly not impermeable to cultural influences. And while the timing and impact varied, it is generally agreed that a number of exogenous innovations attributable to steppe culture—Indo-European languages and word groups, storm gods, and certain items of technology—were gradually introduced along the western littoral, to include the the Middle East, in a number of waves from the fourth to the second millennium B.C.[13] Clearly, this process was complex and is not yet fully understood. But certain mechanisms can be inferred. Conquest, as noted earlier, was one—particularly toward the end of the period. Alternately, as groups were pushed off the plains, some appear to have settled down, mixing with local pedestrian pastoralists (a process bound to quickly strip them individually of their specialized capabilities) and subsequently interacting on a regular basis with developing urban centers.[14] This would better explain the presence of outlander mercenaries such as Enkidu, the transfer of horse-related equipment without riding skills, and the continuing ignorance of true steppe culture.

But while they did play a significant role, external factors ultimately only reinforced the process that was already taking place within the urban pressure cooker. It was here that the future of humankind was being determined, and this was largely a matter of factors inherent to agriculture and sedentary existence.

III

Historically, this first took place among a group of communities that corporately would come to be known as Sumer, located on the alluvial plain formed by the Tigris and Euphrates Rivers. While well watered by these riverine sources, the plain was essentially rainless and subject to searing heat—a fragile but potentially productive environment.[15] This was important because the Sumerians appear to have originated in the northern highlands, where they had depended on rainfall-based agriculture, only moving down on the plain sometime after 5000 B.C.[16] Yet survival here would demand irrigated cultivation, which entailed the digging and maintenance of ditch networks capable of delivering life-giving water. This process probably began modestly, encouraged by the braided pattern of the Euphrates itself.[17] But one almost immediate consequence, attainable at even simple levels of agricultural technology, would have been dramatic increases in yield—approaching an order of magnitude under ideal conditions but almost certainly less here.[18] Nevertheless, given this sort of carrying capacity, it is not surprising that populations shot up, a welcome development since irrigated agriculture was inherently labor-consumptive.[19] It was this primary feedback loop of human fecundity and agricultural intensity, spinning its way though the fourth millennium B.C., that formed the basis for a new type of society.

And as the momentum of the plant trap built, it energized still other elements. Inhabiting flatlands devoid of natural defenses, while possessing large surpluses of grain, created obvious incentives to continue the construction of fortifications— and not just curtain walls around village-sized communities but massive ramparts dedicated to the protection of much bigger "central places." As these became the rule during the late fourth millennium, life in large numbers at close quarters generated dramatic transformations in demographic, social, and organizational patterns. Much of this was nascent in Neolithic life, but the acceleration of change that took place within the intensified urban environment of Sumeria constituted a true difference in kind—and not necessarily for the better.

Populations did not just grow, they took on an entirely different cast. Easily large enough now to support true endogamy, urban breeding pools were no longer obliged genetically to cooperate with neighboring communities. This in turn tended to marginalize the social safety valve inherent in the "escape option"—a development symbolically ratified by circumvallation, which held people in at the same time as it kept them out. And although the loose and pragmatic associations of the band seem to have found analogues in the growth of occupational guilds and quasi-governmental assemblies, the agricultural bias toward the nuclear family was clear in Sumerian communities.[20] Meanwhile, as Peter Wilson has explained, the impact of alternately encountering so many people on a daily basis and then retreating behind the opaque barriers of domestic privacy created a psychological environment absolutely unprecedented in our evolutionary heritage.[21] Given our extreme capacity for social stimulation, it is not surprising that the net effect appears to have been the liberation of vast and potentially dangerous amounts of energy.

But living at close quarters also had biological consequences, the introduction of an invisible threat that would have powerfully reinforced the darker side of urban existence—epidemic disease. Sedentary life of any sort exposed people to unsanitary conditions, such as garbage accumulation and water contamination, largely avoided by populations on the move. But, as historian William McNeill explains, it is only in communities of several thousand that human contacts become frequent enough to allow diseases to spread unceasingly from individual to individual.[22] A still larger population approaching five hundred thousand is required for an epidemic disease to truly perpetuate itself by bouncing from community to community. It appears that the total number of people living in the urban nodes that constituted greater Sumer was the first to reach the requisite figure, likely giving the society the dubious distinction of having become the founder of pestilence.[23]

If this was the case and Sumer did become prey to what one supplicant called "the encompassing sickness-demon, which had spread wide its wings," then the impact should not be underestimated.[24] Disease-inexperienced populations are notoriously vulnerable to the ravages of epidemic, and indirect evidence suggests that significant die-offs probably began in Sumer sometime after 3000 B.C. This would not simply have been a matter of pathology, but instead a culmination of factors inherent in the way of life that had been developing here.

For the cause of epidemic is typically advanced by famine and its handmaiden, malnutrition. And Sumer's reliance on a few cereal crops made less drought-resistant by irrigation, its vulnerability to flooding and low river water levels, and its utter dependence on a well-maintained ditch network practically ensured periodic crop failures.[25] Meanwhile, the steady intensification of cultivation combined with irrigation-induced salinization apparently led to significant and growing environmental degradation.[26] The result would have been a self-reinforcing recipe for recurrent demographic reverses—population increases leading to overcultivation, leading to famine, leading to epidemic and die-offs, leading to chronic shortages of the labor necessary to restore food production. Of course, the survival of Sumer down to the second half of the third millennium B.C. testifies to the system's capacity for recovery, but, as we shall soon see, this was not simply a matter of demographic resilience. Something new and entirely ominous had been interjected into human existence.

Mortality in such numbers and with such regularity had never before been experienced, and it would create instabilities of the most fundamental sort in large agricultural societies, particularly those focused on urban concentrations. As McNeill reminds us, cities from this point until only a bit more than a century ago were to be chronic consumers of humanity, places where deaths always outnumbered births, and fresh waves of people were required to keep the urban mechanism running.[27] Yet the implications went beyond economics and demographics; they struck at the very heart of such societies, necessitating the most profound sort of self-justification in the face of conditions that the majority might well interpret as not just thoroughly exploitative but fundamentally malign.

Quite clearly these circumstances cried out not simply for effective explanation but for organization and control. And institutionalized responses were already well under way. It is probably not accidental that the first elites to emerge in Sumer were priests, or at least those who sought to validate their claims to leadership primarily in religious terms.[28] For human beings are unique in their awareness of personal mortality, and any society that demands the subordination of individual genetic fitness in favor of primary cultural loyalty must address this issue. In an environment where mass death had become a regular occurrence, it is only logical that efforts at sanctification would have become not just concerted but institutionalized. For without an effective means of assuaging public anxiety or channeling it in constructive directions, the intensity of urban existence could easily degenerate into mass hysteria and collapse.

Not surprisingly, Sumerian ritual focused on the assurance of fertility and forestalling natural disasters.[29] Nevertheless, what seems to have differentiated Sumerian religion from previous Neolithic cults boils down to a matter of intensity and a preoccupation with death.[30] Indeed, these elements may have been responsible for the attraction in this context of sky gods and other religious inclusions that appear to have been derived from pastoral existence. For city dwellers could well have found that their lives made better sense from what amounts to a sheep's eye view—a world in which superior beings might nurture and lead mortals, only to strike them down later for reasons beyond comprehension. This, of course, must remain a matter of speculation. Yet the sheer energy devoted to religion in Sumer is far more easily demonstrated.

For the fruits of their devotion remain, even today, heaped in huge piles of rubble. Once, however, they stood tall in stepped pyramids known as *ziggurats*, giant stairways to heaven that were the dominant architectural features of the urban entities growing up around them. Ziggurats were remarkable for several reasons. First, they were prototypes of analogous nonutilitarian structures that appear with puzzling regularity in widely separated irrigation-based societies—so far separated in the case of Meso-america and South America that the possibility of imitation has been dismissed by contemporary archaeological opinion.[31] With the exception of China's boundary fortifications (shaped by sustained nomadic aggression), these structures took the general form of pyramids, and all represented fantastic outpourings of labor. So much labor—an estimated seventy-five hundred man-years in the case of the relatively modest Anu Ziggurat at Uruk[32]—that conventional attempts to explain them crumble in the face of their sheer bulk.

But, as with a number of features of early mass human societies, analogous practices among ants—in this case the penchant of certain species to build large mounds above their nests—seem, at least in some respects, relevant. For not only does the presence of so many ants living in close proximity appear to energize them, both individually and as a group, but the apparent function of the mounds as cooling mechanisms serves to underline the importance of homeostasis in such societies. In the human context, then, ziggurats can be interpreted as prototypical shrines to the vast amounts of energy stimulated by mass existence, and also to the vital importance of dispelling it before explosive levels are reached. And while the implications would have been lost on the builders, ziggurats also could have served as model organizational charts for the kind of societies that large-scale agriculture would eventually produce, locking each individual firmly in place and stacking them all in narrowing layers reaching toward a pinnacle. Although this pattern would not fully evolve in Sumer, several other key innovations in social control would originate here.

Writing was the most important. Once again, the source was organized religion, but the first application, late in the fourth millennium B.C., was not theological but economic, the recording of temple accounts and transactions.[33] It began modestly, just a series of pictographs representing various tangibles, but over some centuries the Sumerians increased the information-carrying capacity of their script through abstraction until it had evolved into true, though still prealphabetic, writing.[34] Known as *cuneiform* (Latin for wedge-shaped) it was produced by impressing wet clay with the end of a reed, leaving marks with this characteristic contour—a medium of extraordinary durability if fired, which was the fate of many of the palaces and temples where such documents were stored. Not only did this provide archaeologists thousands of years in the future with a detailed glimpse of Sumer, but it would constitute a milestone in communications and information management second only to the development of speech itself. The mass cultures of the New World demonstrated that nonliterate civilization was possible, but only at a severe cost. For writing not only transcended time and space by crystallizing thought and enabling it to be passed over substantial distances; it dramatically expanded the capabilities of the human mind by introducing a source of nonvolatile memory.

Transactions and transactors could be tracked in ways never before possible, not only validating reciprocation but opening new vistas in social control destined to be almost compulsively exploited in Sumer, first by the temples and later generally. Literally thousands of lists, labels, payrolls, and vouchers remain from the archives of Lagash and Shuruppak, revealing that virtually the whole working population would eventually find itself mustered into hierarchically administered guilds so minutely subdivided that even the snake charmers were organized.[35] Meanwhile, the keys to this invisible ziggurat would be sequestered in the minds of a select few with the time to learn the meaning and application of what amounted to between six and seven hundred unique abstractions, further segregating humanity into the fundamental ranks of the "enlightened" and "unenlightened."[36]

But even assuming that pen is more powerful than sword, regimentation of this order required the application of both. Although there is no inherent contradiction between religion and coercion, the necessity of maintaining a sense of mystery and otherworldliness generated a natural tendency toward withdrawal on the part of sacerdotal authority (witness the physical detachment of the temple complex at Uruk),[37] thereby opening the way for more purely secular forms of control. In Sumer these took essentially two forms, symbolic poles that would largely define institutional possibility in the subsequent evolution of temporal politics.

The first, the assembly, gave vent to our heritage of reaching decisions through talk and consensus. While unambiguous physical evidence of such institutions stretches back to only around 3000 B.C.,[38] they are probably much older, having developed and been formalized so slowly as to be almost undetectable. As we have seen in the tale of "Gilgamesh and Agga," assemblies at this point appear to have evolved into two separate elements—a gathering of elders and a more widely based body. In both cases membership seems to have been limited to arms-bearing males, and while their full purview remains unclear in this context, it obviously included the use of force.[39]

It is also obvious that Gilgamesh and his henchmen had little trouble overruling and manipulating the respective assemblies. This was probably not coincidental, but speaks to the very essence of governance in the type of society that was emerging in Sumer. For life in Kish, or Uruk, or Lagash, or Shuruppak required too many people to spend too much time doing too much work for government by consensus to work effectively. To keep this kind of economy going, people had to be told what to do, and punished if they refused. While assemblies were certainly manifestations of deep-seated and inexpungible qualities in human nature, the plant trap demanded something else—rulers and enforcers. James Lewthwaite probably comes close to capturing the spirit of this takeover in describing "an industrious but downtrodden peasantry unable to refuse protection money to a mob of flashily dressed racketeers for fear of having their plough oxen kneecapped, pirogues pirated, and olive trees set in cement overshoes."[40]

How was this possible, particularly in an environment where significant portions of the male population were armed and organized? In part it was a matter of will and certain economies of scale having to do with the application of force. For as Stanislav Andreski explains, it is a great deal easier for a body of one hundred to control a population of ten thousand than it is for one man to control one hundred—especially if that body knows what it wants and is willing to obey orders immediately and without question.[41] But the emergence of unitary rulers and military coteries was clearly not just a matter of leverage and determination; it very much reflected the kinds of places the city-states of Sumer had become, the social structure that was building within their ramparts.

It would be a mistake, however, to assume that Sumer was a bedraggled, entirely class-ridden society where status was a matter of birth and little else. By all accounts Sumerians were a lusty, outspoken, venturesome, even entrepreneurial people. At the top of the social hierarchy were a few great families and at the bottom slaves, but between those extremes was a very broad middle group of "freeman" whose well-being was probably largely dependent on energy and acumen.[42]

Nonetheless, this was also very definitely a place of winners and losers, and the nature of the game and its stakes are best illustrated by the fate of the majority who worked in the fields. With the exception of temple holdings, land was alienable and freely bought and sold, both in large tracts by state officeholders and in modest parcels by small agricultural producers.[43] The risks attending such transfers, combined with the hazards of raising crops in what Robert McC. Adams calls "at best an unstable eco-system,"[44] appear to have engendered something akin to a socioeconomic wheel of fortune, with some reaping great rewards and a larger number being driven to ruin and the ranks of a kind of rural proletariat made up of landless farmers and impoverished shepherds.[45] And Sumerian proverbs like "The poor man is better dead than alive" indicate that existence among the dispossessed was not kind.[46] Indeed, the life of the farmer, any farmer, was filled with toil—an endless cycle of field preparation, tilling, planting, weeding, and canal digging—a backbreaking regimen reflected in the lament "I am a thoroughbred steed, / But I am hitched to a mule / And must draw a cart, / And carry reeds and stubble."[47] Few statements better epitomize the ultimate effect of the plant trap on the human spirit. For life, from the vantage point of an irrigation ditch, was not to be enjoyed but endured.

If anything, the lot of women was worse, and certainly more ironic. For barely had the curse of infanticide been lifted before women found themselves confronted by a combination of factors that would bid to reduce them to sex objects and baby machines.

On occasion, women's rights to inherit land, hold property, and practice certain professions have been cited as evidence that their status in Sumer was, if not exactly exalted, then at least satisfactory.[48] This is deceptive, and probably meant little except at the upper echelons of society. In reality, the kind of existence that first developed among Sumerians spelled a precipitous decline in the well-being of most women. Whereas in foraging cultures and even early Neolithic communities females had played a key part in the central economic functions—either gathering or sowing and harvesting—the introduction of the plow, which placed a premium on physical strength, very much undermined the role of women in the fields.[49] Instead, they were largely relegated to domestic duties, a turn of events that was only exaggerated by nearly endless reproductivity.

For women living largely on carbohydrates rather quickly achieve and regain after pregnancy the 20 to 25 percent body fat threshold necessary for ovulation, thereby basically doubling their potential fertility over equivalent hunter-gatherers with high-protein diets.[50] The net effect, virtually continuous pregnancy, not only undermined the health of the mothers but also had a deleterious effect on child care, further contributing to the wild swings of the population profile.

To make matters still worse, the very demographic and agroeconomic conditions that were victimizing women also prompted the introduction of organized male-dominated coercion into human relations. Men had always been stronger than women and basically armed, but this was different. As Marvin Harris explains: "By dominating the armed forces, men gained control over the highest administrative branches of government, including state religions. And the continuing need to recruit male warriors made the social construction of aggressive manhood a focus of national policy in every known state and empire."[51] Indeed, the very notion that force should become the central arbiter of human affairs put women at a distinct disadvantage in nearly all spheres, forcing them to rely on sexual attraction and fertility as the basis of their influence and respect. In the Akkadian version of the Gilgamesh epic, the whore sent to lure Enkidu to Uruk is advised to bare her breasts and "treat him to a women's task."[52] This is exactly right; women were reduced to the point that the sexual act became the fulcrum of what little leverage they retained. This in turn left them captive to the system's basic demographic eccentricities.

But they were far from the only captives in Sumer. Inscriptional evidence, particularly from the archives of the Bau temple community in Lagash, indicates that slaves made up a significant element of the workforce.[53] Moreover, it is unlikely that these were locals who had slipped into this status. Rather, it appears that they were war captives, since the word for "slave" is derived from an ideogram for "foreign country."[54] This is further supported by the sex distribution of the slave population—overwhelmingly female, with the relatively few males being made up of so-called blind ones.[55] On the basis of the evidence, it is reasonable to conclude that male war prisoners—naturally the harder to control—were either slaughtered at the battle site or blinded to work as water bearers while females were all retained to be employed under semi-industrialized conditions, mostly as weavers.[56] It is also apparent that slave girls were exploited sexually and that a certain number bore children.[57]

What does this say about Sumer? Most fundamentally, it seems indicative of a system falling out of balance, one which the internal mechanisms of control—religion, administration, and governance—are incapable of righting. More specifically, there seems to have been significant labor and demographic requirements that were not being satisfied from within. As with ants and dulosis, the most potent means of addressing them was externally through mass aggression and enslavement. Nor was this an aberration, but rather a basic manifestation of the relationship between population dynamics and activity patterns characteristic of a certain level of ecological exploitation. In this respect Sumer and its network of urban pressure cookers represented a transitional stage, but one in which the necessary ignition point had been reached. And, as with certain ant species and also later, larger human empires, because the resulting societies were dynamically unstable along a number of dimensions more amenable to external than internal modulation, the practice of war became generalized. It was certainly never a precise compensator, but in addition to its obvious purposes of aggrandizement and the expropriation of scarce resources, war did find a useful place on a number of societal balance wheels. When famine struck, the stores of others could be taken. When populations rose, new territory could be conquered and excess males eliminated. When pestilence reigned, fresh laborers and breeders could be captured. And so the Second Horseman became a true team player, riding in his grandest days, as just one of four mechanics of apocalypse.

IV

The implosion-ignition-explosion process taking place within the walled cauldrons of Sumer appears to have worked in concert with the rule of "central places" to generate a multinodal structure of approximately thirteen independent city-states,[58] the dynamics of which would not only define diplomacy and war making at this time but also prove characteristic of similar systems in the future, exemplified by the ancient mainland Greeks and the Swiss cantons of the thirteenth century A.D. More plainly than these, however, the Sumerian version operated on two separate levels, featuring two sets of actors, wed by and large to unique styles of warfare— a duplex structure reflecting the political tensions and power relationships evolving within the city-states themselves. These dynamics would, in turn, largely define the geographic focus of the respective camps.

At one end of the spectrum, the emerging ruling class, the so-called lugals, or great men, and their retainers appear to have been particularly concerned with outer borders and areas external to Sumer. While the machinations of the various elites are difficult to sort out, in general they seem to have had both a defensive and an offensive component. In the case of the former, the efforts of the kings of Kish during the Early Dynastic period to suppress raiding from the East appear to have won them considerable prestige and allowed them to dominate locally.[59]

Yet there were ample motives for more aggressive operations. For besides supporting agriculture and animal husbandry, the alluvium of Sumer was largely devoid of natural resources, particularly those favored by the upper echelons of an increasingly stratified urban society—timber, base and rare metals, precious and semiprecious stones, and extra labor to work their growing landholdings.[60] In the main, these were to be found in the highlands to the north, an area inhabited by peoples living at considerably lower levels of economic and political integration. This led to a rather predictable pattern of aggression on the part of Sumerian military elites, the benefits of which served, at least in the short term, to strengthen their positions at home.[61] Consequently, it made sense, not just logistically but politically, to limit participation in such forays. Thus, as Gilgamesh planned a timbering expedition, he was careful to recruit only a small force of those personally loyal to him and without strong family ties: "Let [only] single males who would do as I, fifty, stand at my side."[62]

Despite its size, such a contingent would have to be considered a potent instrument of aggrandizement, combining superior will and organization with the almost magical edge the disease-experienced inevitably held over epidemic-naive peoples. The resulting depopulation and attendant demoralization[63] among highlanders serve, in turn, to explain how cities like Uruk, employing very small forces, were able to establish and maintain networks of enclaves capable of dominating the northern region—a military feat paralleled by the Hispanic conquest of Mesoamerica and South America. From such bases local resources and captives could be appropriated and delivery expedited through Sumerized clients, subject to intimidation of the sort exerted on the lord of Aratta by Enmerkar of Uruk: "Let the people of Aratta fashion artfully gold and silver, / Let them bring down pure lapis lazuli from the slab. . . . Let Aratta submit to Uruk, / [and] . . . bend the knee before you like highland sheep."[64] This archetype of imperial "trade"—the uneven flow of natural resources and humanity—very likely had some effect on power relationships on the alluvium, allowing recipients to consolidate both within and without. But Guillermo Algaze's notion that this was the primary factor leading to the agricultural overintensification that undermined the entire system, ecologically and then politically, seems overstated.[65] Excessive cultivation and its consequences in Sumer were largely a function of the system itself, while political collapse came about as a result of the manner in which the various urban components competed among themselves for power and resources.

It is apparent that warfare was endemic in the lowlands of Mesopotamia by the beginning of the third millennium B.C. And the logic behind it must have been direct and compelling.

As touched on earlier, aggression could be dealt with in one of three ways—flight, submission, or resistance—with the choice depending to a considerable degree on population size and the value, or at least replaceability, of the territory being occupied. In the case of foragers, the obvious course was retreat, since almost any alternative range could support their typically sparse inhabitation densities. Alternately, small agricultural centers farming at low levels of productivity, when confronted by similar but more determined groups, might either accept accommodation and amalgamation or fall back on the inconvenient but still possible option of going elsewhere. Yet large populations dependent on the very limited quantities of land suitable for intensive cultivation were faced with the stark choice between submission and the probable loss of at least some of this vital resource—or resistance using an equivalent level of violence.[66]

Given the implosion phenomenon, the latter case best approximated conditions in Mesopotamia. So it was that among the densely nucleated settlements of Sumer the pragmatic determinants of territorial behavior would have been maximized, and warfare must have jumped from town to town much like the progress of an infectious disease. While this pattern of expansion would prove typical of warfare's spread throughout most agriculture-based societies, it would have been particularly potent in such intensified environments.

Yet the form it took was interesting, for if war conquered all, this did not necessarily imply universal conquest. Quite the contrary in Sumer—once a certain level of consolidation was reached, organized conflict assumed a primary role in enforcing a balance among the system's competitors. So while each resorted to war as a consequence of internal instability, the net result at a higher level was rough equilibrium—or, put negatively, universal frustration. For military advantage could be quickly countered through alliance and opportunism—imparting a characteristic cast of amorality and perceived futility to the entire process. "You go and carry off the enemy's land; / The enemy comes and carries off your land."[67] But if diplomacy and adroitness in changing sides played an important role, sheer force remained the ultimate arbiter. Wars in Sumer were plainly frequent and bloody enough to prompt history's first articulation of deterrence theory ("The state weak in armaments— / The enemy will not be driven from its gates") and also our first public longings for peace and "a day when man abuses not man."[68]

Sumerians may have blamed each other, but the endless ebb and flow of violence would prove emblematic of balance-of-power schemes, both at the city-state level and, secondarily, among the medium-sized monarchies that would tend to form out of the feudal remnants of collapsed empires. These focal points of war would become enduring features of the international environment, casting a long shadow across history well into the modern era. But individually, multinodal political systems balanced by organized violence would prove only metastable, since each was basically composed of a pack of potential hegemonists, one of whom, in time, might succeed in outweighing all the others. Yet the alternative, forbearance and moderation, was really no alternative at all. For ultimately the participants were driven to the balance by internal distortions and were compelled to try to right them with what corporately would amount to perpetually selfish behavior.

Sumer was no different, and its martial history was as much a reflection of conditions within its cities' walls as without. The very composition of the armies and even their tactics were in key respects extensions of the social milieu. And this is not simply a matter of speculation. For a remarkable victory monument carved around 2500 B.C. and known as the Stele of Vultures has preserved for us a limestone snapshot of the Sumerian order of battle—and it reveals a fundamental distinction.

The stele depicts Eannatum, ruler of Lagash, set apart and armed for single combat, leading what amounts to an armored hedgehog.[69] For the infantry column at his back—advancing shoulder to shoulder behind a barrier of locked rectangular shields and a profusion of spears extending forward from several rows to the rear—constitutes nothing less than a full-fledged phalanx. The implications of this rather indisputable bit of evidence frequently have been overlooked by military historians, who prefer to attribute the development of this "advanced" formation to the ancient Greeks, who took it up almost two thousand years later. But, in fact, the phalanx would prove characteristic of city-state-based balance-of-power systems over a broad span of time. This indicates that it is better viewed less as a military innovation than as a very natural manifestation of its constituents' willingness to confront their enemies at close quarters and face danger in a cooperative fashion. These are qualities virtually impossible to elicit in any but highly moti-vated troops, characteristically those who perceived themselves as having a clear stake in charting the course of their polities. Thus phalanx members, besides probably sharing some kinship affinities, typically also belonged to some form of assembly, for these institutions represented two sides of the same ancient coin. Indeed, if the phalanx stood for anything psychologically, it was a willingness to fight as well as govern by consensus—a kind of lethal extension of social solidarity.

What, then, can we say of Eannatum, out front and apparently willing to take on all comers alone? Quite plainly, this is no accidental or symbolic representation. For the more fragmentary and less frequently displayed bottom panel of the Stele of Vultures once again depicts Eannatum well ahead of his massed troops, this time aboard a javelin-packed and presumably onager-drawn protochariot, holding a long spear and a sickle sword. Common sense indicates that this one-man arsenal was not intending to throw himself on the opposing phalanx from Umma—an impressive though suicidal gesture—but rather to seek out and kill their leadership in individualized combat. This is important not only because it reflected the tensions between emerging military elites and those representing broader consen-sus-based politics but also since it prefigured the future of war in the kind of society Sumer gradually was becoming. For although the phalanx was a much more powerful military instrument, it demanded a style of participatory politics that was already waning on the alluvium. Once this reached a certain point, common men would no longer be willing to fight aggressively at close quarters, leaving only leaders and their henchmen ready to wade into the fray.

Meanwhile, all indications are that the Sumerian battlefield remained a lethal and thoroughly frightening place. Victor Davis Hanson has recently reconstructed an unforgettable portrait of life and death within a Greek phalanx, which likely applies to Sumer.[70] For in any such formation, battle must be a short but vicious spasm of terror and brutality, a collision between two compressed human hedge-hogs bound to leave many in the front ranks skewered on the spears of their opposites. This was combat tailored for amateurs, more demanding of courage and the determination to push ever forward than of specialist's skills with weap-onry, a style of war guaranteed to deliver its participants back home in short order—either dead or alive.[71] And available records indicate that there were plenty of the former. For the Stele of Vultures records three thousand Ummaite battle deaths in this single episode. Nor was the carnage over. "The survivors turned to Eannatum, they prostrated themselves for life, they wept. . . . " But to no avail—he marched the Ummaites beneath their city wall and slaughtered them.[72]

Not surprisingly, the war did not end here but sputtered on for a total of 150 years, terminating only when the men of Umma avenged themselves by thoroughly destroying Lagash.[73] But the kinds of acts committed during hostilities—the con-tinuing tit-for-tat confiscation of small parcels of border land, the repeated im-position of huge grain indemnities (in one case 10,800 metric tons), and the gratuitous slaughter or enslavement of combatants and noncombatants[74]—all point to the underlying causes of conflict and the role war came to play in this environment. The logic of these events is the best evidence that Sumer was indeed caught in an eccentric sociodemographic groove, alternately generating excess mouths and the pressure to feed or eliminate them, followed by famine, pestilence, and the compulsion to alleviate the resulting labor shortages. War did all these things, not elegantly but directly and effectively. Yet any social mechanism based ultimately on pure self-interest and violence was bound to have unanticipated results. In this case—a system based on the perpetuation of the balance among constituents—it was hegemony.

It is highly symptomatic that, repeatedly during the war between Umma and Lagash, one or another of the respective cities' rulers, upon winning a significant victory, declared himself suzerain of all Sumer.[75] This was more than mere bravado. For as the succession of bit players acted out their fleeting roles in Sumer's continuing soap opera, they attached to themselves epithets—"warrior from the womb," "lion on the prowl that has no rival," and "torrent thundering against the rebellious land"[76]—leaving little doubt as to their motives and intent. This was not a group to praise or even understand the balance of power. It worked in spite of them, playing on their foibles, but ever vulnerable to a master of the game.

V

"Sargon, the mighty king, king of Agade, am I."[77] But he also wanted it remembered that he began life more humbly—the illegitimate son of a Semite girl who abandoned him, Moses-like, in a cradle of reeds, only to be fished out of the Euphrates and reared in Kish, where he quickly rose to become royal cupbearer to King Ur-Zababa. But Sumerians might well have been wary of the talented young foundling. One day he would rule them all.

Few sources dispute that the transcendence of Sargon transformed Sumerian politics, but there is considerable disagreement as to what exactly was behind it. Some tend to focus on the ethnic dimension,[78] accounting for Sargon largely in terms of a usurpation of city dwellers by Semitic Akkadians, one of those semi-nomadic "outsider" groups inhabiting the hills and fringe areas of the alluvium in what amounted to a dependent symbiosis with the towns. While the evidence certainly supports some fairly sharp distinctions between Sumerian and Akkadian—they spoke two very different languages, for instance—it is also important to remember that Sargon initially was more arriviste than revolutionary; urbanite and functionary by training if not by birth, he exploited the system from within.

His career as ruler began obscurely. But upon gaining control of Kish, Sargon marched on Uruk, where he defeated and captured the ambitious Lugal-zagge-si, who was then brought back in a dog collar and exposed at the Enlil gate.[79] This may have been insulting, but likely of far greater consequence to the citizens of

Gilgamesh's ancient home was the fact that he also destroyed their walls. Sargon did this repeatedly as he picked off town after town—Ur, then E-Ninmar, then Lagash, then Umma—leaving them naked until all the alluvium was his, an achievement he celebrated by washing his weapons in the Persian Gulf.[80]

Others before Sargon had enjoyed strings of victories, but it was the tearing down of Sumer's ramparts that truly signaled revolutionary intent. For the walls did not just represent protection from outsiders, they were the bedrock of the balance, a last refuge in defeat. By smashing them Sargon, the master player, demonstrated his determination to change the rules and also the game.

It was the middle of the twenty-fourth century B.C.,[81] and a new day had arrived in Mesopotamia. For Sargon proceeded to implement a blueprint for tyranny that would provide the basic floor plan of empires across the ages. Indeed, if there was a Despot's Hall of Fame, Sargon would have been its charter member and inspiration. His agents fanned out across the alluvium, framing the structure with tax lists, trustworthy locals, garrisons, and royal governors. But all the beams and cross members pointed inward toward Agade, the capital he built from the ground up as a monument to the new order and an apparent reminder that this was not business as usual.[82] Here he surrounded himself with imperial officials, who established residence around his huge palace complex and administered the stream of exactions that flowed into the capital.

But more critical, given the true nature of Sargon's dominion, was the standing army he kept, quite literally, at his side. For inscriptions report that fifty-four hundred soldiers ate daily at his palace.[83] Gone was the notion of warriors-by-consensus, replaced by the bloated offspring of the "lugal's" band of retainers. This was an army representative of no one but Sargon, his personal instrument of control—both internal and external. For having shorn the alluvium of its ramparts, he was compelled to pacify the hinterlands. But there was more to it than simply securing borders; Sargon waged a concerted campaign of foreign extortion that was at once a fulfillment of Sumerian protoimperialism and a model for future despots. In a relentless series of expeditions he moved both east and west, along the Tigris toward Iran and up the Euphrates in the direction of Syria and Lebanon.[84] Enemies were met and defeated; towns were sacked; the rulers of Elam, Barahshe, Mari, Iarmuti, Ibla, and the Hittites all became his vassals.[85] For a time the uplands were pacified to the point that barges laden with booty—cedar, silver, and other goods—could be floated, safe and secure, down to Agade.[86] Yet appearances deceive, and none more than the stability of empire.

For in old age, after a lifetime of consolidation, Sargon discovered his vassals all suddenly in revolt and himself under siege at Agade.[87] Once again the aging lion "made an armed sortie and defeated them, knocked them over, and crushed their vast army."[88] Still there would be other rebels to suppress before he finally died, counting thirty-four victorious campaigns to his credit but never having found substantial peace.[89]

His son and successor, Rimush, was greeted with more of the same, a general revolt, both in Sumer and among the highlanders of Elam and Warahshi.[90] Rimush would spend much of his reign traversing the rugged hills and valleys to the north and east, reestablishing his father's dominion until, after nine years, he fell prey to a conspiracy from within—assassinated either by his scribes wielding their clay tablets (a prototypical example of bureaucratic peevishness if there ever was one) or by his brother, Manishtusu.[91] He would last longer, a total of fourteen years, but spent his time similarly preoccupied with upland conquests and the acquisition of sources for silver and other metals.

Manishtusu, in turn, would be followed by his son, Naram-Sin, a ruler bearing the true stamp of Sargon. Calling himself "King of the Universe" and aspiring to deification, he filled his sixty-four years on the throne with almost nothing but military operations, all of them on the periphery. Back and forth he marched, winning victories in all directions, until he scorned even sacred oracles: "Has a lion ever performed extispicy, has a wolf ever asked [advice] from a female dream-interpreter? Like a robber I shall proceed according to my own will!"[92]

There is a famous stele, now in the Louvre, depicting Naram-Sin, armed with a composite bow and wearing a horned helmet, treading upon the bodies of the vanquished, one of whom has an arrow protruding grotesquely from his throat. Doubtless, this is exactly how he would have wanted to be remembered, and it is a telling monument indeed, for it captures the true spirit behind despotic consolidation—the arrogant application of naked force in any and all circumstances. The inclusion of horny headgear and particularly the composite bow (the first such representation) is interesting, since it not only points to the psychological and technological influence of steppe culture but makes a point about the availability of force in the imperial context. Ancient-armaments expert Yigael Yadin theorizes that the Akkadians made heavy use of such bows,[93] and this seems quite plausible. For rank-and-file infantries fielded by what Wittfogel calls "oriental despotism" (transnational empires based on irrigated agriculture) were proverbially unwilling to fight at close quarters. Unlike Sumerian phalangites, they simply lacked the motivation and sense of common purpose to advance onto what amounted to ground zero. About the best that could be done with such troops was the provision of a good long-range weapon to support the leadership as they fought it out hand to hand. And all too often these bowmen, even under slight pressure, simply melted away.

This was the fate of the Akkadian Empire. Like his grandfather, Naram-Sin in old age still faced invasion and rebellion. One document of somewhat dubious origins, the "Cuthean Legend of Naram-Sin," has him "bewildered, confused, sunk in gloom, sorrowful, exhausted," but finally still victorious.[94] It would not last. Recently discovered evidence indicates that a three-centuries-long drought commenced around this time, possibly compounding the political instability and pushing the empire beyond the point of recovery.[95] Naram-Sin's successor, Sharkali-sharri, would hang on for a few years only to disappear in a palace conspiracy, along with Agade—the only royal capital in ancient Iraq yet to be located[96]—and the political structure Sargon had doubtless meant to last forever.

It might have, had we been ants. For there is good reason to believe that from a purely materialist perspective, imperial consolidation was not only sound but, in regions suitable for massive agricultural intensification, probably inevitable. Marvin Harris, in particular, has emphasized the basic feedback loop operating between population growth and irrigation, with economies of scale driving each up to and beyond the environment's carrying capacity—more people, digging more and bigger ditches, to grow more grain, until natural disaster or simply salinization, crop failure, and disease reversed the spiral and dictated retrenchment. Short of radical and unavailable advances in agricultural technology, there was no fundamental means of stabilizing this gyrating Möbius strip.

But, as we have seen, state religions, pyramids, hierarchical bureaucracies, political unification, and, most particularly, war were all steps in this direction, a despotic apparatus so logical and necessary that it would reinvent itself with only moderate variation again and again across the face of time and geography. Cybernetically speaking, it was hardly a final solution, but it gave off an aura of permanence and under certain circumstances was capable of self-perpetuation for considerable stretches. But collapse was inevitable, frequently with shattering suddenness. And this was not simply a matter of misfortune pushing us onto the negative slope of the demographic helix, although this frequently was a contributing factor.

The key problem was with the participants. Automata with a genetic stake in the system could have kept it running and running—but we were neither. Instead, we were the most complex and intelligent of creatures, with an evolutionary history utterly contrary to the spirit of the plant trap in its imperial florescence. Self-preservation, culture, and compulsion kept us aboard the spiraling treadmill, but we didn't have to like it. Indeed, the message of our most intimate human contacts likely told us to abhor it. And divided loyalties were a shaky basis for social monoliths. So empires persisted in falling in a day, or a week, or a month, or a year—almost always violently. For war was not only an equilibrator but a terminator.

If there is a lesson here, it is that despite the importance of ecological and institutional dynamics, in human society all still rests on the shoulders of the individual. If we are really to understand from whence war sprung, we must refocus our attentions at this level.

{8}

ANATOMY
OF THE BEAST

I

Fear and curiosity had finally overcome the chills and the diarrhea, as he strained to see what was happening far across the sandy bowl. Others did the same, and their ranks involuntarily lurched forward until the overseers beat them back. Still, standing in the second row, he had a fairly good view of the unfolding panorama—lines of chariots wheeling back and forth, maneuvering for position and throwing up long, dusty rooster tails. It was nearly impossible to tell who was who, but when one of the vehicles pitched over at full speed, slinging its inhabitants forward with bone-shattering velocity, it was assumed they were enemies: "That's two less we'll have to deal with. A good omen indeed!" A wave of guffaws could be heard in the ranks.

But an hour later, any thoughts of victory had evaporated. For something had gone horribly wrong out front. The chariots eventually had come together, struggling at close range in a swirling, nearly stationary mass. Then, without warning, one group had turned and fled, racing off toward the river until they were out of sight. Ours! Now the soldiers found themselves surrounded by enemy chariots, which continued to circle them ominously. Recoiling from the orbiting vehicles and their fearsome steeds, the infantry gradually, unconsciously drew together into an ever-tightening mass. From within the overseers struggled to maintain order, alternately whipping them and urging them to use their bows. But it was impossible to see anything and soon even to move.

As he stood transfixed, thoughts of home and family drew a succession of sobs from his fever-racked chest. He never wanted to be a soldier but could not bring himself to run away when the press-gang had descended on the village. It seemed that it was his duty—to the king and to the gods, who, after all, existed to protect him. Now, instead, they had abandoned him, words so bitter they caused him to spit involuntarily. But the rage of betrayal was soon replaced by the shrieks of his comrades and a blind urge to escape. Pushed irresistibly forward, he staggered and tripped over trampled bodies. Then suddenly the stampeding mass parted, and he saw daylight! Propelled by panic and adrenaline, he dropped his bow and charged off in the direction of safety. It was a mirage. In less than a half hour he would be brought down by a charioteer's arrow and then stabbed repeatedly until the bronze spear point found his heart. The next day his head would find a place in a neat pile erected to mark the victory and to amuse his enemies.

This unfortunate fellow's end would have been neither unusual nor particularly difficult to predict. Dragooned into service, barely trained, and poorly armed, he likely would have reached the battlefield through a series of brutal forced marches, destined to undermine his vitality and conspire with a general state of malnutrition to render him prey to a range of contagious diseases, which must have thrived on the conditions under which ancient armies existed. And even if our hypothetical Bronze Age trooper had been fit as a fiddle, he still would have been little more than a tactical pawn, meanwhile risking his life and genetic future for a politico-economic system in which he had virtually no say and from which he derived relatively little benefit. This rather unremitting set of disincentives raises a question that begs to be answered not just in terms of these miserable circumstances; in many respects it underlies this entire study: Why would anyone in his right mind ever become a soldier?

II

Typically, social paradigms focus either on the institutional or on the individual, but not on both. While the integration of the two existential levels is usually either assumed or at least implicitly endorsed, it is not normally considered crucial to the credibility of the analysis. Due to the reasons outlined earlier, however, an explanation of warfare cannot be considered similarly exempt.

For as we have seen, participation in massed combat is, according to strictly biological logic, simply too risky to a nonhaplodiploid's reproductive possibilities to expect that a prospective soldier would find institutional imperatives sufficiently compelling to join the fray. In order for war to "work" among humans, there must be some source of motivation capable of transcending genetic self-interest in a manner that would lead an individual to face personal and hereditary annihilation for the sake of an organization manned largely by those who are not close relatives. Of course, the very fact that true warfare did become a persistent human institution is an obvious indication that such a source exists. And as suggested earlier, the only driving force that plausibly can be assigned such a role is culture—that revolutionary combination of hardware and software that grew so powerful among humans that it could dispute the tyranny of the gene and pose an alternative evolutionary path. While the nature and operation of the constituent cultural elements, together with their degree of autonomy from genetic equivalents, remain topics of considerable disagreement,[1] the general notion of "coevolution"—quite literally a double or at least braided path of development—remains the most promising avenue to explaining truly eusocial behavior among humans. Such a "dual-inheritance" model posits not only separate mechanisms for genetic and cultural evolution but a subtle interplay of both to produce the most efficient behavioral patterns.

How this works is illustrated by the curiously incomplete spread of adult lactose tolerance among pastoralists. For, except in low-sunlight areas where attendant vitamin D–calcium deficiencies made the drinking of fresh milk necessary, pastoralists did not need to undergo genetic changes allowing adults to absorb lactose; they simply learned how to make yogurt and other predigested milk products.[2]

While warfare does not appear to have necessitated changes in even a portion of the human gene pool, it did take shape out of an analogous but more complicated mix of behavioral preadaptations, symbiotic relationships (horses and bacterial), and purely cultural innovations. This chapter will concentrate on these elements and explore their functional relationships in an effort to better reveal this killer beast's intricate anatomy.

But before we can proceed with any degree of confidence, there remains the necessity of bridging the gap between the institutional and the individual—to provide some plausible means by which solitary humans came to believe and act on the assumption that their culture was more important than their genes. This is critical for social cohesion, since genes are essentially an individual possession to be shared with only one other person at a time, while culture is both individual and corporate, being at once internalized and capable of horizontal proliferation among any number of potential human receptors.[3] While there may be other mechanisms, it does appear that three elements—speech, religion, and the human inclination to reciprocate—played key roles in grafting the individual to his or her culture and spanning the chasm between the solitary and the corporate. Significantly, each of these critical mechanisms is based on a uniquely human capability, a situation that serves at once to explain and emphasize the singularity of our advanced sociality.

The importance of speech to our species' emergence has already been considered. But the centrality of communications, both verbal and written, to the subsequent evolution of culture deserves to be reemphasized, particularly since authorities in the area such as Boyd and Richerson and Durham have rather inexplicably overlooked it.[4] Nonetheless, it is the ability to articulate and transfer vast quantities of information that is the basis for social specialization, which in turn left ever-greater numbers of individual humans physiologically dependent upon cultural support structures. Speech knit us together in a web of chatter, enabling us to rely on one another in truly complicated ways.

And it was reciprocation that came to orchestrate these dependencies. Hymenoptera exhibit certain altruistic behaviors that resemble reciprocity but are actually tightly programmed manifestations of genetic affinity and immediate reinforcement.[5] What differentiates reciprocal acts among humans is the ability to calculate relative gain and loss with a high degree of precision and then to commit these judgments to long-term memory. While innovations such as writing and money would certainly amplify the possibilities, this fundamental human capability would delineate and regulate networks of obligation that not only were the cultural equivalents of the social insects' genetics-based associational superstructures but also opened new and entirely more complex possibilities, such as long-range trade.

But although human reciprocation is capable of almost endless elaboration and abstraction, it nonetheless appears that our commitment to the mechanism is both deep and emotional, as exemplified by the universal tendency to react with exaggerated harshness when obligations are seen as being circumvented or ignored.[6] (Witness the high mortality among card cheats in Western films.) For in a dependency-based society, the implications of untrustworthiness are quite literally life-threatening.

So, ironically, is reciprocation's negative face—revenge—which is similarly subject to iteration, in this case, cycles of violence that not only are at the heart of blood feuds among simple horticultural societies but also act as a considerable force in the prolongation and intensification of true warfare. This tendency of human conflict to persist long past the point of political rationality is one of the most difficult and dangerous aspects of warfare, since it is propelled by the very basic human urge to give back what is received. This takes place at the tactical level, with major battles emerging out of chains of retribution begun by fairly trivial acts of aggression, and on the plane of strategy and high politics, with whole armies and even countries continuing to pummel each other even when the potential gains have been reduced solely to the infliction of harm on the adversary. Indeed, at times the situation degenerates to the point that Clausewitz's famous description of war as a continuation of politics by other means is rendered entirely irrelevant, and combat becomes simply a continuation of retaliation and revenge.

Clearly, this is a problem at all levels of association, one of the law's most basic functions being the removal of retribution from the hands of individuals and the depersonalization of punishment. But with larger groups, magisterial power sufficient to short-circuit spirals of retaliation is frequently missing, so they are left to build upon themselves, generating a terrible momentum for which there is no obvious antidote. Finally, true civil war poses a special problem in this context— virtually a compounding of the syndrome—with conflict typically being initiated through an exaggerated response to a perceived failure to reciprocate and then sustained by an intensifying sequence of vengeful acts.

Religion, the third of culture's fundamental instrumentalities, would prove similarly applicable to advanced sociality and warfare. Once again, it appears that this key institution is based on a distinctive human capability—a unique understanding of death and the inevitability of individual mortality.[7] And if reciprocation serves to join the individual to the society in terms of sustenance, religion does so on the psychological plane. For it can alleviate the angst and isolation of our personal consciousness in part by addressing it at a corporate level and posing alternatives such as an afterlife. Because religion is essentially a matter of ideas, it was particularly prone to manipulation—subject only to the moral boundaries that appear to be perpetuated in our species' most basic and intimate personal interactions. Even these, however, could be circumvented by a variety of artifices such as partitioning existence into the ideal and real, the sacred and profane. So it was that religion became extraordinarily useful in legitimatizing social institutions and generating emotional dependencies on them.

This was particularly important in the military context, since warfare was preeminently a matter of life and death. Consequently, it would prove characteristic of advanced human societies that armies became closely associated with religion, to the point of waging war over matters of faith alone and, as we shall see, rationalizing particularly brutal patterns of aggression. While there was certainly some historical variation, the natural affinity between church and host provides perhaps the most transparent example of culture's power to overcome very powerful disincentives to participation and to weld a mass of individuals into a fighting whole ready to face death without question or hesitation. Of course, assuming the mantle of warrior implied a great deal more than willingness; it demanded the integration of a variety of skills, preexistent behavioral patterns, and innovations to create a Universal Soldier, perhaps more neatly stitched than Mary Shelley's mythic creation, but a composite nonetheless.

III

At the root of human warfare are a bundle of behavioral predispositions—innate patterns liable to activation with relatively little stimulation. They are still subject to very considerable cultural modification,[8] but they remain based on a behavioral package that gives every appearance of being an evolutionary product of our prior experiences as a species.

Yet the relative significance of the various components of this package is not only highly varied but in certain instances counterintuitive. For example, there is experimental evidence for an innate human fear of strangers in both infants and adults,[9] a factor that might logically be assumed to play a key role in both ethnocentrism and warfare. However, this mechanism actually operates at the individual, not corporate, level and is therefore primarily significant in group formation and maintenance[10]—likely rendering it entirely more important in holding military units together than in encouraging them to fight.

Of much more fundamental importance, in terms of behavioral raw materials for war, is the human capacity for aggression. As noted earlier, these aggressive proclivities are most comprehensible when they are simply described in terms of two separate patterns emerging from our experience as hunters and also our tendency to compete among ourselves. Because we are just one of many animals participating in such activities, these two clusters or poles of aggressive behaviors—intraspecific and predatory—are characteristic of numerous species. But in the specifically human context, the development of warfare appears to have enlisted both forms of aggression, leaving them to exist in a state of dynamic tension, so that one or the other would manifest itself to a greater or lesser degree depending on the circumstances surrounding a particular conflict.[11] Thus, the more desperate the war, the more evident activities resembling predation. Conversely, when ends were limited, means mirroring competition within the species would predominate.

As noted previously, intraspecific aggression tends to be ritualized, individualized, confrontational, male-only, and oriented around issues of reproduction, dominance and territory—not killing.[12] Indeed, these themes are reflected in the warlike behavior of a very divergent range of cultures, particularly when hostilities take place in a fairly stable political context. Thus a range of mitigating features—single combat, the adherence (rhetorically at least) to restrictive rules of engagement, general forbearance toward noncombatants and the specific exclusion of women (except as prizes), the conduct of battle on ground segregated from civilian habitation, tactics focused on the gaining and holding of territory, an emphasis on posturing and bluff, and the symmetrical (like-versus-like) use of traditional and sometimes elaborate weaponry—all find very broad and persistent application during periods of what has come to be known as limited warfare.[13] This is not meant to imply that the nature and outcome of this brand of conflict are predetermined, only that its conduct is sufficiently similar to a fairly well defined pattern of aggression to assume that an underlying relationship does exist.

Similarly, when the stakes and desperation climbed and the social and economic gulf between opponents loomed wider, warfare frequently came to be conducted in a fashion highlighting practices analogous to those associated with predatory violence. Norms of combat were ignored, surprise emphasized, living areas became subject to attack, noncombatants, including females, were regularly abused and sometimes drawn into the fighting, weapons use was asymmetrical and promiscuous, the destruction of enemy forces rather than territory was paramount, and mass slaughter was frequently practiced.[14] Here again, the orientation toward killing and the ruthless opportunism of this brand of warfare are strong indicators that the appropriate motivation was supplied in large part through the stimulation of those aggressive proclivities derived from our experience as hunters.

In particular, the excitation of this form of aggression was useful in counteracting a strong and innate human inhibition against killing members of our own species.[15] In this case a cultural rationale would intervene to prompt the aggressor that the victim was not actually human and was, therefore, subject to predation. This sort of species distancing, or "pseudospeciation," appears frequently in this brand of warfare, with racial and religious differences being particularly effective as justifications for indiscriminate death dealing. For, as Pliny the Elder recognized over nineteen hundred years ago, "A foreigner scarcely counts as a human being for someone of another race."[16] The fact that this mechanism is capable of overwhelming not only our inhibitions against killing but the reporting of our senses as to the true nature of the victims only serves to emphasize the power of cultural mediation. For it enables us to manipulate and direct behavioral routines that would lock other animals into a limited and fairly predictable range of responses. We, on the other hand, are capable of true social invention, not only redirecting and creating alternatives to innate behavior but also building upon it.

Indeed, it is just such a compound process that is responsible for the manner in which armies are structured. Groundbreaking research conducted after World War II into the factors influencing cohesion during combat showed clearly that it was founded upon bonding among small groups.[17] Soldiers consistently indicated that their primary motive in assuming risks during battle was a deeply felt desire to support and protect their "buddies." Meanwhile, the anthropologically well documented tendency of human males to join together in small groups, along with its reflection in the primary units of so many armies (even those with very large and homogeneous tactical groupings), points to an innate behavioral foundation, with the logical origin being the hunting of big game by our prehistoric ancestors.[18] For along with weapons use and speech, the kind of teamwork and unity engendered by such banding—plainly a feature of other pack animals—would have provided us with a significant competitive advantage. It makes sense that this capacity would have been selected for and passed forward, making it available for exploitation in later dangerous pursuits such as warfare. In both hunting bands and the small units of ancient armies, this cohesiveness would have been further reinforced by the likelihood of kinship ties and the limited though still powerful force of genetic altruism known technically as *inclusive fitness*.

But if this affinity within small military units can be analogized to the "strong force" operating at the subatomic level to bind nuclei, then the assembly of army-sized groupings still demanded an alternate source of attractive energy, equivalent to electromagnetism at the molecular level, to knit pods of combatants together at successively higher tiers. This force was the cultural medium known broadly as regimentation, but it operated through several instrumentalities of execution and reinforcement. For however much innate behavioral repertoires might be manipulated, armies were still highly unnatural social entities and ones whose functioning inevitably subjected them to unusually powerful entropic forces. Therefore, military cohesion always implied a healthy measure of coercion, but not exclusively so.

How this works on an individual level is most apparent during the initial transformation of a raw recruit into something approaching a fighting man, a basic training process encompassing not only the stimulation of latent propensities but also the teaching of new skills and the calculated repression of certain unwanted inhibitions. It seems likely that the widespread practice of such activities is nearly as old as organized conflict itself,[19] and if the precepts reflected in the writings of the Roman Flavius Vegetius Renatus and the so-called Seven Military Classics of ancient China are any guide,[20] military training was always based on a fundamentally shrewd and pragmatic understanding of our motives and limitations in war. At a higher level, attempts at organizational and institutional manipulation are clearly less consistent and more contingent on economic and social variables, including those of personal whim, but they remain based on a fairly uniform view of what might be useful in welding an individual to a military organization.

Judging by the limited evidence available, the intensity of training in ancient armies apparently varied widely, with most clustering at the rudimentary end of the scale. But it seems probable that drill—weapons-related exercises and, particularly, marching—played an almost universal role in welding-military forces. Indeed, William McNeill attributes a large measure of drilling's effectiveness to the primeval human penchant for dance, with shared patterns of movement performed in unison stirring a deep sense of corporate identity.[21] The evidence indicates that these rhythmic associations were further reinforced by the very widespread use of musical instruments in ancient armies, particularly drums. So widespread, in fact, that Marshal de Saxe, one of the eighteenth century's most famous commanders and military pundits, would theorize that marching in cadence literally prompted the drum's invention.[22] While this is unlikely, it serves to illustrate the significance of rhythmic movement among military practitioners.

But if drumbeats stirred the corporate soul of an army, food could be used to unify it in an even more visceral sense. To state that an army travels on its stomach is perhaps the most basic of military clichés. But it remains overwhelmingly true, and this is why, as the military saying goes, "Lieutenants discuss strategy, while generals talk logistics." For the very act of gathering together thousands of men renders them vulnerable to starvation—an obvious threat in the longer term but also a powerful lever of control.

There are several ways of provisioning an army. On one hand, troops can be left responsible for feeding themselves—either bringing food from home or, more probably, by living off the countryside. The latter, however, demands considerable mobility and presumes that the forager will, in fact, return—both conditions that presumably would have been met by bands of pastoral marauders.

But this was clearly not the case with the armies forming in the agricultural sphere. While the precise organization of military provisioning during the protohistoric period is not well understood, circumstantial evidence such as fortified storehouses along lines of advance—the Incas were particularly scrupulous in this regard—indicates that soldiers normally were fed from centralized supplies closely controlled by the leadership.[23] For if nothing else, experimental psychology has shown food to be a handy and effective reinforcer. With soldiers this was not a matter of classical conditioning but rather focused on the generation of reciprocation-based dependencies that could be used to elicit a certain degree of desired behavior. Regular rations may not have made an army brave and aggressive, but as much as anything they probably held it together.

In an agricultural environment armies could also offer at least a temporary escape from the plant trap—release from the drudgery of the fields, movement, adventure even—a package tour likely to appeal to creatures with a heritage of a far freer existence. Indeed, for those pinned to the land, unlikely to stray more than a few miles from their birthplace during the course of a lifetime, the notion of a campaign must have been heady stuff. More than anything, this probably accounts for the persistently romantic aura surrounding military life. And if armies were relentless in their regimentation of the individual and particularly unforgiving of deserters, they nevertheless did carry their constituents far and wide. If for no other reason, this must have left a place in the ranks with at least a certain measure of attractiveness.

The temporary nature of such sabbaticals should be emphasized, however. For agriculture-based armies as well as their constituents remained thralls to the broader regimen of the plant trap, free to campaign only during seasons when their labor was not demanded in the fields. Subject to variations of climate and geography, this generally meant after the harvest, when provisions were most available, and during a few other slack periods during the year. Among the agrarian, then, war's pulse was dictated by the cycle of vegetation, and its participants loaned out on a speculative basis, which served to underline the motivational ambiguity of the entire enterprise.

Once a soldier was deposited within the ranks, however, his more active participation and risk assumption could be encouraged through decorations for bravery, monetary rewards, and differential access to spoils and captured women. Unless carefully administered, however, this could easily undermine discipline and in certain instances—the battles of Megiddo and Kadesh are examples—transform armies in a matter of moments into uncontrollable bands of looters.[24] More basic still, such rewards were largely contingent upon at least some measure of victory, whereas losers, proverbially those most in need of bravery, were frequently left with little tangible to inspire them.

The limitations of food, travel, booty, and indeed all such positive reinforcers lead us back to the original observation that military regimentation ultimately demanded coercion. While there were exceptions, such as the Greek and possibly Sumerian phalanxes, ancient agriculture-based armies—particularly those representing imperial despotisms—were largely composed of unwilling participants. Retaining them and making them fight often could be accomplished only through harshness, causing individual soldiers "to fear their officers more than the perils to which they are exposed," as Frederick the Great would one day write.[25] In practical terms this must have meant frequent applications of corporal punishment and summary executions, if Herodotus' description of the Persian army and the Chinese codes of discipline propounded by Sun-tzu and Wei Liao-tzu are to be considered typical.[26] And there is little doubt that they were. Such armies remained essentially passive and notoriously brittle, precisely the kind of performance that might be expected from entities driven largely by compulsion.

Forces like these were in fact microcosms of the societies they represented, portable pyramids of power with the minority at the top possessing basically all the advantages the system could provide. In military terms this did not simply entail wide differences in treatment and rewards; it had considerable implications for a combatant's chances of survival. It has long been known that elite members of such forces were better armed and protected than their peasant counterparts. But skeletal evidence from stratified societies in both the Old and New Worlds has also shown that high-ranking individuals tended to be markedly taller (perhaps six inches) and more robust than members of the lower classes.[27] Calling to mind the larger and better-armed "soldier" castes of certain ant species,[28] these human analogues—almost certainly the product of better nourishment, particularly access to meat—would have had a similarly great advantage in battle. Not only would their superior strength and vigor have provided a tangible edge, but they also would have held the upper hand psychologically. Epics such as *Gilgamesh*, the Indian *Mahabharata*, and the *Iliad* all employ elaborate imagery emphasizing the size, strength, aggressiveness, and terrifying demeanor of elite warriors, to make the point that ordinary soldiers stood little chance of survival against them.[29] Even allowing for literary license, there is every reason to believe that this was basically the case, particularly so long as everyone believed it to be so.

This was no level playing field; indeed, it was so tilted that the fighting took place virtually on two separate planes. And although most early agriculture-based armies probably ranged from the low thousands up to around twenty thousand,[30] battles normally turned on the efforts of several hundred elite warriors. Typically, a combat environment would have consisted of opposing lines of archers (perhaps supplemented by slingers and javelin throwers) exchanging desultory fire, while champions on both sides sought each other out to wage what amounted to individualized combat. But prior to and particularly after disposing of the leadership, ordinary soldiers would be subject to promiscuous killing by the elites. Combat, then, could be seen as a highly segmented event, multiple phases combining in very different proportions the basic poles of aggression.

On one hand, the symmetrical matching of archers and of champions took place under fairly strict rules of engagement. While the passivity of the former somewhat obscured the issue, the fighting of the elite clearly reflected the broader characteristics of intraspecific aggression. Thus, in virtually all the martial epics, combatants carefully identified each other, resorted to bluff and posturing, fought in some formalized sequence with weapons that were basically identical, and periodically engaged in acts of chivalrous forbearance.[31] (Indeed, by acting as a repository for acceptable elite combat behaviors, the epics not only defined heroism but effectively transmitted its essence to future generations of aspirants.) Certainly, a case could be made that real combat was far less structured and contrived than its literary equivalent. But not only do other forms of battle reporting, chronicles and iconography, for example, tend to support the epic renditions, but the popularity and authority of these poems with the very classes that did the fighting (albeit far removed in time from the depicted events) indicate that they contained a substantial measure of truth. It also should be emphasized that epic descriptions of elite combat were frequently very gory, featuring all manner of mutilation and dismemberment.[32] On the whole, there is little pretense that members of the warrior elite were anything less than violently disposed toward one another; they were simply polite about it.

This was decidedly not the case with the elite's behavior toward social inferiors. There is every indication that common soldiers were treated virtually as prey, being killed promiscuously and in the most cold-blooded fashion. In literary terms this characteristically took the form of a kind of martial warm-up, during which a swarm of underlings were mowed down as a prelude to a meeting of major combatants. Of more historical significance, however, were the bloodlettings that appear to have taken place with some regularity in the aftermath of major engagements. There are numerous recorded examples of such slaughter, the most notorious being the purported execution of four hundred thousand after the battle of Ch'ang-p'ing (the number seems clearly exaggerated).[33] For unlike the elite, who had better means of escape and retained some value as ransom, the defeated rank and file presented the victors with few alternatives; they could be enslaved, recruited, or killed. In many instances this spelled the end for the common soldiers, for they were little more than sword and spear fodder, the designated tactical victims, whose life or death was of no more obvious consequence than their impact on the battlefield.

But if this was the case, then logically it can be asked why masses of unwilling and weakly motivated combatants were included at all. The perceived psychological impact of such a host probably played some role, but it does not appear to have been crucial given the numerous examples of small but capable forces defeating much larger ones. Rather, this question seems to strike at the heart of war's purpose among centralized, highly populated entities supported by the intensive cultivation of basically fertile environments. Essentially, it was about stabilizing the demography of societies existing at the outer edges of possibility. In times of famine and disease military forces could be used to capture new sources of labor through the expropriation of conquered manpower. In periods of overpopulation the army's role was to gain new land or to self-destruct. And systemically, these objectives were most logically addressed through large numbers of basically expendable soldiers. They were the mediums of exchange, and battles—orchestrated to produce their death, capture, and enslavement, or return with more labor and land—were key mechanisms by which energy was transferred from one political entity to another. This was the plant trap's legacy at the highest levels of intensification—human beings reduced to the level of tokens, destined to be traded freely on warfare's version of the marketplace. Of course, this was a matter of degrees. In less centralized polities, not so dependent upon economies of scale, armies tended to be smaller and soldiers better treated. Meanwhile, on the other end of the spectrum, it is notable that in political environments where power was either highly fragmented or the ecology very fragile—in medieval Europe or among the Mayas during the Classic period, for example—warfare became utterly ritualized and almost entirely a leadership activity.[34]

Clearly, there were other variables, such as the greed and ambition of rulers. But in early societies dependent upon agriculture there does seem to have been a fundamental relationship between economic and political centralization, population, and the general cast of war. And this interaction helps to explain the military fecklessness and frequency of disaster among empires with access to enormous pools of manpower and equipment. For it appears that their armies really did march to the beat of a different drummer.

IV

In some ways the profession of arms is exactly that. And so it is that an understanding of warfare's dynamics demands a grounding in the role played by the tools of the trade—weaponry. This is particularly important in the preindustrial era, since the levels of skill and equipment necessary to engage in certain kinds of combat constituted a much greater proportion of the total time and resources available for pursuits not directly related to subsistence. At a societal level, this fostered—though it did not assure—wide disparities in the type, degree, and effectiveness of armament. Add to this a very significant psychological dimension, for some classes of arms were decidedly, even decisively, more suitable for combatants exhibiting characteristic levels of motivation. Thus prudence dictated that the timid and the unwilling be armed primarily with bows or other long-range weaponry, whereas more aggressive participants could be expected to move in and fight at close quarters, where they would require instruments optimized for such purposes. As noted earlier, this dichotomy of arms and divergent degrees of commitment stretch back to Paleolithic hunters and reflects the necessity of close confrontation with heavy weapons to kill really big animals, whereas safe and ingenious devices could be used from afar, but only against smaller and less lucrative prey.

This does not mean that there was no change. Actually, the early period of warfare saw a good deal of arms innovation, more than at any time until the introduction of firearms. But it was not pursued relentlessly and in a premeditated fashion. Weapons research and development of the scale and continuity calculated to explore every avenue of destructive possibility is strictly a phenomenon of the industrial era.[35] Prior to this point, the evolution of arms better resembled the biological theory of "punctuated equilibria"—long periods of stability interrupted by interludes of dramatic change, which then prompted a spate of compensatory responses until a new systemic balance was reached.[36]

The sudden appearance of warfare among humans had just such an effect on the hitherto static nature of weaponry, and the developments that followed are best seen as moving in the direction of establishing a new equilibrium. Due, however, to the presence of an economically useful assemblage of grazing animals, particularly horses, the process in the Old World would be accelerated and complicated well beyond that which took place in the Western Hemisphere. But the result would be the same, a return to something approaching stability—in this case a stasis that would last from approximately 900 B.C. until the beginning of the Renaissance.[37] This initial period of innovation in the Old World would occur largely during three phases: one coinciding with the earliest depredations of pastoral nomads, a second taking place mostly during the third millennium B.C. as warfare spread across the face of agriculture, and a final short interlude just after 1000 B.C. conducted by the proverbially militaristic Assyrians. These phases should not necessarily be viewed as discrete, however, for each set the stage for its successor, and subsequent interactions among and between settled and nomadic folk proved critical in diffusing arms innovations and transmitting them rapidly over vast areas.

As we have seen, the pattern of violence that begins to become evident in the ruins of Neolithic communities after 5500 B.C. plausibly can be attributed to pastoralists. Initially, the reaction of the sedentary would have been almost purely defensive—the circumvallation of homesites and a reflexive reliance upon weapons traditionally used in the hunt, the most prominent being the spear and one-piece bow. Yet there was also a noticeable pattern of innovation, with more emphasis being placed on weapons particularly effective against humans—our large brains, thin skulls, and erect posture rendering us highly vulnerable to head wounds.[38] Thus caches of clay and stone sling balls repeatedly have been uncovered at Neolithic sites,[39] and it seems probable that raiders would have been met by a hail of projectiles aimed at driving them to cover and depriving them of their mobility. Still more suggestive was the appearance of the mace, which, unlike its lethal competitors, had no hunting heritage. Composed of a stone spheroid drilled though to accept a short handle, the mace was the first weapon created with humans specifically in mind, good at inflicting skull fractures and little else. Eventually it would appear among horse-riding nomads,[40] but the mace was essentially a terrestrial weapon requiring close and sustained contact. This, too, is of some significance, since it suggests that agriculturalists were beginning to fight among themselves.

Yet by the time economic and political consolidation gave birth to true armies, the mace was already disappearing—history's first example of military obsolescence.[41] Human skulls were hardly growing thicker, but they were being better protected. For both the Stele of Vultures and its companion piece, the Standard of Ur, depict massed Sumerian infantrymen wearing helmets.[42] The mace had met its match. For it has been shown empirically that a helmet fashioned of copper two millimeters thick backed by four millimeters of leather was capable of neutralizing the mace's killing power, distributing the force of a blow over an area large enough to render it relatively harmless.[43] The helmet, not the cranium, would be dented.

But the appearance of such helmets among human combatants was symptomatic of something entirely more fundamental in military technology. Beginning around 3000 B.C., the power and reach of political entities arrived at the point of becoming able to generate metal in sufficient quantities to equip forces numbering in the thousands—not totally or, except in unusual circumstances, homogeneously, but enough to initiate a significant period in the evolution of arms, what amounted to the second phase of their adaptation to the emergence of warfare. While copper was used at first, the metallurgical basis for this period of accelerated development was bronze—an alloy not only capable of being cast into complex shapes but, after cold-working, yielding weapons of a hardness and tensile strength rivaling those of iron until Roman times, when tempering came to be understood.[44]

The possibilities, though not endless, were still very considerable. For instance, the basic form of the stone dagger could be lengthened dramatically to produce a true sword, capable of delivering the slash and puncture wounds ideal for the kind of close-in combat that was to be the specialty of the elite warrior class. By the same token, the superior cutting and penetrating qualities of metal also called forth defensive reciprocals, not simply helmets but also metal and metal-reinforced torso protection, leg guards, and shields—culturally generated equivalents to the ant's exoskeleton, destined to remain virtual prerequisites of honorific combat among humans until very recent times.

Yet the influence of metal was hardly limited to the stimulation of new types of armament. Bronze points significantly increased the penetration potential of holdovers such as the spear, the bow, and the battle-ax, thereby encouraging still more metallic protection. So war's appetite for metal fed upon itself, and this in turn further encouraged political centralization and the dominance of military elites intent upon controlling the sources of supply.

Ironically, however, it was the rise of the Sumerian phalanx, an apparently homogeneously armed force, that marked the earliest example of an armywide reliance upon metal-based weaponry.[45] Slow and difficult to control, but entirely formidable once engaged, the phalanx would remain among the most dangerous and persistent of infantry formations, being resurrected as late as the fourteenth century A.D. by the Swiss.[46] But, as we have seen in the previous chapter, it was not destined to last in Mesopotamia, where the rise of despotism made foot soldiers with the requisite motivation to engage in so aggressive a fashion a thing of the past.

Instead, agriculture-based tyranny, here and elsewhere, would come to depend on armies founded on inequality and armed accordingly. This, in turn, was a matter of both tangibles and perceptions. In the case of the former, archaeological finds and particularly iconography make it clear that as a general proposition elite accoutrements were greatly superior to those of common soldiers.[47] For until the proliferation of iron, narrow access to bronze virtually assured that possession of high-metal-content implements, such as swords and body armor, remained the prerogative of the few. There was some variation, however. In Europe, major combatants tended to be equipped almost exclusively for close combat.[48] Heroic fighters in Egypt and Asia, on the other hand, would place a much greater reliance on the bow, but they too persisted in being highly aggressive and dressed in a manner that would protect them at close quarters.

But in the Old World, at least, one weapon became emblematic of practically all elites, and that was the chariot. Among the Achaean Greeks, Assyrians, Egyptians, Vedic Indians, and Chinese of the Shang and Chou dynasties—virtually across the gamut of agriculture-based militarism—the horse-drawn chariot marked the focal point of aristocratic warfare, the stage upon which its champions entered the fray, elevated above other men and propelled by the equine equivalents of conspicuous consumption. For not only were chariots themselves costly and difficult to produce, but they were inherently horse-consumptive, normally requiring a team of at least two, plus a spare.[49] Thus even a relatively modest fleet must have necessitated a very considerable investment in facilities, as attested to by the 450-stall stables unearthed at the Israelite fortifications at Megiddo.[50] For larger chariot corps, such as the Assyrians', the requisite equine establishment must have grown to truly massive proportions—all this for animals that, with rare exceptions, were never alternately employed at agricultural tasks such as plowing.[51]

But contrast this undeniable expense with the actual military capabilities of the vehicle. For detailed iconography and actual examples recovered from tombs have facilitated accurate replication, revealing a rickety, unstable vehicle, barely controllable on all but the smoothest ground. And while Egyptian and Assyrian reliefs typically depict charioteers thunderously bearing down on their intended victims, bows drawn, aim presumably unerring, further consideration raises some disconcerting questions. Just how was a warrior expected to fight from so precarious a perch—to let go with both hands, while standing on this erratically bouncing platform, and fire an arrow with any degree of accuracy? Quite probably he wasn't. For further study reveals a number of less heroic poses with chariots at rest as their occupants use their bows.[52] Or consider the circumstances in the *Iliad*, where the heroes mostly fight on foot, using their chariots simply to ferry them to and from the scene of battle. When chariots did charge, we can assume their occupants were holding on for dear life.

Obviously, this limited military utility is not easily squared with the vehicle's cost and ubiquity. Yet to understand the chariot's true importance is to understand a good deal about ancient warfare. It is interesting that only a few centuries after the invention of the wheel and well before real horses were available, the Sumerians were already experimenting with protochariots hitched to asses—the Standard of Ur (c. 2500 B.C.) depicting a line of four-wheeled battle wagons drawn by teams of quadruple onagers. But, as ancient-transportation authorities Mary Littauer and J. H. Crouwel point out, the primitive design of both vehicles and tack, combined with the notorious obstinacy of the beasts, would have ensured a slow, unmaneuverable, and highly unreliable fighting platform.[53] Once again: Why would the Sumerians have gone to so much trouble to field such a technically faulted weapons system?

Some hints are provided by evidence of the importation and introduction of true horses for warlike purposes into the agricultural sphere. Very recent information recovered from Tell Es-Sweyhat in northern Syria—a forty-three-hundred-year-old clay sculpture of a bitted domesticated stallion, along with accompanying finds of model chariots—indicates that the horse was not ridden but was used to pull such vehicles.[54] The care taken in modeling the figurine, along with the costs attached to importing and breeding animals with no obvious economic function, indicates that they were not only prized but considered militarily valuable largely in combination with chariots. If this weapon was a dud, then it could not have been obvious at the time.

For the accumulated centuries of raiding by pastoral nomads would have conditioned the agricultural world to view horses as terrifying in and of themselves, the consummate symbols of military potency. It was on this basis that chariots would have been perceived—dangerous precisely because they employed horses or, in the earlier period, animals that looked like horses. And Bronze Age battlefields would have been no place for skeptics. Picture a crowd of agricultural laborers-turned-soldiers, each armed only with a lightly strung bow and a few reed arrows, awaiting the charge of a line of chariots. Imagine the thoughts of this half-trained infantry—minions of a power structure that did little more than tyrannize them—as the chariots drew close and the horses' pitiless hooves threatened to overwhelm them. Was it any wonder that in battle after battle so many turned and ran? It was this power to terrorize, to appear to be what it wasn't, that enabled the chariot to sweep triumphant across Eurasia, embraced by far-flung aristocracies as the ideal instrument of their dominance.

In the case of the distant Chinese, however, the means of transmission were more than a little ironic. For while it remains possible that the chariot was independently developed in China, there is good evidence—rock drawings clearly depicting the vehicles at six principal sites strung across the vast expanses of central Asia—that they were brought east by pastoral nomads. Indeed, David Anthony and Nikolai Vinogradov, on the basis of chariot remains recovered from the steppe along the Russia-Kazakhstan border and dated around the same time as the Tell Es-Sweyhat model, have suggested that horse nomads actually may have invented the chariot, which then spread very quickly to the agricultural Middle East.[55] Whatever the direction of the initial diffusion, the notion of the ultimate horsemen taking up a device that existed in most places as a surrogate for horsemanship is not as implausible as it seems. For who would be more apt to appreciate a device that after all, survived on its ability to apotheosize equinity? Chariots were light and easily carried or towed, and this was a people already dependent upon wheeled transport to allow them to penetrate the deep steppe. So it seems logical that the chariot would have found a place in the portable lifestyle of the wandering nomads—an affordable luxury for leadership transport, racing, and ceremonial purposes but not as a serious weapon. For no chariot could have challenged the awesome military efficiency of the mounted archers.

But this would not have been the case in agricultural China. The evidence indicates that sometime during the Shang (founded c. 1720 B.C.), and perhaps even before, dynasts began to field chariots, which would dominate Chinese warfare until they were replaced in the third century B.C.[56] There does not seem to be any precedent for these vehicles in the form of other wheeled conveyances such as wagons or carts.[57] And not only do these chariots bear a strong resemblance to Near Eastern types, but they seem to have appeared first in the northwest, precisely where one might expect contact with pastoral nomads.[58]

While this vignette of cultural transmission is somewhat counterintuitive, it is highly symptomatic of the limits imposed upon ancient weapons technology, and the manner in which the worlds of pastoralism and agriculture interacted without real and substantial integration. For unlike today, the margins of surplus in subsistence patterns were insufficient to allow for promiscuous adoption of superior armaments. Within limits of time and space they were available, but they had to either match or at least reflect socioeconomic reality. Thus the chariot could be passed between two very separate agricultural environments by an intermediary using it basically as a novelty, yet still could end up being employed in a nearly identical fashion.

A more straightforward example of this convoluted process is provided by the very belated adoption of true mounted cavalry in the agricultural sphere. While isolated evidence of riding among the sedentary extends back to the early second millennium B.C. the Assyrians around 900 B.C. appear to have been the first agricultural people to employ mounted horsemen for purely military purposes.[59] This time lag stands as a testimony to the difficulties involved. For the requisite skills were the natural products of a nomadic existence on the steppe and hard to approximate elsewhere. But, as we shall see, no one in the ancient world was more willing than the Assyrians to invest in arms innovation, and, as prodigal charioteers themselves, they must have long been aware of the vehicle's shortcomings, especially in rough terrain.

Yet their progress with cavalry was painfully slow, as chronicled by a series of revealing reliefs. Assyrian troopers are first shown riding bareback (sitting precariously is a better description) in pairs, with one rider holding the reins of both horses while the other attempts to use his weapons.[60] Next we find them walking their mounts through rugged territory.[61] Finally, around 750 B.C. we start to see them poised more comfortably on saddlecloths above the horses' withers, operating independently with some degree of confidence in hilly terrain.[62] Yet it is interesting that Assyrian cavalry is almost never shown in true battle scenes, only in pursuit on ground inhospitable to chariots. This was to be the primary role of cavalry in ancient warfare among agricultural peoples. With some exceptions—the true professional mercenaries employed by Hellenistic monarchs and Carthage, along with the Chinese, who sent their mounted forces to live on the steppe only to see them "go native" and desert—a fundamental lack of horsemanship would always place limits on the effectiveness of this kind of force in pitched battle. Yet mounted troops would still prove extremely useful hanging at the edges, reconnoitering and harrying the survivors of broken formations to prevent them from rallying, and, even more importantly, running them down and conducting the after-battle slaughter. Isolating victims, leaving them no avenue of escape, and then killing them as prey—this was the ruthlessness of the hunt applied in practically its purest form. And its initial application by the Assyrians provides a deeper insight into their martial ethic, or lack of it.

It seems hardly accidental that the Assyrians were also the first true masters of siegecraft and perfecters of its most serviceable implement, the battering ram. For what was at this point many thousands of years, sedentary peoples in the Near East had taken refuge behind walls built around their homes and families. These ramparts provided not only physical security but a profound psychological line of demarcation between the dangers of the outside world and what amounted to inviolable territory. For walls stopped the most terrifying of outsiders, the pastoral nomads, and subsequently provided a reliable barrier against all other threats. There is some evidence that, prior to 900 B.C., fortified communities periodically fell to starvation, treachery, undermining, or even direct assault with scaling ladders and proto–battering rams—really only long poles with sharp, probably metal-covered tips.[63] But generally, attacking forces lacked reliable means of reducing masonry and stone defenses and therefore tended to bypass them. War then remained practically circumscribed—intraspecific inhibitions against the inclusion of noncombatants being buttressed by the sheer solidity of rubble and brick.

The Assyrians would have none of it. They descended upon previously safe havens "like a wolf on the fold," as Byron put it, bearing truly formidable siege engines capable of pulverizing defenses that had only frustrated others. The Assyrian battering ram would prove to be one of the most ruthless and devastating weapons in history. Introduced during or shortly before the reign of Ashurnasirpal—the same military genie who fostered mounted troops—this device was already elaborate and well developed when it first appeared in reliefs.[64] Built on massive six-wheeled wooden frames up to eighteen feet long and covered with wicker protective plates, the engine's heavy projecting beam was internally suspended like a pendulum and in a fashion that allowed it to be directed either horizontally or upward as the situation demanded. Different bits were fitted for brick and conglomerate, and there was even a turret for defensive fire.

Yet it is important to realize that these battering rams were aimed asymmetrically against noncombatants and what had previously been, for the most part, sacrosanct ground. Siege engines such as these implied something akin to total war, calculated, predatory, and utterly without mercy. Thus, after sacking one town Ashurnasirpal would exult: "Men I impaled on stakes. The city I destroyed, devastated. I turned it into mounds and ruin heaps, the young men and maidens in the fire I burned."[65] Not every power would practice siege warfare with the enthusiasm of the Assyrians, and some, like the Greeks, would even abstain for long periods. But the notion of sanctuary had been breached in a fundamental way by the very agriculturalists who had once huddled behind ramparts to escape war. Weapons like the battering ram opened the way to a day long in the future when civilians would be bombed relentlessly and targeted with nuclear weapons.

V

For the individual soldier, however, the crux of the matter was always the possibility of his own injury or death. This is what makes war among humans such a difficult proposition to explain. Yet the magnitude of that difficulty logically must turn on the nature and degree of the danger. It has been assumed here that warfare among large, highly organized agriculture-based societies was generally a very risky activity. Given ample indications of high casualties, combined with the well-documented propensity of peasants to avoid military service,[66] this appears to be a reasonable assumption. Nevertheless, it certainly bears some substantiation.

This is not easily done in absolute terms. For the scarcity and unreliability of records, the confusion of battle, and the very elusiveness of disease among the ancients all militates against arriving at estimates of war-induced mortality professing any degree of statistical precision. But the issue is clearly not unassailable; it can be approached along several avenues provided the objective remains within reach of what amount to subjective vehicles. And on these grounds it does seem that war's bad name was no accident.

From an actuarial perspective several things could happen to a newly recruited common soldier, most of them bad—he could get sick, he could be injured, he could arrive sick and spread his illness, he could be killed in battle, or he could be wounded. In all cases there was little that could be done for him, though, with the notable exception of the Roman army, this would not have been a matter of great concern for those in charge. The elite was better treated—not to mention better fed and protected—but they took greater risks and probably were wounded more often, hardly a comforting thought given the quality of care available. Thus it is possible to say that soldiering at all levels was a calling hedged in by lethal prospects.

An ancient army on the move was virtually a medical disaster in progress. And while its sources of misery were many, disease would have posed the greatest threat, both to the force itself and frequently to its adversaries. For until the arrival of modern military health care early in the twentieth century, far more soldiers were lost to contagion and infection than were ever felled by enemy weapons.[67] This would have been particularly true prior to around 500 B.C., when regional patterns of immunities finally began to be established—initially in the Near East and somewhat later in Mediterranean Europe, India, and China.[68] Until this point, however, vulnerabilies to infection would have continued as a geographic patchwork, ensuring that contagious disease remained, for the most part, epidemic, not endemic.

Pinning down exactly what were the primary killers remains problematic, however, since contagion is characterized by a continuous pattern of adjustment between host and parasite, with symptoms being subject to considerable change over time.[69] Thus paleopathologists, working almost exclusively with ancient bone samples and literary sources, continue to have difficulty equating past outbreaks with contemporary diseases.[70] Nevertheless, it is generally agreed that not only are the basic protozoan, bacterial, and viral pathogens very old—far older than humanity— but that certain conditions of existence (close association with the major domesticated animals, continual exposure to standing water in irrigation ditches, and urban crowding) were bound to have fostered characteristic classes of infections among local populations, which then were subject to further transmission.[71]

An army would have provided an almost ideal instrument: a large, compact mass of hosts under nearly continual stress, steadily infecting and reinfecting itself as it wandered about the countryside in search of physical and biological conquests—a veritable van of contagion. Repeatedly, this mechanism appears to have had an important, even catastrophic, effect on political history, with invading armies bringing about horrific losses among the bacteriologically naive. Of course, the process could be a mutual one given the condition of the aggressors. For the rigors encountered on a campaign were almost guaranteed to have steadily eroded the vitality of a fighting force, not simply ensuring the perpetuation of prevailing pathogens but also rendering it increasingly susceptible to new infections.

Regardless of its psychological compensations, the necessity to move was fundamental to the degenerative process. For marches were conducted not at the leisurely pace of lightly burdened hunter-gatherers but at a rate demanding purposeful treks exceeding ten miles a day—all the while loaded down with weapons, equipment, and even food if sufficient pack animals were unavailable. Jammed into columns compact enough to preserve some semblance of order, while traversing only an approximation of roads, soldiers would have been assaulted by clouds of dust and chronic thirst. For the Middle Eastern lowlands are nothing if not hot and dry, and in such an environment an average person must consume a minimum of three gallons of water a day just to keep going.[72] Failure to meet this requirement for even a few days in succession could literally cause an army to wither away. A related problem, almost as dangerous, was sunstroke, witnessed by the fate of Aelius Gallus' Roman force, the greater part of which perished in the Arabian Desert of heat stress in 24 B.C.[73] While disasters quite this stark were clearly the exception, they are nonetheless illustrative of the kind and severity of factors that would have led to an army's steady deterioration after more than several days on the march.

Hygienically, however, immobility was probably worse. For outside of the Romans (who were scrupulous in this regard) and the Egyptians, it was a rare ancient army that practiced even rudimentary sanitary procedures while camped,[74] thereby subjecting water and eventually food supplies to probable contamination. But these heretofore transient nests of pestilence would grow more enduring with the spread of siege warfare, immobilizing by its very nature and often devolving into stalemates determined by which force, attacking or defending, first succumbed to the relentless wastage of manpower. Indeed, from a purely medical standpoint, a battering ram might be seen as a positive factor, akin to a lancet used in draining an abscess. Of course, from a broader perspective, this is misleading and points to the necessity of considering lethality in more direct and conventional terms—those for which fighting forces were consciously intended.

For when all is said and done, ancient armies were still made up of sharp and otherwise harmful instruments along with agents more or less ready to employ them. In recent times there have been numerous attempts to determine experimentally the effectiveness of early weaponry.[75] Although the results lack consistency and would, at any rate, be highly dependent upon the skill of the users, one conclusion seems beyond dispute: all of them, even the hapless chariot, were easily capable of inflicting lethal injuries under the right circumstances.[76] As noted earlier, opportunity and the protection worn by recipients probably largely defined these circumstances.

But hard data on the nature and extent of wounds suffered in battle remain rare. One mass grave from around 2000 B.C. yielded the bodies of sixty Egyptian soldiers sufficiently preserved to reveal arrows still lodged, gaping wounds, and severe injuries from maces.[77] Literary sources, specifically martial epics, can also be used to provide a statistical glimpse of wound profiles, and from this perspective, at least, combat appears deadly indeed. For example, of the 147 wounds inflicted in the *Iliad*, all of those to the head and 77.6 percent of the total prove lethal, with swords and spears being the prime causes of mortality.[78] While these figures seem high, they are not unreasonable considering that the ancients had no effective means of dealing with either severe hemorrhaging or trauma-induced shock.[79]

Should the victim survive the immediate repercussions of his injury, the chances of subsequent infection and death were still substantial, even in the case of minor wounds. For there is good evidence that two of the major causes of death from gash-and penetration-related infection, gas gangrene and tetanus, were extant in the Bronze Age.[80] And, although the ancients did have several effective techniques for dressing wounds—honey and lint is one example—practices such as packing them in dung and binding them so as to cut off the blood supply would have been more dangerous than the original injuries in many instances. While the equivocal nature of available treatment makes it difficult to determine the true efficacy of military medicine, one thing does seem clear: in most ancient armies the supply of battle surgeons and related personnel was so limited that only the elite could have expected any care at all.

Whether this was good or bad, it does seem that the overall profile of risk that can be deduced from the conditions and contingencies outlined in this section must be considered a high one. It would be futile and misleading to reduce these indicators to some numerical estimate of attrition among ancient armies; the data are simply too limited and imprecise. But all appears consistent with the elevated levels of mortality appropriate to warfare's postulated function among very large agriculture-based societies. In conditions of overpopulation, armies act essentially as vessels of death, either conquering new land and eliminating its autochthonous population or self-destructing. Soldiers in this context are simply pawns driven to participate through cultural loyalty and coercion. But in times when numbers at home are forced below a certain threshold, their role must necessarily shift to that of procurers—finding and collecting fresh sources of humanity and herding them into the jaws of the plant trap.

VI

The Israelites were certainly the most famous captive people of antiquity, but they were hardly the only ones. In fact, large agricultural despotisms routinely used their armed forces to practice human transplantation—removing entire populations from their homelands and resettling them in far-off districts. Thus Esarhaddon would write to the god Assur of his newly acquired human resources "I filled Assyria to her borders like a quiver. . . . I distributed (them) like sheep and goats, among my palaces, my courtiers . . . and the men of Nineveh, Calah, Kakzu, [and] Arbela."[81] While it is true that the Assyrians were inveterate people-snatchers, the practice was so common among agrarian imperial tyrannies that it points to a connection to the system's basic functioning. For by this time social control and regimentation were sufficiently developed that large numbers of war captives, males and not just females, could be incorporated and put to productive use. Conquest would continue to include the periodic slaughter of the men and impregnation of the remaining women. But among very large agricultural societies, it appears that organizational capacity and demographic necessity had reached a point that war captives became more valuable for their labor and other uses than for their biological potential. Under such circumstances, the removal of large numbers and even entire populations began to make sense.

It is probably true that most prisoners of war eventually found themselves employed in the fields tending crops. Nevertheless, uses and intensity of application varied considerably. It has been suggested that in certain societies, the growth of the bureaucracy and a resulting demand for labor to work on public works projects led to the raiding of neighboring peoples.[82] Also, as an alternative to civilian applications or slaughter, Egyptian armies of the New Kingdom and others absorbed male prisoners of war into their own ranks to compensate for wastage during campaigns. Far to the east, the victorious Ch'in sought to reinforce their dynasty's imperial consolidation of China in 221 B.C. by moving 120,000 families of former enemies to their capital Hsien-yang, along with sending around 300,000 laborers to stabilize the northern border by constructing linear fortifications—both activities being supervised by the army.[83]

In the Western Hemisphere most of the major imperial entities appear to have engaged in the forceable acquisition of peoples, with some exhibiting highly exaggerated tendencies. The mysterious pyramid builders of Teotihuacán moved 80 to 90 percent of the entire Basin of Mexico's population to their capital, in what apparently amounts to history's most drastic relocation.[84] The Incas were also habitual resettlers and plainly engaged in conquest in order to expand and rationalize the labor pool.[85]

But the most bizarre and troubling case of captive taking was that of the Aztecs, whose religious ideology drove them to what amounted to endless warfare so as to obtain on the order of fifteen thousand sacrificial victims a year.[86] Obviously, they were no contribution to the labor pool, but there is good evidence that their bodies were systematically butchered and eaten, thereby entailing at least some energy transfer.[87] In any case, it is clear that captive taking was absolutely fundamental to Aztec war aims and, at least in their minds, to society's very survival; for if the gods went hungry, the cosmos was sure to collapse.[88]

If the case of the Aztecs tells us anything, it is that captive taking could be applied to a range of societal deficiencies. Conditions of servitude also varied considerably. In some cases they were quite tolerable. Among the Chinese, slavery never gained a foothold, and, after the initial imperial consolidation, forced resettlements were peopled from within, not without.[89] In other instances, however, predatory enslavement was the norm and the subjugated were worked to death.

Indeed, it is safe to say that literally the worst examples of human exploitation in history arose out of this institution. Yet its very ubiquity among large societies based on intensive agriculture indicates that this was essentially a functional phenomenon and not one of human perversity or simply "bad ideas."

If this is the case, then the practice of organized slave raiding among various species of ants is potentially relevant, and suggestive of a fundamental impulse among large societies that directly exploit the environment to seek homeostasis through the coercive acquisition of captives. A number of scholars looking at the roots of human slavery have noted similar patterns, with fundamental population instabilities and resultant labor shortages being addressed through the use of armies to acquire foreign hostages.[90] At best, however, it was only a temporary solution, a societal bandage. For, statistically at least, the captives had the last laugh by refusing to reproduce, so slave-dependent human societies were faced with the same necessity as their ant counterparts, a continuing quest for fresh prisoners. William McNeill has even suggested that the very act of military campaigning abroad further disrupted reproductive patterns at home, thereby exacerbating the population problems.[91] This, in turn, could well have the effect of locking particularly aggressive societies into a perpetual feedback loop of imperialism. Whatever the case, the quality of the labor acquired was bound to be low and the overall strategy corrosive to the emotional and psychological well-being of all participants. For humans were not ants, and the concept of slavery was antithetical to our evolutionary heritage. This "peculiar institution" would linger throughout the history of agriculture, marginally useful though ultimately self-defeating. But as soon as industrialism began to transform the very basis for our ecological dependencies, slavery would be cast from our midst in a flurry of moral indignation. For the refractory nature of our heritage as a species would never allow us to forget that we were born to be free.

VII

While it certainly appears that war at this level of ecological adaptation is not simply a matter of territorial ambition, it clearly should not be discounted as a key motivator. In general, intensive agriculture-based despotisms did tend to expand, and they did so almost exclusively through the use of armed force. Nonetheless, territorial aggrandizement had its limitations, particularly in the Bronze Age. Historian John Keegan makes the point that much of the earth's surface is in one way or another prohibitive to large-scale operations, helping to explain why so much military history transpired on the same ground.[92] Meanwhile, land suitable for the kind of agriculture upon which such regimes were based was not only scarce but usually widely separated by far less useful tracts. Essentially, these were societies dependent upon warm climates, copious amounts of water, and fertile soil—conditions found only in a small number of river valleys and lacustrine environments. Without such an environment the economic engine that ran on human fecundity, irrigation, and elevated food production simply would not run.

The conquest of marginal territory, when feasible, still may have made sense in terms of resources, captives, the establishment of buffer zones, and the potential for colonization at lower levels of cultivation, but it did not directly address the fundamental demographic disequilibriums to which such societies were subject. Nor did it imply an equivalent degree of control. Maps that display the expansion and extent of ancient empires are misleading in this regard. For much of the territory encompassed in such representations was held only in the most tenuous fashion, and frequently this was as much a matter of intent as of necessity. Often campaigns were undertaken simply to acquire tribute, with the local rulers being left in place so long as they proved reliable in handing over the requisite goods and services. Indeed, in the case of the Assyrians and Aztecs, there are strong indications that conquered territories were purposely held so loosely as to encourage revolt, thereby opening the way for further campaigns and still higher exactions. There were other variations, but the point is that expansion frequently masked other intent.

Thus territorial ambitions remained relative, a matter of local geography, demography, and various other factors relating to societal competition—an amalgam that serves to explain the rather considerable variation in the aggressiveness and appetites of various imperial entities. In the case of Egypt, for example, factors promoting isolation and internal preoccupation would predominate. Among the Chinese an initial expansion along the Yellow River, followed by a momentous leap to a second river valley, the Yangtze, would be succeeded by quiescence, punctuated only by a string of futile expansions into the steppe in hopes of precluding nomadic attack. On the other hand, Assyria would remain perpetually aggressive. Born surrounded by enemies, it would grow into a state dedicated to and eventually consumed by war and expansion. Somewhat later, improved transportation and the spread of more effective dry farming techniques along the Mediterranean littoral would open the way for still greater imperial consolidation, epitomized by the hegemony of Rome. In the New World ecological and technological peculiarities would promote its own set of territorial imperatives.

All of these societies would use war according to their own circumstances, and while the applications were clearly analogous, their very substantial variety serves to drive home the point that this was no compulsion but an institution and a finite one at that. Still, it will serve us well to look still closer at these imperial entities. For in terms of the numbers of people engaged, they constituted warfare's center of gravity for thousands of years. Meanwhile, it appears that the object of our inquiry, this mount that has spirited us across the face of recorded history, is as much beast of burden as charger—blessed with a protean anatomy, but a limited one nonetheless.

{9}

GARDEN OF
OTHERWORLDLY
DELIGHTS

I

"Be a scribe. It saves you from toil, it protects you from all manner of labor," read the characters neatly inscribed on the base of the broken pot. "Be a scribe. Your limbs will be sleek, your hands will grow soft. You will go forth in white clothes, honored, with courtiers saluting you."[1] The boy looked at the symbols—phrases he had scratched out by rote hundreds, perhaps thousands, of times—and thought seriously about their meaning for the first time within memory. For his teachers at the temple there could be no other way. "I have never seen a sculptor sent on an embassy, nor a bronze founder leading a mission,"[2] another of the hortatory practice phrases reminded him, as if he needed prompting. Only scribes were worthy of elevation—true responsibility and the good things of life.

But gradually events outside of school were teaching him otherwise. He had been with his friends when the charioteers swept into the village, on their way to fight the Asiatic dogs. As the soldiers watered their horses, he was transfixed by their splendor—their bows, their hard bodies and obvious bravery. These were Pharaoh's men, the country's protectors. They, rather than the scribes, extended the realm and gave vent to the king's desires throughout the foreign lands.[3]

Nothing made his teachers angrier. "The gods, not these popinjays in chariots, will defend Egypt, now and forever!" And for what had seemed months, the boys labored over tales of military misery—bedraggled infantrymen more dead than alive and charioteers driven into poverty by the cost of steeds and equipment.[4] But at night his grandfather told him of the Hyksos and their cruelty; how they had infiltrated the delta and gradually taken it over right under our noses. Soon the foreign princes of Avaris and the Nubians were conspiring to rule us all, and would have, had it not been for Pharaoh and our brave warriors.

"You remain unconvinced, young brother." In the midst of his musing he felt a priestly hand upon his shoulder. "I know that you still find the ways of the soldiers attractive. But it is wrong for you—and for all of us. I believe that with all of my heart."

Why did the priests and the scribes hate them so, he wondered later. It was more than jealousy, he sensed. But was not the threat of invasion real? Had it not always been so? Strangely, the boy felt their misgivings as he gazed across the river at the red cliffs in the distance. But he could see no other way, and at that moment he vowed to defend his beloved homeland, though it might cost him his life.

Although the temple officials would have had only the most skeletal knowledge of their country's origins and social history, their reservations were well founded. It was 1455 B.C. and Egypt was in the midst of change, the system's most basic readjustment in its three-thousand-year life span. For the land along the Nile was no longer insulated from war and invasion, and the necessity of dealing with this reality would lead to a substantial militarization of the regime that would alter power relationships fundamentally.

Traditionally, Egypt had been quite a different place, a venue where the role of organized conflict, though hardly nonexistent, was decidedly muted. This factor, combined with an unusually fortuitous ecology, would produce a distinctive version of agricultural despotism, a social equation in which the variables arranged themselves to compensate for war's diminished function. The result, an entirely more relaxed version of the plant trap, not only provides a useful historical counterpoint but also serves to illustrate the relativity of war and its true identity as nothing more than a social mechanism.

II

"Egypt is the gift of the Nile," pronounced Herodotus of Halicarnassus.[5] Seldom has a historian had so penetrating an insight. Virtually everything distinctive about Egyptian history can be attributed, at least in part, to the singularity of this river and its location. For the Nile's generosity transcended basic geography and extended to unparalleled offerings of fertility, sustenance, and transport. Flowing through 625 miles of desert, acting as the country's sole source of water, and carving a valley through territory sufficiently hostile to shield the country from outsiders, the Nile literally defined the boundaries of Egyptian history—not just laterally across the narrow green strip extending outward from its banks but to the south and north, the barrier rocks of the First Cataract marking the boundary with Nubia, and a point just below Cairo, where the river fans out and the delta begins, defining the traditional seam between Upper and Lower Egypt.

It was a land shaped like the long-stemmed papyrus and fed by a liquid conveyor belt. For as paleoecologist Karl Butzer notes, none of the world's rivers is more predictable and reliable[6]—or as beneficent, it should be added. Each year, beginning in mid-August, the monsoon waters of the Ethiopian highlands would reach Aswan and continue rolling gently northward, progressively breaching the river's banks, inundating the floodplain, and depositing a rich layer of silt upon it.[7] The Nile irrigated. The Nile fertilized. It even made it unnecessary to plow furrows in the fields.[8] Just plant and wait. If there could be such a thing as agriculture for the lazy, then this was it. And there was more. For the river would also double as highway and beast of burden, its current providing a free ride downstream and prevailing winds making it possible to sail effortlessly home.[9]

In short, the valley of the Nile provided the most perfect setting on earth to practice the kind of agriculture upon which early civilization was based, and by and large the long and stable existence of its inhabitants would reflect these happy conditions. This was particularly true of the first 950 or so years—the Archaic period followed by the Old Kingdom. Indeed, the very inception of Egyptian civilization appears to have taken place under circumstances considerably less intense than was the case in Mesopotamia. Quite clearly, the mechanisms of consolidation were analogous, but the pressures that drove people together were more moderate and their requirement for protection palpably less urgent.

The conventional wisdom has it that, in comparison to other ancient civilizations, a great deal is known about Egypt. Indeed, the extreme aridity of the setting and its impact as a preservative has allowed archaeologists to accumulate an unprecedented body of artifacts and information over the past two centuries of digging. But this grand heap of data is deceptive, for the great bulk of it has been recovered from tombs, whereas materials derived from actual living areas remain in short supply. In part this can be attributed to the likelihood that most occupation sites remain buried irretrievably beneath thick layers of alluvial mud and contemporary population centers.[10] But there is an alternative interpretation—that many never existed in the first place, or at least that urbanization on the scale of, say, Sumeria or China was the exception, not the rule. This possibility strikes at the heart of how civilization in Egypt got started, and the role of violence in that process.

Recent work has clarified somewhat the picture of early settlement patterns and the process by which agriculture came to the valley of the Nile. Paradoxically—considering the suitability of the environment—it appears to have taken place in fits and starts. Surprising evidence indicates that the plant-human symbiosis may have been kindled as far back as 12,000 B.C. at Isna and Naqada—sites located around the fated point where the upper Nile bows out to the east—and at Tushka in Nubia, only to fizzle out prematurely about two millennia later.[11] Not until around 5500 B.C. does hard evidence begin to indicate that agriculture had finally moved into the valley for good, a hiatus Butzer attributes to the almost profligate lushness of the environment and the resultant long-term success of hunting and gathering.[12]

At any rate, the conditions that emerged give every indication of having been a product of the same basic domestication processes that gave rise to the Neolithic elsewhere in the Old World—a deepening human-cereal symbiosis, followed by increased sedentism and then a neoteny-driven accumulation of the basic herd ruminants (sheep, goats, and cattle) plus pigs. Where exactly this occurred, locally or more probably transplanted from North Africa or the Sudan, remains somewhat unclear.[13] But in either case, the Nile and its surroundings would impose a unique cast on the results.

This would be particularly true of the pastoral component. For within the valley the pressures that normally pushed farmers and herders apart were likely mitigated somewhat by the river itself. As Butzer points out, the single annual inundation of the Nile made for a short three-to four-month growing season, leaving the fields and their surroundings free for grazing during the remainder of the year.[14] While the possibility of herd animals trampling and eating crops certainly must have existed, it would have been a relatively short-lived concern and unlikely to drive local animal keepers far afield and toward independence. It is important to re-member that, at least throughout the historical period, the valley was surrounded by desert. It would be possible for small numbers of herders to eke out a living on the fringes and among the scattered oases, but the valley itself and the delta remained congenial environments for grazing animals, particularly cattle.[15] Given these circumstances, a more relaxed and cooperative relationship between local farmers and herders can be inferred, a supposition that seems to be supported by low-density settlement patterns and a persistent absence of defensive walls in most areas.

There was an important exception, however. At the point in the river mentioned earlier, where the Nile bows out to the east, unmistakable evidence of walled fortifications was unearthed at Abadiyeh and Hierakonpolis.[16] There was even a model showing sentries peering out over a stronghold. Plainly, those inside were afraid of something. What was it? We can quickly rule out horse-borne raiders. This was a very isolated area, over 450 miles up the Nile, and the first indications of any such depredations in Egypt are arrowheads of the Scythian type dated no earlier than the sixth or seventh century B.C.[17] Still, it is difficult to dismiss the possibility of outside pressure. But from where, and by whom?

To address these questions we must step back in time and broaden our field of view. For the desert had not always been as dry as Egyptians of dynastic times knew it. During a period known as the Neolithic Subpluvial, beginning sometime between 6000 and 7000 B.C., portions of the Sahara grew sufficiently moist to support farming in certain oases, along with a relatively continuous blanket of savanna vegetation suitable for pastoralism out on the tablelands.[18] In particular, these conditions predominated in an area extending to the west and southwest of the Nile's bow. Here, a line of Neolithic sites at places such as Nabta, Bir Terfawi, and Bir Sihara have been documented, along with extensive rock drawings in the surrounding territory that depicted the herding of plainly domesticated cattle.[19]

But this Subpluvial was not destined to last, and during the fourth millennium the area gradually reverted to extreme aridity.[20] As the bloom came off the desert, survival pressures on its inhabitants must have risen, logically driving them toward the most reliable source of water remaining, the valley of the Nile—to raid and even to settle. A similar process may well have taken place among the people of the Red Sea Hills to the east. At any rate, historical times would find a penumbra of nomads hanging at the edges of the upper valley, linked to the Nilotic inhabitants in a trading symbiosis but also viewed with exaggerated fear and hatred.[21]

For it is plausible that they, or rather their ancestors, were the source of the outside pressure that drove the residents of the Nile's bow to construct fortifications and embark on a path that would lead to the political consolidation of the whole valley. This is not meant to imply that the other forces of state formation inherent in settled agriculture—task specialization, social differentiation, control of surpluses, the manipulation of religion, and so forth—were missing here, or not taking place elsewhere along the Nile. It simply seems logical that nomadic pressure in and around the river bow was more intense and would have accelerated the process, leaving the local statelets—Nagada, This, and, most important, Hierakonpolis ("Falconville")—in a more mature and militarized condition.

But there is also reason to believe that the implosive forces experienced here were basically milder than was the case in the Tigris-Euphrates basin—the kind that might be expected from pedestrian rather than horse-borne raiders. Modern Egyptologists such as Barry Kemp and Michael Hoffman persistently speak of "walled towns," using representations in stone palettes such as those of Narmer and Tjehenu, along with traces of mud brick ramparts as evidence.[22] Yet these predynastic structures are uniformly very small in appearance and can easily be interpreted as fortresses—keeps for the military elite and temporary sanctuaries during emergencies but not roomy enough to provide a home for anything like the substantial numbers thought to be living in the area at this time.[23] In other words, it seems likely that most residents were not drawn into the equivalent of an urban crucible, even in the relatively advanced politico-military environment in and around the Nile's bow.[24] It follows that this would have had an important impact on the subsequent social and political development of dynastic Egypt.

Meanwhile, between around 3500 and 3100 B.C., during a period known as the Gerzean, the process of consolidation ground forward.[25] It was plainly not peaceful, but the intermural violence was presumably confined mostly to elites. The action among the little states of the bow region would have been driven by the logic of power but rationalized in terms of order and protection from the spectral "outsiders" who appear to be depicted on the wall painting in tomb 100 at Hier-akonpolis.[26] At any rate, it was here at "Falconville" that the power struggle was apparently settled and a victor emerged, now the de facto ruler of a serpent-shaped kingdom stretching south along the river toward Nubia and north almost to Akhmim. As Kemp points out, the logic of geography did not favor the per-petuation of a multinodal "city-state" model but rather continued consolidation until the entire valley and the delta fell under one hand.[27] For the Nile's availability as an avenue of transport and communications, along with the differential matu-ration of the communities along its banks, made the notion of unifications tempting to the point of inevitability.

The impetus behind this terminal phase of growth is and will continue to be debated.[28] For there are also signs in the north of economic development and perhaps contact with nomads, particularly at Maadi. But even here it is not pos-sible to detect any significant accumulation of wealth or prestige.[29] Probably there was some northern political evolution and, by inference, even fortifications; but the weight of evidence argues that the primary force behind national amalgamation came from the south.[30] It is difficult to imagine this endgame transpiring without considerable organized violence. Indeed, the very names of the earliest Pharaohs—Scorpion, Catfish, Fighting Hawk, Serpent—bear a highly predatorial cast.[31] But subsequent history yields few signs of deep and abiding animosity. Instead, the almost instantaneous flowering of what amounted to the first truly national culture along with the people's enduring loyalty to the notion of a universal god-king argue that this was a state whose time had come, and not simply one imposed by naked force.

III

While the relatively modest role of warfare in Pharaonic Egypt down to around 2200 B.C. is frequently noted, it has been largely overlooked as a significant influ-encing factor on the type of society that emerged and thrived during this time. Nonetheless, a case can be made that the opportunities and requirements imposed by not having to regularly prepare for and wage large-scale war had an important effect upon the institutional development of civilization here. These adaptations, however, were inextricably linked to the country's location and unique environ-ment.

Gradually, it has become clearer that generalized models such as those of Witt-fogel and Marvin Harris, which assume an invariably taut relationship between demographic pressure, intensive irrigated agriculture, and political development, do not fit well with conditions in early dynastic Egypt as they are now understood to have existed. Despite the well-known representation of the Scorpion King cer-emonially cutting a watercourse, the Nile's gentle gradient was not suitable for radial canalization, such as was practiced in Mesopotamia.[32] Rather, the available evidence indicates that agricultural land use, at least until the New Kingdom, was limited to winter cropping, essentially confined to the floodplain, and remained utterly dependent upon the annual inundations of the Nile.[33]

While yields from this crude but effective system were normally bountiful, it nonetheless worked to automatically control population growth—albeit brutally.[34] For the regularity of the Nile was not absolute, but instead was punctuated by abnormally high waters and, still worse, unusually low or abbreviated crests. For the gradual rise in elevation of the land away from the banks ensured that during times of deficient flooding, substantial portions of the convex plain would remain dry and without a fresh layer of fertile sediment. The distribution of stored surpluses might carry the population through one or even two cycles, but a string of failed floods inevitably resulted in mass starvation and depopulation.[35] But it stands to reason that this periodic culling process would not normally have been compounded by epidemic disease. For the low-density settlement patterns typical of residents along the Nile would not have provided a favorable environment for the wildfire spread of contagion among those weakened by hunger. This, in turn, would have moderated the swings of the demographic curve and left Egypt with a somewhat more stable population base than more urbanized societies—a judgment supported by Butzer's estimates, which show a near-doubling of the people in the valley and the delta in the five hundred years after 3000 B.C. followed by perhaps a 20 percent drop-off down to 2000 B.C.[36] Finally, historian John Riddle points to later papyrus sources that document the use of agents to prevent conception and induce abortion. Just how effective, how widely used, or how far back such practices might have extended remains unknown, but they do at least raise the possibility of some conscious population control at this point.[37] So, if warfare was less frequent along the Nile, it was also likely to have been less called for as an agent of demographic readjustment.

Nevertheless, we are left with a very substantial population—in excess of one million—whose members, in turn, had a great deal of time on their hands. This is significant, for the Nile's gift of freedom from field preparation combined with the single-crop regimen meant that only about a third of the year had to be reserved for agricultural work. Nor were the herds that would have moved into the fields during the fallow season inherently very labor-consumptive. So what would have remained was a massive number of idle hands—hands available to do the Pharaoh's work.

The evidence indicates that the Egyptians of the upper Nile long had a fondness for richly furnished tombs,[38] but nothing on the order of what began to take place after unification. During the first two dynasties, Pharaohs favored mud brick mastabas—massive slablike superstructures built over a subterranean gallery of rooms crammed full of burial goods and sacrificed retainers.[39] While mastabas were already big—some measuring over a hundred feet on each side—they provided just a hint of future developments.

Shortly after 2700 B.C. Djoser, the first or second king of the Third Dynasty, was buried in a huge tomb constructed along the Sakkara ridge near Memphis. It was the first structure in Egypt built on truly monumental proportions and not of mud brick but stone throughout.[40] Standing in the midst of a temple complex bounded by a wall six hundred yards long and three hundred yards wide, it was a sight to behold. That, in fact, was the intent. For the original plan had called for simply another mastaba, very large but only about 25 feet high. At some point, however, the architect, Imhotep, must have realized that this structure would be rendered invisible by the surrounding walls and reacted accordingly—a radical enlargement of the project, piling mastaba atop mastaba until the result reached six steps and 180 feet tall, a giant white pyramid looming over the Memphis skyline for all to see, forever. It was a groundbreaking scheme calculated to gladden the hearts of bureaucrats, and future generations of scribes would worship Imhotep for his wisdom and foresight, remembering to sprinkle a few drops of water from their ink pads before beginning to write.[41]

For it was organization—the scribe's pen, not the overseer's whip—that was central to this and the twenty-five or so other megamortuaries that would follow during the Old Kingdom: planning, measurement, the mastery of detail, the acquisition of materials, and, most of all, the ability to gather and coordinate the vast outpourings of labor necessary to undertake construction on this scale. It was in this accumulated expenditure of effort that we can find the ultimate meaning and motivation behind these tombs—the key not necessarily to their existence but to their size.

The pinnacle would be reached at Giza with the construction of the great pyramid of Khufu, the second king of the Fourth Dynasty. It was and remains the biggest stone building ever constructed on earth. A smooth-sided pile of 2.3 million stone blocks, each averaging 2.5 tons, according to Herodotus it required the labor of at least one hundred thousand men for twenty years to construct.[42] Many think this is not a bad guess. But an analysis of the site plan is even more telling. For example, most scholars now agree that for pyramids the size of those at Giza the constituent blocks must have been dragged up huge ramps winding around the outside of the structure. These rubble inclines may have equaled two-thirds of the volume of the pyramid, all of which would have to have been removed when the job was done.[43] Add to this the necessary quarrying, grading, and site preparation, plus worker housing and support structure, and the true magnitude of the project comes more clearly into focus. For if the composition of temple gangs provides a reasonable model, then the operational code was job sharing—multiple crews brought in for very limited periods.[44] Not very efficient, but it did spread the work as widely as possible, involve as many idle hands as could be managed.

Another symptom was the treatment of the tombs once completed. For their interior decoration was as lavish as their exterior dimensions; indeed, they amounted to Bronze Age Bloomingdales, filled with the best that Egypt's craftsmen could provide.[45] And despite halfhearted efforts by the priests to protect the contents, one gets the distinct impression that grave robbery was treated with a wink and a nod. Pharaohs themselves had the embarrassing habit of stripping their predecessors' tombs of stone and decoration to build their own eternal dwellings.[46]

Tomb building during the Old Kingdom was no casual activity; next to agriculture it was probably the country's biggest industry, with not only rulers but a host of notables being interred in more modest though still elaborate circumstances. Yet the manner in which it was approached makes it appear that the key was as much in the doing as the result. The unavoidable conclusion is that these were giant make-work programs, designed to consume as much labor as possible, to bleed off, in an ultimate sense, the excess energies of so many people living under one polity—energies that under different circumstances would have been devoted to war and national survival. Indeed, much later during the New Kingdom, when Egypt was far less free from attack, tombs would remain elaborate but far less grand in scale and were constructed by a permanent cadre of workers only a fraction of the size of the labor pool that had once done the Pharaoh's work.[47]

Tombs also provide another perspective, a picture window into life along the Nile. For the whole intent behind grave goods and associated representational art was to reproduce the earthly environments of the deceased, so that they could continue their respective lives for eternity. Subject to certain religious conventions and an understandable idealization (this was supposed to be paradise), the picture that emerges appears both consistent and accurate.[48]

Egyptians of the Old Kingdom were represented as zesty, nature loving, and given to the pleasures of the flesh. And although careful examinations of mummies indicate that in reality parasites such as worms and blood flukes wrought considerable misery and brought an early end to most, it was not death but life and its prolongation that were their obsession.[49] Various forms of labor are certainly represented, but the impression conveyed is that the environment was more to be enjoyed than exploited.[50] This outlook extends to human relations, particularly intimate ones. There was an obvious joy in family and procreation. Early Egyptians were clearly intrigued by the aesthetic dimension of the human body, and usually depicted it as trim and attractive, though more modestly clad than later. And while there is no overt sex play in tomb art, symbolism indicates that this was a major preoccupation.[51] Moreover, there is the appearance of considerable equality between the sexes, with husbands and wives frequently portrayed as true partners. Other evidence indicates that this was no aberration but, in fact, reflected the status of women. This was a hierarchical society where various forms of exploitation were common, but within groupings women seem neither repressed nor oppressed, having the same legal rights and expectations of life after death as their male counterparts.[52]

All of these elements represent very long-term trends in tomb art, and are not necessarily restricted to the Old Kingdom. But there is one major discontinuity. In comparison to the New Kingdom, the vistas were remarkably pacific. Indeed, as John Romer notes, "There are very few military scenes in the tombs and temples before the Sixth Dynasty, and there is nothing about war or its 'splendours' in Old Kingdom literature.[53] Not unexpectedly, little is known about the nature and organization of the military in early Egypt. But the available evidence points to a kind of national militia raised on an ad hoc basis and commanded by what amounted to amateurs—courtiers and priests appointed for specific missions.[54] As late as the Middle Kingdom weaponry remained notably crude—little more than flint-tipped spears, axes, and lightly strung bows—and besides an occasional shield or helmet, there was nothing in the way of protection.[55]

Such a force would have been most useful in preserving internal order—putting down local rebellions and recalcitrant nobles, along with protecting, or at least trying to protect, grain stores during times of famine.[56] For as much as the Pharaonic propaganda mill might shower the people with mythological and ceremonial justifications for its legitimacy, such regimes are ultimately based on at least some ability to cow the skeptics. Then there was the lingering nomad problem, with the continued fortification of the Nile's bow indicating that this remained a hot spot to be periodically quenched.[57] Finally, there was even some expansion, the initial steps of the conquest of Nubia—a halting process that would gain momentum later, but a start nonetheless.

Taken together, however, this does not weigh very heavily on the grand scale of militarism. And if it is possible to say that early Egypt was substantially free of war and its immediate consequences, it could well follow that a number of this civilization's more profound and divergent tendencies are directly related to this happy fact. For example, the equation of peace with the manifest sensuality of early Egyptians and their respect for the female as well as the male side of human nature appears to take on real causal significance when Nilotic society is compared with more militaristic and decidedly more prudish equivalents. Similarly, there does appear to be a relationship between the early Egyptians' perception of relative safety from external attack and their looser settlement patterns. As noted previously, this would have had the effect of at least blunting the impact of epidemic disease and moderating the swings of the population pendulum, thereby reducing the usefulness of aggressive war as a demographic stabilizer. Thus the normal operation of this feedback loop would have been in the direction of peace and associated activities. Pyramid building, in this context, can be seen as an archetype of William James's "moral equivalent of war," providing not only an outlet for excess energy but also a national rallying point and a means of mass participation in a unifying activity.

If, indeed, this is the case, then the plant trap demonstrated considerable flexibility during the Old Kingdom. Organized force certainly played a role, but a relatively minor one with compensations taking place elsewhere. Given access to the basic cereal crops and domesticated animals (excepting the horse), early Egypt provided almost a laboratory for the development of hydraulic civilization under near-optimal circumstances: a rich environment, only minor nomadic pressure, and relative isolation from other settled polities. Unfortunately for the sons and daughters of Osiris, once stasis is reached, experimental conditions are generally altered. And so, near the end of the third millennium B.C. their garden of otherworldly delights would be deluged by change.

IV

Already dry, the climate grew dryer, culminating in a series of disastrously low floods (c. 2250–1950 B.C.), which many scholars now believe plunged the country into a time of instability known as the First Intermediate Period.[58] Politically, this apparently translated into a loss of equilibrium between the central court and the provinces, a condition accompanied by something approaching civil war.[59] Dynastic stability would be regained with the establishment of the Middle Kingdom, but nearly two hundred years would be required to restore the country's prosperity.[60] Even then, some argue, national self-confidence remained unhinged, almost imperceptibly so, but never to be regained.[61] At any rate, the river would again intervene, this time with a series of freakishly high floods (c. 1840–1770 B.C.) that would undermine the agricultural base of the Middle Kingdom and bring on the so-called Second Intermediate Period.[62] This time, however, the political consequences were to be more serious and persistent.

For into the resulting power vacuum was drawn a new and alien element. The Pharaohs of the Middle Kingdom had long employed foreign workers, and this influx gradually developed into a peaceful but persistent migration from Syria-Palestine into the thinly settled northeast delta.[63] With the devolution of central authority, the leadership of the newcomers—Hyksos, or "chiefs of foreign lands"—became increasingly aggressive and militant, expanding through a series of fortified towns, the most important of which was Avaris, taken over around 1725 B.C. During the next half century the Hyksos worked their way up the eastern edge of the delta toward Memphis until they established themselves as the predominant power in the lower Nile basin and set up their own dynasty, the Fifteenth.[64] Subsequently, the Hyksos would conspire with the Nubians, who also seized their independence during this confused period, urging a joint invasion. But they never succeeded in eradicating Egyptian authority in the upper Nile; nor was their spread likely to have been as cruel and violent as later portrayed. Nonetheless, the Hyksos' intervention was a deeply humiliating, even traumatizing, event in Egyptian history. They would be expelled, but never forgotten.

How had the intruders managed it? Even considering the natural disruptions in the valley, how could what must have been a very small number of active invaders cause such sustained and serious problems for a society of this size and capability for mass action? Essentially, the Hyksos' advantage must have been military, for they would have brought with them the weapons and tactics of the much more intensely warlike environment of western Asia. Egyptians clearly had had some earlier contacts with similarly armed forces, but the experience with the Hyksos was likely their first extended exposure to what amounted to a state-of-the-art Bronze Age military technology—true swords, metallic armor, and, perhaps most unnerving, chariots—all synchronized in a manner calculated to strike terror into the kind of lightly armed levies upon which the Egyptians relied.[65] Very probably the result was a string of military disasters, and in short order the Egyptians would turn to similar arms,[66] a factor that could help explain the new Theban dynasty's success in eliminating the Hyksos and reasserting Egyptian authority along traditional dimensions running from the delta into Nubia.

At any rate, the subsequent New Kingdom was a far different place than the Egypt of the Old and Middle Kingdoms. In terms of both real policy and ideology, warfare and empire became preeminent,[67] thereby providing a prime illustration of the ability of organized conflict, through contact, to jump from society to society, elevating the violence and bearing with it all the impedimenta characteristic of militarized polities at similar levels of ecological adaptation. The Pharaoh himself was now as much warrior as icon, and to join him on the battlefield there arose a professional army and the infrastructure to support it.[68] Indeed, much of the internal politics of the era probably revolved around the machinations of this new class, as witnessed by the scribes' bitter denunciations of soldiers and particularly charioteers.

A good deal of this was obviously self-serving, but behind what Kemp calls "the shadowy outline of two opposed interest groups in the New Kingdom" is a real sense that militarism was not well suited to Egyptian civilization.[69] While expansion was clearly the order of the day, most would probably agree with O'Connor that it was a "logical necessity to prevent further invasion."[70] Even in the best of times, there is good reason to believe that the legions of the Nile were never the military juggernaut the warrior Pharaohs' inscriptions would have us believe. Nubia was periodically secured, and an unstable presence was established along the west Asian littoral, but there was always an element of self-delusion attached to the New Kingdom's records of military success. As with imperial China, defeats were marked victories, fictions of conquests maintained for ideological purposes; even material received in trade was referred to as tribute.[71]

This is understandable. For looming behind the facade was the frightening fact that the country was now irrevocably caught up in the Darwinian landscape of Middle Eastern power politics. And this was a society that had evolved and thrived for fifteen hundred years under a set of conditions that did not require a heavy emphasis on the ability to wage war effectively. Besides war's infectious nature, Egypt demonstrated that hydraulic despotisms were capable of a modicum of adjustment. But there were also clear limits.

In the end Egypt's interlude as predator state was little more than a masquerade, and after a relatively brief spasm of energy the slide toward foreign domination began in earnest. Even during the New Kingdom, settlements remained essentially rural and unfortified, and therefore easy to conquer and control.[72] And when Egyptian troops were unable to match the military prowess of outsiders, shelter was sought behind the ranks of mercenary foreigners, a trend that would culminate with the ill-starred Saite dynasts. Meanwhile, the pace of the invaders picked up, punctuated by brief periods of flimsy and anachronistic national self-assertion, until finally Ptolemy and the Greeks turned the place into a giant food factory destined one day to feed Rome. It was a sad fate awaiting the inhabitants of this garden on the Nile. But in a world where war had come to balance the scales, it was the lot of those who would depend upon the kindness of strangers.

{10}

LORDS OF
EXTORTION

I

Ashurbanipal's face was an impassive mask as he dictated the letter; only the trembling in his right hand betrayed him. Kudurranu, his agent in Babylonia, was commanded to gather the scribal experts of Borsippa and to scour the city for tablets about battle, about prayer, and about "what is good for kingship." "Seek out and send to me any rare tablets which are known to you and are lacking in Assyria. . . . No one is to withhold tablets from you. And if there is any tablet or ritual which I have not mentioned to you, and you find by examination that it is good for my palace, then get it and send it to me."[1] Just as they had delivered their gold, now they would pay tribute with their rites and their memories. For these were grim times. The omens were all bad—the livers misshapen, the stars misaligned. Somehow the proper formula had to be found, before all came to ruin.

When the scribes left, Ashurbanipal, the great king, the mighty king, the king of the four quarters, the king of the universe, began to pace.[2] "O Nabu," muttered this man most feared among men. "Take my hand, guard my steps continually."[3] Yet no incantation could soothe his sense of dread. Already Egypt had slipped through his fingers, thanks to the treachery of Psammetichus, whose father he had treated as a son. And when Assyria finally abandoned Gyges, the fool tried to stand up to the Cimmerians alone and had been killed. Now Lydia, too, was lost, and still the Cimmerians roamed, lethal intruders from the steppe. There was no more frightening thought. For three generations they had ridden their mares unchecked, mocking the power of Assyria like none before. For three generations the death of Sargon, Ashurbanipal's great-grandfather,[4] had gone unavenged. Yet they could not be caught, and they would not stand and fight. Now they had been followed by Scythians, who, it was said, drank from their enemies' skulls and had driven the Cimmerians from the steppe in the first place. And this was not the end of it. For the Medes, settled perhaps and half civilized, remained marauders at heart and almost as dangerous.

"Ashur, why hast thou abandoned me?" Had he not sought and obtained a favorable oracle from Ishtar of Erbil to crush Assyria's ancient enemy, Elam? Had he not hung the head of their usurper king Teumman in his garden, and then, when they continued to disobey him, ravaged the whole country, pulling up the very bones of their former rulers and bringing them as captives to Assyria?[5] Had he not besieged the rebellious Babylonians, starving them until they ate the flesh of their sons and daughters?[6] These were things in which Ashurbanipal took pride. For he had never failed to obey Ashur and take up the sword against his enemies. But somehow the ancient rituals had become polluted, and now he found himself forsaken by the gods. This had happened before, but always Assyria had risen again. He brightened at the thought of his agents methodically harvesting the wisdom of the ancients. Surely they would locate the correct interpretations, the key to purging the corruption that had crept into the divine ceremonies. "Then the sun will shine upon us," thought the great king, "and we will be free to do Ashur's work again."

It was not to be. In short order Ashurbanipal would die, still burdened by his troubles. And fifteen years later, in 612 B.C., Nineveh, his capital, would fall to an alliance of Chaldean Babylonians, Medes, and probably Scythians. This effectively marked the end to the Assyrian Empire. Few would mourn its passing. Barely two centuries later, when Xenophon the Athenian came upon the ruins, not even the locals could tell him what it once had been.[7]

Assyria's reputation has not improved. It is little studied today, and its few defenders maintain only that it was hardly worse than its adversaries. This remains questionable, however. For the suddenness and entirety of its end, along with the intensity of the flames that consumed its cities and baked its clay tablets, have preserved for us an extraordinary record of what the Assyrians thought were their accomplishments. They left an audit trail of accelerating aggression, which reached a kind of crescendo near its end. Thus during the period between 890 and 640 B.C., corresponding roughly to the last fifth of the twelve-century span of Assyrian history, wars were fought during 180 of those years.[8] And it was not just a matter of frequency. Belligerence pervaded even the arts. Thus, the most striking and well-known examples of Assyrian sculpture, the bas reliefs with which the rulers decorated their palaces, are an almost uninterrupted cavalcade of combat, death dealing, and mutilation. Even while conceding that martial affairs have a disproportionately high profile in the physical records of such societies, Assyria must be considered an extraordinarily militaristic example. Not coincidentally, the later Assyrian kings appear to have perceived themselves as agents of an especially voracious god, Ashur, whose claims to universal rule it was their duty to enforce.[9]

Indeed, it is hard to escape the conclusion that this was a polity that came to be driven by war; that in the final stages of empire armed aggression was pursued virtually as an end in itself—primarily as a means of acquiring the tribute and human resources necessary to undertake still more campaigns.

This is highly suggestive at the theoretical level, since war alone, among the basic variables governing societies dependent on intensified agriculture, seems relatively exempt from absolute ceilings imposed by diminishing returns. And if war is less directly subject to biologically and technologically imposed upper limits than the spread of disease, food production, or population growth, it is also more under conscious human control--at least from the perspective of the aggressor. So the very factors which permitted warfare to operate as a serviceable societal equilibrator—that it could be initiated as a matter of choice—also allowed it to become an all-consuming pursuit. This was the fate of Assyria.

Yet continual aggression did not lead to greater stability. Quite the contrary, exceedingly militaristic states, such as Assyria and Mexico under the Aztecs, were chronically volatile, a habitual motive for campaigning being the frequent revolts among the subjugated. While a certain amount of this may actually have been elicited, it remains true that continual rebellion was systemically harmful and often motivated by sheer, blind hatred. For the oppressed were not simply factors in some sociopolitical model but human beings with long memories and a refractory sense of justice. Therefore, while macroparasitism[10] based primarily on mass aggression may provide a stable foundation for certain insect societies, among humans it must always remain subject to sudden and complete collapse. At the systemic level, however, it was powerful and self-reinforcing and could be perpetuated for considerable periods, as the long, if proverbially oscillating, history of Assyria testifies. In this sense war not only paid but tended to feed on itself. Yet it did not maroon us on an endless feedback loop, nor could it ever when applied to a species such as ourselves.

II

The case of Assyria is instructive, in part because it started quite differently. It began early in the second millennium B.C. as a narrow strip of territory along the Tigris River, tucked away in northern Mesopotamia. Unlike Egypt, however, Assyria was anything but isolated, and, as in all matters of real estate, location was critical. This spelled both danger and opportunity. To the north and east the country was surrounded by high mountains, the haunts of predatory elements such as the Lullubi and Guti and a conduit for raiders spilling off the Iranian plateau and the Russian steppe. West of Assyria, the plains of Jazirah stretched out for hundreds of miles, leaving the country wide-open to invading armies and foot nomads from the Syrian desert.[11]

But if Assyria was hemmed in by real and potential enemies, it also lay athwart the long-distance trade routes connecting northern Syria and particularly Anatolia with Mesopotamia. It was along these arterials that Assyrians first began to move outward, largely at the initiative of private merchants who set up trading colonies, such as the one found at Kultepe in eastern Asia Minor.[12] Thus, textual data from this earliest period deal almost exclusively with trade.[13] But while the exchange of goods might have been the impetus for this initial expansion, transmontane entrepôts and ass caravans moving tin and textiles back and forth from Ashur to Cappadocia also required protection. And this would come to be the province of government.[14]

For the geography of the homeland—its physical insecurity along with factors of climate and terrain—served to encourage political consolidation and the development of a strong military. Unlike the hot flat griddle of southern Mesopotamia, where irrigation was absolutely necessary, Assyria was a land of rolling hills and some rainfall that could support limited dry farming. And while the records clearly indicate that ditching was practiced,[15] agriculture in general would remain less intensive and productive than in the south. The area was and continued to be largely rural, with its few great cities—Erbil, Nineveh, and Ashur—being based almost purely on strategic and political considerations rather than the necessity to concentrate labor for vast hydraulic projects. From this perspective, probably the countryside's most important product was its tough, tenacious soldiers who had learned to fight while defending their villages against wave after wave of invaders.

These troops, along with the country as a whole, appear to have been united for the first time by a usurper named Shamshi-Adad in the late eighteenth century B.C., perhaps taking advantage of the end of a three-centuries-long drought, which helped to topple the Akkadian Empire and left a power vacuum in the area.[16] With characteristic Assyrian braggadocio Shamshi-Adad called himself "king of the whole world," claiming not only to have collected tribute in "the upper land" of north Syria but also to have set up a stele in "Lab'an [Lebanon] on the coast of the great sea [Mediterranean]."[17] More to the point, it is apparent from inscriptions found there that he annexed and stationed his younger son, Yasmakh-Adad, in the city of Mari along the road from southern Mesopotamia to northern Syria.[18] Quite clearly, his aim was to keep such routes open, and to do so by force. So, if it is possible to maintain that great merchant families did indeed dominate the most lucrative aspect of Assyria's international affairs at this point,[19] it is also true that the monarchy and the army rather quickly grabbed a piece of the action, establishing themselves as the enforcers—but not for long.

For, in addition to trying to control the trade routes, Shamshi-Adad and his sons did what all Assyrian kings had to do—dabble in Babylonian politics and cow the dangerous elements to the north and east. Ultimately, they failed on all counts. Not only did Hammurabi of Babylon come to dominate southern Mesopotamia, but he eventually held sway even over Mari and, presumably, the trade that went with it. Even worse, however, was the failure to stem the flow of infiltrators, particularly Hurrians, who appear to have originated in the highlands of Armenia.[20] By the time of Shamshi-Adad these Hurrians were already a strong element in northern Mesopotamia, and they would grow stronger. But exactly how remains unclear. For very soon the lights went out on the Assyrian kingdom and would stay out for the next three centuries.

When the story resumes around 1470 B.C., it found the Assyrians being dominated by a Hurrian-based aristocracy known as the Mittanni. Precisely who they were and how they got there remain issues of considerable scholarly debate, but they had horses, drove chariots, and one of their kings, Saustatar, had just looted the city of Ashur of "a door of silver and gold."[21] This was perhaps instructive, since the Assyrians themselves were destined to become among history's all-time great looters. But better times would have to wait for upwards of a century, when good fortune arrived in the form of the Hittite king Suppiluliumas, who swept through Syria and into the land of the Mittanni, sacking the city of Wassukanni and dealing the kingdom a near-fatal blow.[22] In the confusion, Ashur-uballit I successfully asserted his independence and pushed Assyria back on the imperial fast track.

Recent scholarship indicates that the process that would culminate in what has come to be known as Middle Assyria—the period down to 911 B.C.—was still, to some degree, a matter of aristocratic families and trade.[23] Yet it is also apparent that imperial authority and the role of sheer force is considerably more evident than before.[24] The most successful monarchs—Shalmaneser I (1274–1245), Tukulti-Ninurta I (1244–1208), and Tiglath-Pileser I (1115–1077)—were all essentially soldiers and their rhetoric that of the conqueror: "His royal neck I trod on with my feet, like a stool."[25] More pertinent still, the fate of Assyria rose and fell with the martial fortunes of its rulers. Thus, immediately after the father-and-son team of Shalmaneser I and Tukulti-Ninurta I managed to stabilize the restless northern and eastern highlands and then push south all the way to the Persian Gulf, their empire imploded and remained basically inert under a string of weak successors for much of the twelfth century.[26] Similarly, Tiglath-Pileser I would preside over an imperial renewal, beating back the nomads—he fought the Aramaeans twenty-eight times in fourteen years[27]—and eventually reached the Mediterranean, only to have it all go to ruin shortly after his death. So by the end of the tenth century—at the very brink of that time of hegemonic florescence known as Neo-Assyria—the country had sunk to what Georges Roux calls its "lowest ebb," surrounded, near economic collapse, and reduced to a patch barely a hundred miles long and fifty miles wide.[28]

Nonetheless, in spite of its subdued status, it also appears that the elements of transcendence were now in place, that Assyria was ready to demonstrate what war could do if given a free rein. Indeed, it is possible to go further and maintain that the forces that would cause conquest to assume the proportions of an addiction were already evident in the shape of events and society in Middle Assyria. For example, while individual Assyrians may have continued to engage in truly recip-rocal trade, at the governmental level the economic motivation behind military operations was at this point focused not primarily on secure routes but on tribute. Thus Assyriologist H. W. F. Saggs points out that Tiglath-Pileser I's northward expansion was clearly less concerned with strategy than with booty, metal, and animals, which he then sent home with the deliberate aim of increasing produc-tivity.[29]

This is suggestive. And although socioeconomic data are far less available than the comings and goings of kings, consistent signs point to problems of this nature in the Assyrian homeland, including chronic crop failures. In this context tribute can be seen as a means of reversing the domestic slide, the equivalent of restocking the pond.

This brings up another matter. As far back as the reign of Shalmaneser I, Assyria had begun shifting conquered peoples around on a scale that would eventually assume monumental proportions.[30] This generally is explained in terms of control, that the Assyrians were primarily intent on quashing the rebellious tendencies of recently conquered peoples by literally disorienting them through deportation. Saggs questions this logic, however, pointing out that the ever-bloodthirsty As-syrians could have assured themselves peace and quiet a great deal more easily through mass slaughter. Instead, he argues, their aim must have been economic stimulation through repopulation.[31]

Assyrian demographic information is admittedly slim, but what is available points to serious problems maintaining their own population. For example, first-millennium documents indicate that Assyrian villagers in the district of Harran had an average of only 1.43 children, well below the replacement rate. Other, somewhat later, data indicate a notable lack of female offspring, a possible sign of differential rearing by a society with a heavy and continuing requirement for young males to serve as soldiers.[32]

While this apparent want of little girls is based on but a single source, the evidence is much better for concluding that the overall status of women was not good. For a copy has been unearthed of Middle Assyrian laws from the time of Tiglath-Pileser I dealing specifically with relations between the sexes, and it is weighted heavily in favor of males, to say the least. In addition to garden-variety efforts to control and repress female freedom and sexuality—severe penalties for adultery and abortion, the possibility of divorce without compensation, restrictive dress for prostitutes, and so forth—the code contains an interesting conception of restitution allowing the husband or father of a women who had been beaten or raped to seek redress through treating the perpetrator's wife in the same fashion![33]

While this concept of justice basically speaks for itself, it is worth noting that this and other sources seem to reflect a pervasive suspicion of sex in general. Homosexuality, for example, was severely discouraged, with the penalty for sod-omy being castration.[34] By the same token, the frequent representation of beardless figures along with other evidence indicates that eunuchs were quite common, particularly among high officials who might have contact with palace women. Also, there is a notable absence of unclad females or the representation of any sort of sexual license in the bas-reliefs of Assyrian conquests. Thus, while mutilations of the most diverse and horrific sort are depicted in anatomical detail, captive women are shown primly dressed and in rather matronly poses.[35]

This sort of prudishness was not necessarily accidental or superficial. Psychoanalysis is hardly needed to recognize a noticeably inverse relationship between the manifest sexuality of early civilizations and the degree to which they emphasized warfare. Granted, the evidence is limited to iconography and literature, and there were clearly exceptions, but at the extremes—Assyria and the Aztecs compared with the Egyptians, for example—the contrasts are quite stark. Procreation and death are fundamental to human consciousness and are rather naturally schematized as polar opposites. Therefore, it makes sense that they would play a similar role in the deepest motivations of large societies. These themes have been explored before in the historical context, sometimes excessively. Nevertheless, it would be unwise to overlook their fundamental importance.[36] Clear evidence— body counts, if nothing else—indicates that polities can become obsessed with death dealing and that this amounts to a societal pathology. Assyria, particularly in its latter stages, gives every appearance of having fallen into this state.

Why exactly this would have occurred remains elusive, however. Certainly, there were tangible causes, social and economic grounds, along with factors of geography, that would have predisposed Assyria to repeatedly seek solutions through armed aggression. Similarly, matters of ideology and religion—hungry gods to be fed—should not be underestimated.[37] But in the last analysis, it may be that the very act of emphasizing warfare, militarism itself, had the most profound influence. For given the right conditions and in the absence of viable institutional counterweights, the open-ended nature of warfare would have encouraged practically unlimited amplification. Martial success fed upon itself, and as it did so, its central instrument grew into a veritable juggernaut, at once terrorizing the Middle East and gradually sucking the life out of the state and society. This was the Assyrian army.

III

Like much about this insatiable empire, its military forces had roots deep in the past. Yet the army's unique efficiency and functional diversity were largely products of the last three hundred years of imperial history, the era of Neo-Assyria. This is the army portrayed in the Old Testament and described in most historical accounts.[38] As a force structure it marked both a culmination of Bronze Age military trends and something entirely new. For the Assyrian army was history's first embodiment of the principle of combined arms, capitalizing with ruthless pragmatism on all known weapons and adding new ones when the opportunity arose.

This constituted a significant departure. Rather than being guided exclusively by tradition, Neo-Assyria was actively, even consciously, experimental in its approach toward the instruments of war. But this outlook did not simply mirror an unprecedented flexibility and vision; it also reflected the degree to which military power had come to subordinate competing claims on resources. For at this level of economic adaptation, weaponry of the kind introduced by the Assyrians would prove extraordinarily, even prodigally, expensive.

As with other contemporary armies of the region, the basic tactical unit consisted of foot archers, either deployed as skirmishers or massed in formation. But the Assyrians parted company with their rivals in the lengths to which they went to equip these troops, particularly those in the ranks. Theirs was no lightly strung self-bow but a carefully crafted composite weapon of extraordinary power, with its ends characteristically curled forward to resemble the bill of a duck.[39] But still more unusual was the extent of their protection. After the reign of Ashur-nasir-pal II (883–859 B.C.) concentrations of archers are shown as not only screened by shield bearers and heavily armed spearmen but themselves clad in conical helmets and long coats of mail.[40] Such expedients would have been extraordinarily costly in metal, particularly iron, which the Assyrians pioneered using on a large scale for military purposes. Thus the palace arsenal of Sargon II (721–705 B.C.) alone would yield 160 tons of the metal to archaeologists.[41] Yet the result, a stability in these formations necessary to exploit other tactical possibilities, was deemed worthy of the cost.

Meanwhile, as noted earlier, this scale of expenditure would have been exceeded, if anything, by the Assyrians' ever-growing appetite for horses—not simply to pull their progressively more numerous and capacious chariots but also, after around 900 B.C., to mount the cavalry that constituted one of their two major contributions to the arsenals of militant agriculture. Never great riders, Neo-Assyrians were great killers, and their use of the horse to pursue and dispatch survivors is emblematic of their near-total commitment to a brand of warfare aimed at the utter vanquishment of their enemies.

The other prime example is siegecraft. While the social and tactical implications of this Assyrian specialty have already been discussed, the sheer magnitude of their operations against fortified places deserves to be emphasized. Becoming the first true masters of the siege also necessitated that they become virtuosos of logistics. For the construction, delivery, and application of all the necessary impedimenta—battering rams, scaling ladders, earthen approach ramps, and undermining tunnels--implied a massive consignment of men and matériel that could stretch on for years. Yet the price would be paid again and again, since, to paraphrase the bank robber Willie Sutton, this was where the loot was.

The analogy is not necessarily fatuous or inappropriate. The steady growth in the scale and complexity of the Assyrian military was propelled in large part by the problems and possibilities of plunder—reaching the victims, overcoming their resistance, and then devouring not simply their wealth but their very populations. For the force structure grew and grew until Assyria was capable of deploying armies of up to 120,000—virtually the maximum that could be effectively controlled in the field without electronic communications.[42] Such numbers plainly exceeded the homeland's capacity to supply troops, and increasingly the army was peopled with imperial levies, often made up of recently conquered elements, even units of the hated Elamites.[43] So gradually Assyria was transformed from a state possessing an army to just an army living off an empire. And as the imperial beast grew and diversified, so did the target list from which it might be fed, along with the urgency of its appetite.

The role of ill-gotten gains in ancient warfare should not be underestimated. Heretofore, the underlying causal dynamics of population, disease, and food production have been emphasized in order to show how war meshed with the other basic functions of human society. Nevertheless, at the conscious level it does appear that the desire to expropriate was frequently the key motive for possessing and employing military forces, particularly among pastoral nomads and aggressive agricultural regimes. And this was a matter not simply of physical conquest but also of intimidation. For a large and effective military frequently allowed states to cow others and then to milk them. Indeed, if tribute can be conceptualized as a kind of way station between trade and naked theft, then it provides a useful guidepost to the historical path trod by the Assyrians. For they began as traders and ended as lords of extortion, driven to that role by the insatiable hunger of an army that came to exist as virtually an end in itself.

IV

Despite the general impression of Neo-Assyria as a peak period, a good case can be made that it was, in fact, a time of devolution. Seen in this light, the spasmodic enlargement of the area under Assyrian control is best viewed as a desperate and continuing effort to right a polity already fatally out of balance. Indeed, applying the very term *empire* may be somewhat misleading, since it implies to the modern reader an element of coherence and economic integration that was frankly missing for most of the period. Rather, Neo-Assyrian rulers continue to give the impression of having treated the surrounding territories as a kind of hunting preserve stocked with victims perceived as hosts or simply prey, depending upon the circumstances.[44]

Inscriptional evidence at this point comes to be dominated by excruciatingly detailed tribute lists, whereas specific references to commercial transactions virtually disappear.[45] This could be a function of different accounting mediums—durable clay tablets versus parchment, for example—but a more plausible explanation is that the meticulous records of incoming booty now constituted the national balance sheets.

Military expeditions were still undertaken for strategic reasons, but they also resembled the comings and goings of gangsters collecting protection money—payments were made, feet were kissed, warnings were issued to the delinquent, beatings were administered, and worse inflicted upon the rebellious. Considered in terms of management, however, the Assyrian ruling hand was a loose one, so loose, in fact, that the suspicion arises that behind lax procedures was an unspoken intent of encouraging revolt as a pretext for squeezing the perpetrators all the harder. It is true that the problems and dangers engendered by continual insurrection finally forced the Assyrians to begin annexing and directly administering satellite states during their last imperial century and a quarter.[46] But prior to this point and down to the end in outlying districts, the Assyrians give the distinct impression of having subscribed to the notion that in chaos there is profit.

It would have been difficult to argue with the results—a mesmerizing stream of gold, silver, copper, iron, horses, grain, and practically every other commodity under the sun, all of it flooding yearly into the royal centers to be meticulously registered by an army of scribes and disbursed by the bureaucrats, who had displaced the great merchant families in the power structure.[47] In certain instances the quantities were truly breathtaking—Metenna of Tyre was forced to pay Tiglath-Pileser III 150 talents of gold, whereas Sargon II relieved the city of Musasir of more than five tons of silver and over a ton of gold in 714 B.C.[48] Plainly, these were unusually large hauls, but the memorable ten-day housewarming Ashurnasirpal II threw for nearly seventy thousand guests in 879 B.C. to commemorate the completion of his new palace serves to illustrate in still more unmistakable terms the scale and success of Neo-Assyrian imperialism.[49] Yet it also points to its profligacy and fundamental unsoundness.

Beginning around 750 B.C., deportations began to take on truly massive proportions.[50] This accelerating process—a relentless dragnet that would eventually ensnare an estimated total of between four and five million souls during the final three centuries of Assyrian history[51]—was on a scale indicative of a good deal more than political motivation; it resembles instead a desperate attempt to shift the ballast in the floundering, top-heavy ship of state. Ironically, social and political derangement was antithetically reflected by trends in the homeland that reduced freeholders to the status of serfs, thereby tying them to the land and preventing escape from chronically depressed circumstances. But neither deportation nor detention was likely to reverse the slide and revitalize a center that long had been sacrificed in favor of external contingencies and an imperialist superstructure far too heavy to support except by continued conquest.

Yet endless aggression carried with it the seeds of eventual defeat and disaster. For one thing, the Assyrian recipe for victory was no secret to its victims, and centuries of success inevitably led to imitation. Thus, as ferrous metallurgy spread, mail, helmets, and particularly iron-bladed swords became increasingly common throughout the region. Specialization also flourished, as new forms of arms were consciously assimilated by other armies. Consequently, upon the formation of the central kingdom of Israel, Samuel declared that a chariot squadron might be established for the first time.[52] And this was but one example of a process that, weapon by weapon, specialty by specialty, eroded the Assyrian qualitative advantage.

Perhaps of equal concern, the legions of her enemies were growing more numerous. As Nadav Na'aman notes, Neo-Assyrian thrusts westward in the ninth and eighth centuries B.C. were increasingly met by coalitions, which gradually enlisted broader participation and more far-flung powers.[53] Frequently, small Phoenician and Palestinian states had little choice but to join one side or the other—a process that actually brought the Assyrians a certain number of allies.[54] But the net effect was to increase the scale of warfare and render adversaries steadily stronger, while placing still more pressure on the Assyrian resource base and making further lucrative conquests all the more imperative.

This addiction to war would have particularly disastrous consequences when applied to the rival mountain kingdom of Urartu. For this state, which had been formed out of Hurrian elements in the highlands of Armenia, had enjoyed a meteoric rise in the eighth century B.C. to become a counterpoise to Assyrian power and an object of almost obsessive hatred by its rulers. But, in a very real sense, the Assyrians had only themselves to blame for the existence of Urartu.[55] For continuous incursions into the Taurus range and beyond—the inveterate taking of hostages and expropriation of timber and metal resources—not only familiarized the natives with the ways of the Assyrians but gave them a palpable motive for unity. So Urartu thrived, raising livestock and mining copper and iron, while protecting itself with a series of massive stone fortresses, which in combination with the rugged terrain frustrated the Assyrians for over a century.[56]

Finally, in the summer of 714 B.C. Sargon II—an Assyrian's Assyrian if ever there was one—dispensed with border clashes and personally took charge of a major expedition aimed at the heart of Urartu's mountain fastness. We know from a letter to the national god Ashur relating the course of the so-called eighth campaign that progress through the rough highlands was extremely difficult, and at times the army was on the verge of mutiny.[57] But Sargon persevered. He personally defeated the Urartian coalition in the field and then encouraged his troops to systematically loot the whole country, the climax of which was the ultraremunerative sack of Musasir. It was a sweet moment for Assyria's ruler. "I Sargon, guardian of righteousness, . . . reached my heart's desire and stood in triumph over my haughty enemy."[58] He had delivered a blow from which Urartu would never recover, but it was also a blow that would redound upon his own head with unexpected suddenness and catastrophic consequences.

For Urartu had not only been an impediment to Assyrian ambitions; it appears to have served as a bulwark against military power of a different order entirely. Virtually as soon as Sargon departed with his epic haul, a group known as the Cimmerians flooded into the country, further ravaging it and administering the coup de grâce to the remnants of Urartu's political structure. Just who these Cimmerians were has been the subject of some conjecture, but Herodotus' original assessment—highly mobile pastoral nomads driven off the deep steppe by the Scythians—remains the most plausible explanation.[59] There are archaeological indications that they, having established themselves in the mountains, began raiding Assyria itself.[60] What exactly took place subsequently is not entirely clear—the Assyrians were only scrupulous in recording their victories. But the signs point to Sargon, now an old man, having taken the field against them and suffered a disastrous defeat, losing not only his army but his life.[61] For a country that lived and died on its ability to inspire fear, this sort of defeat would have been as damaging in terms of prestige as it was tangibly. Something new and very dangerous was happening.

Much remains shrouded by the past. But sometime after 3000 B.C. the firm establishment of true states and the final proliferation of walled sites across the Middle East may have stemmed or at least reduced the hypothesized first incursions among the horse-borne—who at this point would have been very few in numbers—and sent them to join the nomadic migration gradually filling up the deep steppe. For the historical record indicates that subsequent agriculture-based polities were confronted mostly by two pastorally derived elements: nearby pedestrian groups and descendants of steppe dwellers who began moving in sometime around 2000 B.C. to mix with the locals—losing their riding skills but not their ability to raise and employ horses in combination with chariots. In the Assyrian case these two would have been represented in the first instance by the steady stream of Aramaeans, who began causing trouble after the reign of Tiglath-Pileser I, and in the second by the Mittanni. As conjectured earlier, this latter element—Kassites and Hittites were other examples—seems to represent among the first of those pushed off the now-contiguously populated steppe by a growing ripple effect. But the pressure would have been low and the process fairly gradual. Around the time of the Cimmerians' arrival, however, the internal dynamic—probably propelled by increased population—seems to have grown much stronger, literally thrusting whole groups suddenly and intact off the steppe.

Once loose in the littoral, large bodies of true horse nomads would have been almost impossible to eliminate. In the case of the Cimmerians, they appear to have quickly fanned out into the interior of Asia Minor, where they continued to cause trouble for at least the next half century, overrunning first Phrygia in 676 B.C. and then Lydia in 651 B.C., killing its king, Gyges.[62] Although the Assyrians were plainly worried by this continuing presence, they appeared unable to do much about it. For no matter how well organized, armies such as theirs were simply too plodding to come to grips with an enemy that could outrange and outrun them in virtually every circumstance. Nor was there any well-defined home-land to address, for the Cimmerians were truly representative of a nation on the move. About the best the Assyrians could do was a combination of diplomacy and the reconquest of areas the nomads had already swept through.[63] But it must have been like boxing shadows, and meanwhile their misfortune was compounded by the arrival off the steppe of the even fiercer Scythians, and the rise of the Medes, a recently settled power from the Iranian plateau whose tribal warriors still retained some nomadic military capabilities.

Yet in the face of change, the Assyrians could only offer repetition—the sack of Babylon in 689, Sidon in 677, Memphis in 671, Thebes in 663, Babylon again in 648, and Susa in 646—a string of conquests whose frequency only served to emphasize the imperial hunger and practically endless list of enemies. "The King knows that all lands hate us," a civil servant from Babylonia once wrote to Esar-haddon (680–669), Ashurbanipal's predecessor.[64] It was axiomatic: Assyria was surrounded by wrath, and it was only a matter of time until her foes joined hands in sufficient numbers to bring her down. Ashurbanipal might search his empire for liturgical salvation, but it was not to be found. For in the end Assyria's most heartfelt legacy was a loathing strong enough to unite both nomad and settled in a determination to destroy her.

On balance it does appear that Assyria truly did live and die by the sword. Several other agriculture-based imperial powers—frequently upland newcomers, bearing the cultural standard of much older societies—fit a similarly militaristic mold. Thus, just as Assyria was the aggressive vessel of an earlier Sumerian legacy, so did Macedon push the language and civilization of the Greeks to the verge of India, and the Aztecs forcibly united central Mexico under traditions perceived as inherited from the Toltecs.

In particular, the parallels between the first and last are notable—especially in light of their separation in time and distance. Both the Assyrians and the Aztecs leave the impression of peoples literally driven by war—caught on a kind of treadmill of organized violence. In both cases there is an obsessive, repetitive quality to their manifold acts of cruelty, along with their urge to represent and record them, which is strongly indicative of compulsive behavior. Not illogically, each society sought to justify its serial megasadism in analogous religious terms—the unavoidable commands of jealous, exclusive, and virtually insatiable gods. And just as the Assyrians and Aztecs were inveterate takers of war prisoners, so did they endeavor to capture and return to the homeland the images of their enemies' deities. For if hostile gods were difficult to kill, at least they could be imprisoned. But even this would have brought the aggressors little respite. Ashur and Huit-zilopochtli were always hungry and ever importunate. So perpetual aggression was assured.

All of this seems symptomatic of the open-ended nature of war, discussed earlier. Among both the Assyrians and the Aztecs there was a notable absence of institutional curbs on warfare. Each society began as something else but evolved while surrounded by enemies in geographies that placed a premium on successful war making. Increasingly, state institutions came to reflect this environment. So if war began as the health of the state, there was no real alternative when it became an all-consuming mania. Nor does there seem to have been much conscious awareness of the predicament. Endless war was simply viewed as the normal course of events, everybody's duty, however onerous. So each society marched in lockstep toward destruction.

Such a fate was in no way inevitable. But the case of Assyria indicates that it did require preventative measures to avoid.

{11}

HEAVEN'S MANDATE

I

When the duke's invitation arrived, the Master said nothing. After a time he told Yu to inform the court that he would be delighted to participate in the contest. This surprised us, since the Master had never shown the slightest interest in the drawing of the bow. "Gentlemen never compete," he had once remarked. "You will say that in archery they do so. But even then they bow and make way for one another going up to the archery ground, when they are coming down and at the subsequent drinking bout. Then even when competing, they still remain gentlemen."[1] Only Tzu-lu felt it necessary to elaborate, reminding us that it was entirely appropriate for a *shih* to practice *all* of the disciplines. Later the Master said, "That is Yu indeed! He sets far too much store by feats of physical daring. In archery it is not the hide that counts, for some men have more strength than others."[2]

On the day of the contest the Master arrived promptly, modestly taking his place among the contestants. He greeted only those he knew, questioning them intently on every detail of protocol. Yet when it came time for him to shoot, he approached the ground very much at ease. But there was nothing for him there. Despite all efforts he could not string his bow and had to be helped. His best shots went wide of the mark, and most failed to reach the target. He remained calm, remarking only that arrows, like birds, flew according to their own whims. The knights could barely contain their laughter. If their scorn was obvious, the Master would have none of it.

Later, however, "a villager from Ta-hsiang said, 'Master K'ung is no doubt a very great man and vastly learned. But he does nothing to bear out this reputation.' The Master, hearing of it, said to his disciples, 'What shall I take up? Shall I take

159

up chariot-driving? Or shall it be archery? I think I will take up driving.' "³ But though he ranked next to the Great Officers, and was not permitted to go on foot, no one could remember his driving a chariot either before or since.

Benjamin Schwartz, in his book *The World of Thought in Ancient China*, focuses on this incident, elaborated upon here, as symbolic of a basic conflict between the presumed mastery of both martial and intellectual skills by any *shih*, or nobleman, and the increasing cleavage separating the civilian and military wings of the functionary class destined to dominate government and politics in China for more than two millennia.⁴ The fact that the Master "singles out the military arts as peculiarly ridiculous examples in his own case"⁵ is not to be taken lightly. For this man, Confucius, was no ordinary mortal. Rather, he was the archetypal Chinese culture hero, a figure of towering significance whose influence on his society finds no real parallel in any other ancient civilization.⁶ More to the point, Confucius's rejection of military values—though subtle and admittedly not total— would have a profound impact on the ultimate form and political behavior of the society that is frequently associated with his name. Yet Confucius was no revolutionary; rather, he was a proverbial defender of tradition, a figure whose significance largely rests on his ability to justify submission to hierarchy and authority in terms that were truly powerful and binding. But the process involved a basic redefinition of sovereignty—a transference of emphasis from the ability to coerce to the sanctity of the family, ritual, and individual virtue. This, he would have argued, was the true Mandate of Heaven, and it would lead to a culture that was uniquely enduring—persistent in good measure because it was equipped with the philosophical justification for controlling the exertion of force and limiting the role of warfare. This came at a price; but China would never become Assyria.

Yet to really understand why, it is important to realize that the Confucian subordination of war did not emerge out of a vacuum. It was very much a product not only of the times in which the Master lived but also of the circumstances in which Chinese civilization had originated. And although there is much that was indigenous and original, in one crucial respect it appears that China was derivative. For a good case can be made that war had been brought to China, like a cold wind blowing off the steppe.

II

Seeking to capture the ineffable uniqueness of the society, historians and Sinologists have repeatedly posed the question in one form or another: What is Chinese?⁷ It is a query that is truly Eastern in its simplicity and infinite complexity. But the very fact that vastly learned students of the area could be driven back to so basic a proposition only serves to illustrate the enigmatic nature of the civilization—the haunting awareness that, despite having been subjected to the same basic conditions and limitations imposed on all human existence, China really did grow to be different. The issue is a transcendent one affecting the present as well as the past, but there is a growing body of evidence that a substantial source of this singularity has to do with origins, the initial evolution of Chinese society.

"We are in the midst of a Golden Age of Chinese archaeology," notes Kwang-chih Chang, probably the leading Western synthesizer of physical data on early Sinic cultures.[8] For over the past several decades a dazzling array of new finds, combined with a much better understanding of extant information, has led to a far broader and deeper conception of the societies that would form the foundations of Chinese civilization. The field remains in ferment and there is much to be learned, but certain things already seem clear. Perhaps most surprising are the sheer age and continuity of the Neolithic. Forty years ago many authorities believed not only that agriculture was a relatively recent phenomenon in China but that metallurgy, the idea of script, and ceramics had all been brought there by outsiders.[9]

All of this has changed.[10] Radiocarbon analysis has pushed the cultivation of millet well into the seventh millennium B.C., while rice grains found in the south have been dated to around 5000 B.C.[11] Although it is entirely possible that even older examples will be discovered, it is already apparent that agriculture had a very long and indigenous history in China, and it has now been well established that the fabrication of pottery is virtually as old.[12] But perhaps more surprising, there is now tangible evidence to show that several characteristic elements of ancient Han Chinese culture—silk manufacture, the production of bronze, and writing—all found their roots deep in the Sinic Neolithic. Thus silkworms and a half-cut cocoon have been recovered from a site in southern Shansi, while a copper-tin alloy knife, the earliest bronze artifact yet found in China, was discovered amid the strata of Neolithic Ling-chia.[13] Yet most remarkably, it now seems that the origins of literacy can be traced back to the character-like markings cut into pottery beginning sometime between 4800 and 4300 B.C. While this is clearly not writing, it also seems, as several authorities point out, "that these marks were purposefully and knowingly incised by human beings to represent definite concepts."[14]

The weight of evidence points to a very long and smooth cultural development, what archaeologist Richard Pearson calls "a kind of local continuous evolution of culture."[15] The use of the term *local* is significant because it reflects the hard-won understanding that the Chinese Neolithic was very widespread, not confined to the north and the relatively well-known Yang-shao culture of the middle Yellow River valley but also flowering in the coastal regions, along the Yangtze, and still farther south in societies now identified as Ta-wen-k'ou, Ma-chia-pang, Ta-hsi, and Shi-hsia, among others. All were predominantly agricultural—with slash-and-burn techniques and shifting repetitive occupancy being the characteristic subsistence-settlement pattern.[16] Domesticated animals were confined primarily to the dog and pig, although sheep, goats, and cattle were present in small quantities in the north. Some cultures were more advanced than others, but shared artifacts testify that all were drawn increasingly into patterns of mutual interaction. Yet there are no indications that these were other than peaceful. Certainly, there may have been individual violence, but no evidence has been uncovered to indicate that these cultures conducted or were subject to organized aggression.

Then something happened. Shortly after 3000 B.C. a new cultural horizon appeared, the so-called Lung-shan. While there were some signs of transition in the south, the Lungshanoid was primarily a phenomenon of the north, with sites mostly clustered along the Yellow River and its tributaries, stretching east to west in a great concave arc.[17] Areas of habitation were larger and permanently occupied, reflecting a more settled intensive pattern of agriculture. The material culture was richer, with increasing evidence of copper-bronze metallurgy and the widespread use of the potter's wheel.[18] Also, burial patterns indicated the beginnings of social differentiation. But most striking were the manifest signs of violence. Lungshanoid sites in the north—though not, apparently, in the south—were walled, surrounded by defensive ramparts methodically built up by packing tamped dirt layer upon layer into a tightly stratified mass. This so-called *hang-t'u* ("stamped-earth") technique was extremely labor-intensive and would not have been casually undertaken. The occupants of these sites must have felt seriously threatened. But by whom?

"Each other" is the answer most mainstream archaeologists familiar with the Lung-shan would give us. According to this paradigm, social intensification was solely responsible for starting what must have amounted to organized warfare. As we have seen earlier, societal development undeniably contributed to the evolution and spread of war. But there are problems in the Chinese context with assigning it an exclusive role. Why, for example, had it taken so long? Despite the long duration of sedentary existence here, fortifications appeared in China over two thousand years after they first began to proliferate in the Middle East. In this regard it is worth noting that Lung-shan cultures, like their Neolithic predecessors, seem to have been much more heavily weighted in favor of agriculture, with domestic ruminants present, though almost always in small numbers.[19] So an internally generated farmer-pastoralist split and resulting pattern of raiding do not seem likely to have assumed a primary causal role in originating hostilities. Meanwhile, the question still looms: Why had fortifications appeared in the north and not the south, especially when both appear approximately equivalent in development? This is not an issue easily ignored, as a number of respected authorities concede.[20]

Indeed, clues found in the Lung-shan stratum of a place called Chien-kou do point in another direction. For here was uncovered the first direct evidence of violence in Chinese prehistory, a site of mass slaughter where the decapitated victims—men, women, and children—had been dumped into several abandoned wells.[21] Also recovered were six human skulls with signs of blows and scalping, craniums that one of the excavators, Yen Wen-ming, speculates were intended to be used as drinking cups.[22] This is suggestive, since authorities as far back as Herodotus have observed that drinking from the skulls of their enemies was particularly characteristic of steppe nomads.[23]

The suspect has been under investigation before. Indeed, there is a long tradition of scholars supporting the notion that the eventual rise of civilization in China was deeply affected by "a shock coming from outside" and that the source was Inner Asian pastoralists.[24] Until very recently, however, chronology made this explanation—much less its application to the still-earlier appearance of fortifications—very hard to support.[25] For, as we have seen, it was generally believed that horse riding did not predate 1500 B.C., long after the residents of Lung-shan villages began building walls around themselves. And if it was possible to raise serious questions as to how pastoral nomads could have possibly gotten out to China at this point, then the whole thesis begins to crumble.

However, Anthony, Brown, and Telegin's work, by firmly pushing the origins of riding back to around 4000 B.C., changes things considerably. Although the evidence presently indicates that full penetration of the Inner Asian plateau did not come before the invention of the wagon and the domestication of the camel shortly after 3000 B.C., the timing remains plausible. For the archaeological record reflects a nomadic presence during the same time frame out as far as the Altai Mountains in the form of the Afanasievo culture.[26] A vanguard of nomads still would have had a century or two to reach the fringes of China, ample time given their extreme mobility and the absence of anything in the way of occupied territory to slow them.

As in the Middle East, the population of those first involved in a pattern of raiding was almost certain to have been minuscule, but the terror they inspired would have been out of all proportion to their numbers. Most probably attacks were devastating but sporadic, intended for plunder, not conquest. But the shock and incomprehension they must have elicited could well have been sufficient to prompt the hitherto peaceful cultures of northern China to take desperate measures to protect themselves—very possibly in the form of tightly packed earthen walls, the remains of which we see today. As in Mesopotamia, this act of circumvallation would have had an intensifying effect of its own, helping to speed the development of social complexity and the rise of true civilization in and around the middle Yellow River valley. But this hypothesis is not aimed at discounting the importance of other factors involved in the rise of the city and the state in China[27] so much as it is intended to locate a point of ignition.

This is of particular significance here, because it apparently came considerably later in the process of social evolution than it had in the Middle East. This, in turn, has important implications for the plausibility of the argument, especially since the direct physical evidence is minimal. But if it is true that a distinctive Chinese culture was already well on the way to formation when the first nomads could have arrived, then it follows that the impact of the intruders would have been decidedly more limited and selective than had been the case among Near Eastern Neolithic cultures. In fact, there is little evidence that the development of language, religion, or the familial basis for politics in China was significantly altered. What did change had to do almost exclusively with what nomads did best—fighting, along with its implements and customs. Thus, when the curtain rises on Chinese civilization, the actors appear already equipped with domesticated horses, chariots, and composite bows. Perhaps of more significance, there is reason to believe that they carried with them a distinctive outlook toward warfare—one that combined the nomad's preference for deception and surprise with a fundamental distaste for the entire affair, an aversion that arguably still registered the peaceful farmers' shock and dismay at the depredations that so changed the course of their existence.

III

Looking back on their origins, ancient Chinese historiographers pieced together the episodic remains of their very early past into a sequence of mythical ruler-sages followed by three dynasties: the Hsia, said to have been founded in 2205 B.C. by Yü, conqueror of floods; the Shang, allegedly begun by T'ang in 1766 B.C.; and the Chou, traditionally established in 1122 B.C. by Wu Wang.[28] Excepting the Chou, for whom there was verifiable documentation, this chronology—its dates and subject matter—was long regarded as little more than fabrication. However, the continuing recovery of inscribed bones known to have been used in antiquity for divination eventually led the Academy Sinica to begin formal excavation at An-yang in 1928. What it found was the site of the last royal capital of the Shang.[29] Subsequently, archaeologists also uncovered the probable (though yet to be confirmed) center of the ancient Hsia at a place called Erh-li-t'ou.[30] Still more recently, radiocarbon analysis has provided a rough concordance with traditional dating.[31] Myth, it seems, has been largely vindicated.

But what specifically has archaeology revealed about these societies? First, the Three Dynasties are best viewed less as separate and sequential entities than as what Chang calls "a system of parallel and interrelated regional developments with shifting centers of gravity"[32]—in other words, three manifestations of the same fundamental civilization, which emerged out of the Lung-shan in and around the middle Yellow River basin toward the end of the third millennium B.C.

All were agricultural, based on the cultivation of millets in the traditional Neolithic manner, apparently without the aid of the metallic farming tools, draft animals, and extensive irrigation that would become characteristic of later dynasties.[33] This is somewhat puzzling given the clear signs of rapid population growth (the complex at An-yang was vast), but it may testify to the inherently high carrying capacity of the environment, which was considerably warmer and moister than at present.[34]

The evidence indicates that the Erh-li-t'ou, Shang, and early Chou all shared the same physical culture, language, and a single system of writing that was employed for religious and political purposes.[35] Each was dominated by kings and a ruling elite living in urban centers made up of specialized quarters including palaces and ceremonial centers, workshops, and residential areas--all of it surrounded by massive stamped-earth ramparts. Within these complexes, power was maintained and exercised through several mechanisms. One of these was control over bronze foundry, cast in the form of ceremonial paraphernalia and weapons— the instruments of what the ancient Chinese themselves called "the two major affairs of the State, the ritual and the warfare."[36]

The former, amounting to a monopoly over high shamanism,[37] enabled the leadership to communicate with ancestors, to be knowledgable of future events, and provided them with the legitimacy to rule. This, however, required careful attention to ritual, performed in large part with meticulously designed bronze vessels, the beauty and virtuosity of which increased until reaching a stage that has yet to be surpassed.[38] The other key element was scapulimancy, divination through interpreting the cracks generated by heat on a piece of bone, the results being subsequently inscribed on the surface. While small numbers of oracle bones have been recovered from Erh-li-t'ou and the Western Chou, some one hundred thousand inscribed examples have been collected from An-yang. Corporately, they constitute the best written record of leadership activities in the Shang state and early dynastic civilization. On this basis, it appears that military power and warfare were major preoccupations.

This, however, is not universally conceded. A revisionist view—still linked to the assumption that true horse nomads had not yet arrived—maintains that the Three Dynasties were able to create and rule their far-flung but loosely knit empires essentially through common culture and ritual, not bureaucratic coercion and military power.[39]

In the case of the Shang, at least, the bones demur. The picture that emerges is one of rulers constantly on the move and on the make—hunting, inspecting the realm, collecting tribute, rewarding or punishing subject lords, and fighting off invaders mostly from the north and west—a perpetual odyssey of arm-twisting and instructive violence.[40] In these activities, at least, ritual and cultural affinity were decidedly secondary. On the other hand, the shadow of the steppe nomad can be discerned as hanging over the entire enterprise.

For while the Shang military was clearly representative of an agricultural people, their armament was not. No Shang bows remain extant, but their repeated depiction leaves little doubt that these were highly developed composite weapons, quite conceivably derived from the shorter models carried by horse nomads.[41] Equally suggestive was the appearance of horse-drawn chariots. As explained in an earlier chapter, rock drawings of chariots in central Asia, the absence of any prior wheeled conveyance among the Chinese, and the telling resemblance between early

Sinic examples and counterparts in the Near East all point to the chariot having been brought from without by steppe nomads.[42] How and when this might have happened are very much open to question. But because chariots are vastly inferior to skilled horse archers tactically, it seems likely that these vehicles remained primarily devoted to prestige transport while in the deep steppe. Yet as was the case earlier in the border regions of the Middle East, some of the populations that gathered along the outskirts of China—those quite possibly derived from pastoralists spilling off the Inner Asian plateau—would have found the chariot offered at least a modicum of the horse's military advantages to those no longer really able to ride; this would have constituted the bridge of transmission.

Interaction with Sinic cultures may have gone on for a long, if indeterminate, period (tradition, at least, has the Hsia possessing chariots),[43] but oracle bones do provide us with specific examples in the case of the Shang, a record that offers some insight into the aims and motivation behind their military conduct. For while the Shang behaved aggressively toward a number of peoples pursuing mixed economies to their north and west, one group in particular was treated as implacable enemies. Known collectively as the Ch'iang, they have been linked archaeologically to the Ssu-wa-shan, a non-Han Chinese culture of the upper T'ao Ho valley. Like some of the others deemed barbarian, the Ch'iang demonstrably leaned toward pastoralism, but what distinguished them was the striking importance of horses to their way of life.[44]

This, it seems, was key to their dealings with the Shang. Although far to the west, they are known to have employed chariots and were apparently considered extremely threatening. For the Shang launched repeated and massive expeditions—one numbering thirteen thousand troops—against them. The frequency and duration of these forays indicate less than total military success, but they did produce numerous war captives.[45] A number of these prisoners were systematically recruited, apparently to raise horses and drive Shang chariots.[46] This was significant, since it presages a horse-based symbiotic relationship with the peoples of the steppe that would dominate much of Chinese military history. But as would be the case later, the interaction between the Shang and the Ch'iang had its darker side—one that very clearly reflected the hatred and disdain underlying the relationship. For in addition to being exploited for their equestrian skills, Ch'iang war prisoners either were enslaved and sent to toil in Shang fields or were ritually killed.

Human sacrifice constitutes one of the most distinctive features of Shang archaeology. And while it does appear to have roots in the Lung-shan, the scale is vastly greater—corpses numbering in the thousands buried in leadership graves, below their houses, and beneath their public buildings—literally at the very foundations of Shang civilization.[47] The records indicate that a good many of the victims were Ch'iang; indeed, they were the only people specifically enumerated as having been sacrificed.[48] Yet we can assume there were others, for cranial analysis indicates that many of the sacrificial victims were indistinguishable from those presumably doing the sacrificing. In other words, it is quite possible that the Shang took to killing not just war prisoners but their own in what could well have been a crescendo of bloodshed.[49]

What was behind this? To some extent Shang warfare does reflect a motivational pattern analogous to that of other large ancient agricultural societies—new territories were expropriated and put under cultivation; captives were taken and used to augment the labor supply. However, the absence of large-scale irrigation or any particular emphasis on the technological intensification of agriculture, combined with apparently limited investment in public works (basically stamped-earth ramparts and an array of relatively modest public buildings), raises questions as to just how labor-hungry this society actually was. Thus, it has been suggested that capture, virtually for its own sake, was a central incentive for war[50]—at first perhaps motivated by the fear and hatred engendered through the juxtaposition of two profoundly different ways of life but ultimately taking on a momentum of its own, as the tempo of Shang militarism increased and expeditions grew larger and more draining on the state. Unlike in Mesoamerica, there is no evidence for large-scale purposeful cannibalism. In effect, sacrifice was an end in itself, driven presumably by ritual and ideology, but a consumer of substantial numbers of people nonetheless.

This is significant, because the last ruler of the declining Shang was remembered as particularly cruel, ritually killing even his closest advisers.[51] Indeed, it was this cruelty and the provocation of foreign wars, as much as the regime's perceived decadence, that were remembered as having spurred the confederation formed around Wu Wang and the Chou aimed at overthrowing the Shang.[52] Notably, the Ch'iang would count themselves among those joining the Chou at the pivotal battle of Mu-yeh.[53] As for the Shang, its army, depleted by other campaigns, would put up only feeble resistance before turning on the king, who then fled to perish in his own palace. By dusk King Wu would occupy the capital, beginning a hegemony of three centuries down to 771 B.C., labeled by Chinese historians as the Western Chou.[54]

They are remembered as rough-and-ready westerners with a checkered past, and it does appear that nomadic pressure at one point uprooted them into a wandering pastoral existence until they located suitable farmland in the Wei valley.[55] However, it is unlikely that at the time of the overthrow they were viewed as anything like barbarians.[56] On the contrary, the political agenda they revealed was conciliatory and sophisticated, aimed at assuming the mantle of agricultural civilization and extending its sphere of influence in all directions. Yet the Western Chou were and remained essentially a small military elite, incapable of directly controlling even the substantial dominions inherited from the Shang. Their alternative was quite remarkable; they proposed to rule by example. The new Chou stewards promptly articulated the conditions necessary to obtain and preserve what was termed the Mandate of Heaven and confirmed their readiness to apply a strict moral standard to themselves—a standard that encompassed sincerity, service to the people, diligence in administration, honoring tradition and the worthy, care in the use of force and punishment, and abstention from personal excess.[57] It is notable that, in this context and indeed most others, religion is interpreted and applied as essentially a cultural instrument, a pattern of behavior, rather than as a matter of eschatological compulsion—solicitous gods demanding war and promising an afterlife in return. Prosperity and political stability are offered instead as rewards for the measured pursuit of ethical conduct—literally a ritual of good deeds. It was a credo never to be forgotten, not simply the essence of governance for subsequent dynasties but quite possibly imperial agriculture's earliest comprehensive attempt to come to terms with our behavioral heritage, to square the contradictions between settled hierarchical existence and certain key aspects of human nature evolved during our time as hunter-gatherers.

But accompanying this ideology was also a hardheaded commitment to the practice of power politics. Thus, when the newly established Chou kingdom was almost immediately faced with a great rebellion led in part by recalcitrant Shang, the leadership was quick to send experienced troops and commanders to quell the widespread uprisings.[58] Yet like the initial conquest, this episode brought no orgy of retribution. Instead, the Chou continued their program of co-opting their former enemies, along with erstwhile allies. So not only did the remaining Shang nobility and their subjects become a part of the new coalition, but their way of life and customs were tolerated—even, it appears, the aberrant ones. Symbolizing this rapprochement was a cemetery recently uncovered near the city of Tung-chia-lin, which was found to have been divided into two distinct and contemporaneous sections: one, a Shang quarter, containing both dog and human sacrifices, and the other with no sacrifices at all.[59] Eventually the practice would disappear entirely, demonstrating the pragmatic if not exactly moral fruits of this sort of toleration.

Meanwhile, the Chou were busy expanding the realm, spreading the common conception of Chineseness, or *Hua-Hsia*.[60] Their central instrument was the vassal community, an entity aimed at blending and managing people, not specific territories. These were also portable, not tied to a given locality but movable like pieces on a geographic chess board.[61] Plainly, the intent was not just to control core areas but to colonize and in doing so assimilate new populations. This is reflected in Western Chou archaeological finds far to the south in the Yangtze basin, which frequently appear as isolated examples of high culture in a sea of less advanced communities.[62] But these insertions rather quickly sparked intensified development in the surrounding areas, and the process of Sinification rolled on. Eventually what emerged was a vast, multilayered feudal construct, based essentially on a network of walled cities—both the seats of the Chou-designated military elite and key conduits for cultural diffusion.[63] And stringing everything together was a web of personal bonds and family ties, reinforced by regularized procedures and ritual—the neural net of the Western Chou. But the brain was also evolving. For at the center, administration was growing increasingly complicated. Out of the executive branch sprang department heads with staffs, the royal secretariat, the Vassals of the Four Quarters—throughout government the literate infiltrated, creating functionaries out of what once had been priests and diviners.[64] For better or worse, the process of bureaucratization had begun and was gaining momentum as the Western Chou struggled to keep track of its far-flung dominions.

But the center would not hold. In the end the Chou concept of rulership was simply too abstract not to be pulled apart by the entropic forces of crisis and changing circumstance. In the north it was nomads. The Western Chou brought with them considerable experience in dealing with the pastoral peoples living on the fringes, and apparently there was some optimism that they could be absorbed into the sphere of the Chinese. But whatever success they might have achieved was limited and temporary.[65] For the Chou would find that the line of demarcation between farming and herding was not easily erased—some symbiosis might be possible, but a true integration of the two ways of life was wishful thinking, especially at this point.

For nomadic pressure was building in a way not before experienced. It would appear that the same forces that had driven the Cimmerians off the western steppe late in the eighth century B.C. were also operating at the opposite end of the Inner Asian plateau—giant ripples coursing through the now-contiguous populations of horse nomads and then crashing against the edges. For a time these waves of energy would be registered primarily by increased movement and aggressiveness on the part of the more settled "buffer" pastoralists, the Ti and the Jung. Although a serious threat, these groups primarily fought on foot with what amounted to conventional weaponry,[66] and their depredations leave the distinct impression of acts perpetrated by fugitives fleeing before something entirely more frightening. That "something" would shortly manifest itself in the form of the Hsien-yun, a group first appearing around 800 B.C. "The most striking thing about the Hsien-yun invasion is its concentration, its immensity, and the limited time it took," remarks Jaroslav Prušek, a specialist on the northern tribes. "We get the impression of a new fast-moving warrior . . . coming from far away, and disappearing again."[67] Quite clearly, true horse nomads had penetrated the outer shell of more settled pastoralists and reintroduced themselves into Chinese history, only this time in numbers sufficient not just to terrorize but to threaten and keep on threatening the entire political order. Apropos of their arrival, the Western Chou capital was soon sacked by nomads, forcing the dynasty to move far east, leaving their former homeland in the hands of a tough group of retainers later known as Ch'in. History had plans for the Ch'in; but for the Chou it was the beginning of their death spiral.

Meanwhile, as great as were their external difficulties, the Chou were also beset by internal developments that would just as surely undermine their power. For the seeds of Chinese civilization, the vassal communities which the Western Chou scattered so assiduously, had germinated and were growing into something unexpected. The rearranged populations of the Chou vassaldoms took root and started to behave as permanent political entities—military colonies became city-states. As this occurred the authorities inside the walled enclaves reached out and began taking exclusive control of the land surrounding them.[68] As the vassal states consolidated their territorial control, they became increasingly able to draw resources from their own dominions, growing proportionately less dependent upon Chou suzerainty. Soon the texture of Chinese political geography was transformed—what had previously been a fairly fluid and manageable environment suddenly precipitated into a minutely partitioned landscape of something like 170 separate statelets.[69] Worse yet, they were increasingly disorderly. For as the process of territorialization proceeded, the vassals rather quickly exhausted the unclaimed areas between them and took to encroaching on one another's land, setting off a seemingly endless sequence of wars and violent consolidation. It was a cycle the Chou were helpless to break or even influence, beyond trying to ensure their own titular survival. China's history had entered a new phase.

IV

Confucius and his fellow sages were not sanguine about the future. They viewed their own society as marching toward perdition, and instead looked back fondly to the Western Chou as a sort of golden era of relative tranquillity and goodwill. While the very frequency and persistence of this nostalgic view may indeed reflect some measure of truth, the bad reputation of the following periods, the time of Spring and Autumn (722–481 B.C.) and even the Warring States (403–221 B.C.), seems exaggerated when viewed from our own materialistic "developmental" perspective.[70]

Around 500 B.C., economic intensification had begun to accelerate, with large-scale flood control and irrigation projects greatly increasing the food supply. The introduction and proliferation of iron agricultural implements brought further gains during the fourth century, until the land's enormous productive potential was brought essentially to fruition.[71] All the while, population grew astonishingly. Abetted by the availability of food and the system's capacity to absorb labor, conceivably it could have totaled around twenty-five million by 400 B.C.[72] Central government, although short-circuited at the uppermost level by the demise of the Western Chou, continued to evolve among the component states. And advances in effective organization and control meant more opportunities for the literate element epitomized by the sages.[73] Trade and commerce also thrived, and a prosperous, though officially despised, mercantile class began to emerge. So it seems that economically, technologically, and administratively this was a period of growth and creativity, a time when Chinese civilization was fleshed out into something like its classic proportions. Why, then, were the sages—not just Confucius but also Mo-tzu and Mencius—so pessimistic?

In a word, it was war. For the mosaic of states that made up the Chinese ecumene was undergoing a process of political consolidation, the key instrumentality being organized warfare. And the sages, however philosophical, were men of their times, who could not ignore the rising tide of violence and avoid wondering where it might end. Their fingers of blame pointed in various directions, but all were united in decrying the brutality and destructiveness of the wars that surrounded them. Thus for Confucius, as Schwartz shows, the supreme public goal was "to bring peace to the world."[74] Mencius went farther, proclaiming warmongers criminals.[75] And Mo-tzu saw the problem as sufficiently urgent to take matters into his own hands, making himself an expert in defensive war and urging innovations that would protect cities against sieges.[76] The sages' combined outlook is historically important, not simply because Chinese civilization would one day be guided by the principles of Confucius and Mencius but also because of what it says about the general climate of thought with respect to warfare during the Spring and Autumn and Warring States periods.

This was a society not much taken with war. Having undergone a long cultural evolution in relative peace only to have it shattered, likely through the intercession of terrifying outsiders, Chinese of all casts were predisposed to view war skeptically as a dangerous and somewhat alien phenomenon. Circumstances, however, did not allow for outright pacifism. Thus Confucius, the staunchest defender of legitimacy, would praise a certain minister, though he served a usurper, on the grounds that his actions helped stem a great invasion of Ti: "Were it not for Kuan Chung we might now be wearing our hair loose and folding our clothes to the left [like barbarians]!"[77]

But if survival dictated that the menace from the steppe be countered through force, this did not mean that warfare, especially between Chinese, was cause for much enthusiasm. Thus historian David Keightley, comparing Sinic accounts of war with concurrent Greek martial literature, was struck by the sparseness of the former's combat descriptions and the disinclination to celebrate heroism, particularly of the sanguinary Homeric sort.[78] There is little glory in killing; battle, instead, is portrayed as an exercise in social solidarity, a matter of duty, not an opportunity for personal aggrandizement.[79]

The pervasiveness of this outlook is underlined by the fact that it is not simply characteristic of poetic and historical literature but is very much reflected in compositions specifically devoted to military training and education. Indeed, it is in the writings of professional strategists that the distinctiveness of China's martial vision is most tellingly delineated. This body of thought has come down to us in a series of seven so-called Military Classics, by far the most famous of which is *Sun-tzu's Art of War*. While all but two seem to have been compiled sometime in the fourth century B.C. each draws on a variety of older sources.[80] Yet their outlook is remarkably consistent, presenting what amounts to a unified view of how the Chinese should and would fight.

The contrast with the West could hardly be more stark. As military historian Victor Davis Hanson explains, "Not merely the ability but also the *desire* to deliver fatal blows and then steadfastly to endure, without retreat, . . . have always been the trademark of Western armies."[81] This was far from the case with the Chinese. Thus the strategists as a body endorse Sun-tzu's belief that "subjugating the enemy's army without fighting is the true pinnacle of excellence."[82] Repeatedly, weapons are termed "evil" or "inauspicious implements," and Ssu-ma appears to speak for the others when he warns: "Thus even though a state may be vast, those who love warfare will inevitably perish."[83] Yet this martial diffidence is strongly contrasted by a uniform tactical ruthlessness that supports surprise, deception, sudden attack, and feigned retreat in order to prey on enemies when they are least able to defend themselves[84]—a style of fighting utterly at odds with the blunt confrontational ethic of Western combat.

Our own age, chastened by millions of battle deaths, has rediscovered the Chinese way of war as potentially a more humane and cerebral alternative. This may be. However, it also appears to be a historical composite, arrived at as much through necessity as by reason. For example, the introduction during the fourth century B.C. of the crossbow—a weapon apparently suppressed in the West by the ancient Greeks[85]—made it increasingly difficult for Chinese forces to fight at close quarters, since bolts could easily penetrate body armor. Yet the crossbow was a late development, used extensively only toward the end of the Warring States period.[86] The military outlook reflected by the strategists, on the other hand, seems to stretch back much farther, repeatedly drawing on semilegendary figures associated with the very foundations of civilization in China. If there is even an element of authenticity here, then a good case can be made that Chinese attitudes toward combat basically always existed—a proposition none of the strategists appear to doubt—and that they were the creation of a people who came late to war and were almost immediately subjected to its worst aspects by greatly superior adversaries. Granted, there was an interlude of battlefield chivalry that began and ended during the Spring and Autumn period.[87] But at its core the thinking of the strategists can be said to mirror a very ancient syncretism between the nomads' extreme tactical opportunism and the agriculturalists' reluctance to embrace war in the first place. Seen in this light, the Chinese way of warfare appears less coherent and considerably more contradictory than is generally recognized.

Yet it was these very contradictions that would have prompted the strategists to confront issues fundamental not simply to the military environment of their own times but to the very nature of organized aggression. They worry, for instance, about motivating troops and frankly admit its difficulties. "People do not take pleasure in dying, nor do they hate life, [but] if the commands and orders are clear, and the laws and regulations carefully detailed, you can make them advance," advises Wei Liao-tzu.[88] As a group they are very much aware of war's excesses, and—with the notable exception of Sun-tzu—specifically advise against plunder and the abuse of civilians.[89] They particularly worry about the disruption of agriculture. This is highly suggestive, because China at this point was well on its way to becoming a classic manifestation of Wittfogel's "hydraulic" society—a massive irrigation-based ecosystem that would have been increasingly subject to the very demographic instabilities which organized warfare at this level of subsistence was employed to address. But there is a reluctance to do so. The strategists on the whole are wary of offensive war and do not dwell on territorial ambitions. Moreover, Wei Liao-tzu proposed a specific alternative to capture, after the Shang no longer a primary object of war but still likely to have retained some allure: "One who is enlightened . . . will attract displaced people and bring unworked lands under cultivation."[90] Practically to a man, then, the strategists emerge as thoughtful critics as much as active practitioners of war, arguing in essence that it was as much the wealth of the state that creates military power as the reverse.

Yet on the surface of events, it was military power that appeared to dominate. For the China of both the sages and the strategists was caught in a whirlpool of political amalgamation, dragging it with ever-increasing violence toward the center and an endgame. By 403 B.C. the 170 or so original participants had been reduced to seven powerful survivor states, who then engaged in a grinding free-for-all culminating in the unification of 221 B.C. And as the process of elimination moved forward, combat intensified.

As noted earlier, the unfolding of the Spring and Autumn period saw the conduct of war—though not necessarily its planning or intent—taking place under circumstances best described as chivalrous and characteristic of intraspecific aggression, at least for the leadership. While the emergence of this sort of ritualized combat is understandable, given the aristocratic contenders' shared ties stemming from the Western Chou's hegemony, it should not blind us to the pitiless nature of the larger process, or the active pursuit of tactical advantage during this time. Trickiness is still stressed, and smaller states are warned repeatedly in commentaries to keep up their guard.[91] Soon enough the larger powers' penchant for battlefield etiquette would be swept away by the increasing desperation of the fighting.

The Warring States saw the scope and brutality of combat surge phenomenally. Force structures, even among the weaker powers, were said to number in the hundreds of thousands—probably an exaggeration—but they were still fed by masses of peasants now subject to mandatory training and conscription.[92] The aristocratic chariot-based force became a thing of the past, its vehicles driven from the battlefield by the crossbow. Instead, huge infantry formations would be led (directed, really) by purely professional soldiers specializing solely in military matters, thereby registering the further estrangement of civil administrators from the province of war.

Meanwhile, war's dominions continued to expand. Increasingly, during the fourth and third centuries, cities became subject to sieges, assaulted by a range of engines including battering rams, catapults, and mobile towers—an arsenal that prompted Mo-tzu and his followers to take up the cause of defensive warfare. Yet they could not prevent the spread of suffering and dislocation to what had once been considered safe havens.

And beginning during the closing years of the fourth century, the urge to defend also manifested itself in the construction of border fortifications hundreds of miles long. Although none were likely to have been nearly as massive or elaborate as today's Great Wall, they were still on a sufficient scale to have helped drain off the excess energies of increasingly overheated and populous societies, as well as providing some measure of defense--a persistently Chinese formula for monumental architecture. While some of these barriers were thrown up between warring states, the positioning of others clearly reflected the danger from the steppe.

For at the same point in time, after a hiatus of around a century, barbarians were again causing serious trouble. Now it was Chung-shan horsemen, descendants of the Hsien-yu.[93] The Chinese reaction was interesting and most contradictory. They went on the offensive. In 307 B.C. King Wu-ling of Chao, a northern border state, began training his own force of horse archers, even dressing them in barbarian garb—short jackets and trousers—to facilitate their riding. With time and doubtless a great deal of effort, they grew skillful enough to meet and even defeat true steppe nomads. Indeed, they had virtually become steppe nomads themselves, eating large quantities of meat and living a life that was very similar.[94] But more significant than their capabilities or achievements was the fact that their existence did not go unnoticed by the other Chinese, particularly the Ch'in.

Since the evacuation of the Chou, these former retainers had not only survived in the west but flourished—mostly due to their toughness and ruthless pragmatism. And rather quickly the Ch'in became known for their mounted archers, even going the Chao one better and recruiting actual steppe nomads.[95] Cavalry soon spread to other states, but Ch'in alacrity was characteristic. Despite being thought of as raw outlanders and soft on barbarians, the Ch'in led the rest of China in irrigation and the intensification of agriculture on a larger and larger scale, encouraging peasant immigrants from the east until their economy was among the strongest and most productive of the contenders.[96] Meanwhile, the innovative Ch'in dabbled in bounty of an entirely different sort, offering a price for the head of literally every soldier who opposed them and turning the aftermath of battles into orgies of decapitation.[97] But the wagon loads of heads that trundled back toward Ch'in constituted a harvest no less bountiful in political terms, since it allowed them to obliterate the infrastructure of their former opponents and absorb conquered populations into an expanded and centrally administered state.

Ch'in ruthlessness would never be forgiven, but the momentum it generated flattened adversary after adversary. In 246 B.C. a teenager ascended to the throne of Ch'in, a figure whose talent and cruelty predestined him to preside over a dazzling endgame beginning fifteen years later. After destroying Han, Chao, and Wei in rapid succession, he managed in 223 B.C. to remove the last major obstacle, the great southern protoempire of Ch'u, and two years later establish himself as Ch'in Shih-huang-ti, the first emperor of the new China. Virtually his inaugural act spoke to the nomadic threat. Rather than entirely demobilize his great armies, he sent units north to consolidate the already existing defensive lines—once again, not apparently the forerunner of the present Great Wall[98] but still one of a series of monumental attempts to define what was China and what was not. The framework of imperial politics was in place, the process of consolidation consummated. But the settlement would not last, for it was based essentially on force.

Over the next twelve years he traversed his empire, not just building roads, digging canals, issuing currency, standardizing laws, and creating administrative districts but also moving hundreds of thousands of people, executing literati, burning books, and reputedly even punishing an entire mountain by painting it red—a record that would live on mostly in infamy.[99] Within a year of his death, insurrection would break out, ending only in 206 B.C. with the establishment of a new and far different regime, the Han. Unification triumphed, but the rule of Ch'in and the violence leading up to it had been decisively rejected.

Instead it was the Han dynasty that bequeathed to China a concept of empire which would endure virtually intact for two millennia.[100] Effective government might ultimately depend upon compulsion, but it was to be justified in moral and intellectual terms. Conversely, the mutually reinforcing potential of religion and warfare would be left, for the most part, latent. Civilian administrators, not warlords and priests, would rule, and it followed that the use of capture and territorial aggression as societal equilibrators would remain in abeyance. There was, however, a cost attached. China would suffer famine and plague stoically, and quite possibly its ride along the population roller coaster was steeper and wilder than it might have been. But this was deemed acceptable. The legacy of the sages and the civilization's essential historical experience stood vindicated. War's appetite had been shown to be boundless, and it would be curbed if it could not be eliminated.

For the pastoral threat remained an enduring fact of life. At times China would lash out. In 141 B.C. Han Wu-ti, like the Chao, created his own horse nomads and brought half a century of war to the steppe, but at such a price that the state was forced back on more passive measures.[101] The alternatives were appeasement and fortification, and in toto the three policies would define the dynamics of a continuing, indeed nearly perpetual, symbiosis between imperial agriculture and militant pastoralism. Like all parasitic relationships, it was draining—but less draining, the Chinese understood, than unrestricted war.

China would endure not because it had optimized the variables associated with mass subsistence through intensified farming but because it reached the best possible accommodation between hydraulic tyranny and the human spirit. Plainly, it was not optimal for creatures who had evolved in freedom and equality, but it was bearable. When the covenant was broken, Chinese peasants would rise up in righteous indignation. But so long as mass agriculture remained the sole alternative, they would turn again and again to Heaven's Mandate.

{12}

THE WORLD ANEW

I

Bernal Díaz was a soldier, a practical young man with an excellent eye for detail. But as he marched inland toward the Mexica capital of Tenochtitlán, he and the rest of Hernando Cortés's little band of conquistadors had the repeated sensation of being caught in a dream. There was the same eerie combination of the familiar and the fantastic. He had seen towns and cities with streets and plazas not unlike those of Spain. Everywhere there were carefully cultivated fields, filled with hardworking peasants. Along the way they had encountered a familiar mix of the exalted and the degraded—aristocrats and slaves, priests and bricklayers, soldiers and prostitutes.

But there was also a palpable strangeness to this place, an absence of everyday things that left Díaz with a nagging sense of wonder. Where, for instance, were the animals? Not so much as a single goat, or sheep, or pig, or ox, or cow could be found anywhere. There were no plows, nor were there wagons or carts. The only beasts of burden were the people themselves. Even more bizarre, the Spaniards had gradually come to realize that humans were also a source of meat, that war captives regularly had their hearts ripped out and their limbs delivered to waiting stewpots.

Yet Díaz was already convinced that this would not be his fate, nor that of his fellows, however small their numbers. Native warriors were numerous and brave enough, but in previous skirmishes had shown themselves to be ill armed and curiously without purpose. They fought capriciously, suicidally intent on taking captives, and they were terrified by the Spaniards' guns, often swooning when they were fired. Warhorses so bewildered them that they seemed to believe mount and rider were one. But no military failing was greater than the locals' disregard of fortifications. Amazingly, these warlike people had left their municipalities un-shielded by surrounding walls and therefore vulnerable to assault. Coming from a world where circumvallation was the rule, Bernal Díaz puzzled over why this should not be the case here. But as he marched across the causeway and saw Tenochtitlán shimmering in the distance, he became convinced that it was God's will, that he meant for Spaniards to rule here and deliver this blasphemous people from the clutches of Satan. "Soon they will be ours," he whispered almost in prayer. "Nothing can stop us."[1]

Díaz was right. Within months the realm of those we know today as the Aztecs had crumbled, deserted by its tributaries and beset by a host of new enemies. For the Spaniards were not the only invaders. In the first half of the sixteenth century A.D., Mexica and the other great Amerindian empire, the giant Peruvian "Land of the Four Quarters," Tawantinsuyu, were confronted by what amounted to an ecomilitary onslaught, an alliance of humans, domesticated animals, infec-tious disease, and even weeds—all representing a separate and entirely more com-petitive environment—which would sweep over the Americas with a suddenness that was practically without historical precedent. Although the religious and po-litical significance of this epic conquest has long been appreciated, it is only within the last several decades that anthropological and biological authorities have deci-sively broadened the context, showing this to be a true "war of the worlds"—a clash between two hitherto segregated environments with implications at practi-cally every level of life.

For it is almost universally conceded that the civilizations of the New World were truly "pristine," giving every appearance of having evolved without any knowledge of or influence from the advanced cultures of Eurasia.[2] In this light they can be seen virtually as experimental controls, showing, among other things, that the rise of complex societies was not ultimately a response to local conditions but a more or less pervasive phenomenon addressing problems shared by all human groupings living at a certain level of subsistence. And though the tendency has been to focus on the considerable similarities between Old and New World civilizations (such as intensified agriculture; large, hierarchically organized popu-lations; institutionalized religion; and parallel forms of monumental architecture), there were also substantial differences—variations that can serve as a basis for comparing the relative importance of the respective dynamic factors in the societal equation and assessing their mode of operation. Such an approach is especially useful in understanding war, illustrating just how sensitive it is to key exogenous variables.

As presently understood, the Americas were first populated by humans some-time around twenty-five thousand years ago, the result of small-group migrations across the Siberia-Alaska land bridge, exposed by the dramatically lowered oceanic levels during the last glacial period.[3] At this point only dogs would have been available to accompany them, and as the human bands spread downward through

North and South America they would encounter none of the ruminants that sub-sequently would form the Old World pastoral complex. It is also important to know that these human groups would have been too small to sustain the bacterial and viral parasites that would generate the major infectious diseases in Eurasia. It follows that when New World human populations eventually did grow large enough to perpetuate contagion, there would have been neither the number nor the variety of domesticated herd animals to act as key vectors for transferable infections.[4] So humankind's new home began relatively free of disease and pastoral animals. And when the glaciers melted and the land bridge disappeared, it was destined to remain so, physically isolated from further biological and cultural diffusion.

Thus sequestered, social evolution in the Americas would unfold at a notably leisurely pace. Recent evidence indicates that the domestication of plants began nearly as early in the New World as it had in the Middle East,[5] but the subsequent transition to farming took place more gradually. For a long time agriculture tended to be a supplemental activity, a seasonal interruption in hunting-based migratory patterns.[6] And with the exception of certain very early coastal communities in northern Peru, the rise of fixed population centers also transpired over several millennia. More to the point, settlement patterns tended to be sprawling, not compacted. Fortifications remained the exception, not the norm, and if present at all were likely to be strongholds located well away from areas of habitation.[7] There are occasional signs of violence among the remains of communities, but destruction is generally limited to governmental and religious quarters, not domiciles. More often than not, abandoned settlements betray few if any signs of physical devas-tation. This is a far different archaeological record than has been unearthed in the Old World, and would seem to beg for careful analysis.

Contemporary archaeologists and anthropologists have devoted considerable at-tention to describing and analyzing the role of war in complex Amerindian soci-eties. While often conceding that it was more ritualized and ceremonial than in Eurasia, there has been little inclination to view it as substantially different. Nev-ertheless, by drawing connections between the faunal and epidemiological singu-larity of the New World, its patterns of settlement, and the general paucity of fortifications, it is possible to reach significantly different conclusions.

The only indigenous ruminants with any pastoral potential here were the Pe-ruvian camelids, llamas in particular. Yet these animals were primarily good for wool and transporting light loads, clearly not sufficient to support significant in-dependent populations.[8] Elsewhere, there was nothing. So the decisive break be-tween herders and farmers simply never occurred in the New World. There may have been some trouble with residual hunter-gatherers, but the nature of these groups and, once again, the dearth of fortifications argue that this was not that serious or sustained. And absent the fundamental enmity between the tenders of plants and the keepers of animals, there would have been no general urge to huddle behind walls, thereby initiating the implosion-explosion phenomenon that I argue had accelerated and intensified social evolution in the Old World. Instead, political consolidation and social stratification evolved in a less abrupt fashion, with fewer obvious signs of coercion. War was clearly part of this process,[9] but its role was more exclusively political. Looser settlement patterns and the reduced role of epidemic disease make it likely that the populations of complex New World societies were inherently somewhat more stable than those of Eurasia, and therefore less in need of war as a demographic equilibrator. When combined with the lack of a basic pastoral-agricultural enmity, it appears that this resulted in a considerably different martial tradition—one that largely excluded noncombatants from the consequences of organized violence and kept war sufficiently a matter of elites and armies to generally obviate the need for circumvallation. Warfare in the New World was fundamentally about leadership, not peoples. Unlike in Eurasia, there is little reason to suspect that its roots were tangled with genocidal hatreds.

Of course, no social theory is without inconsistencies. In virtually all the regions of Mesoamerica and Peru where large agriculture-based polities did evolve, there is archaeological evidence that during chaotic periods some elements found it necessary to build walls around themselves.[10] Just as was the case with Jericho, where fortification occurred prior to pastoralism, this illustrates that any settled population can be sufficiently threatened to resort to circumvallation; the danger, not the means of subsistence, is critical here. As we have seen earlier, however, such danger was greater and more prevalent when agriculture and nomadic pastoralism were juxtaposed. Thus in New World centers of social evolution circumvallation remained not only exceptional but temporary, with fortified sites generally being abandoned with the return of more stable conditions.[11]

But war in this context was no toothless game, somehow divorced from a fundamental role in societal dynamics. The civilizations of the Americas had a number of functional shortcomings and vulnerabilities—protein deficiency was a chronic problem particularly in Mesoamerica, and on the whole their ecologies were fragile and climates prone to disastrous fluctuation. These issues were reflected in a preoccupation with capture and, in certain instances, cannibalism, along with the very sudden and surprising collapse of apparently healthy polities. Also, in the Basin of Mexico and on the Andean plateau improvements in agriculture and distribution appear to have culminated in substantial population increases, which, in turn, were paralleled by a decided intensification of warfare and the emergence of more frankly expansionistic political entities during the period just prior to the arrival of the Spanish. So war here, despite its divergent origins, appears to have retained an intimate relationship with imperial agriculture. But to explore further requires that we move from the level of generality and look more closely at the key regions where complex societies evolved and the civilizations they eventually produced.

II

The ancient Mayas were an enigma wrapped in vegetation, a civilization discovered without warning after a thousand years in the shadows. "Architecture, sculpture, and painting, all the arts which embellish life, had flourished in this overgrown forest . . . and none knew that such things had been, or could tell of their past existence. . . . All was mystery; dark, impenetrable mystery," wrote American diplomat and explorer John Lloyd Stephens upon arriving at the ruins of Copán in 1839.[12] Over the next century, finds accumulated and verdure-clogged temple complexes were discovered across the Yucatán and much of Central America. But the Mayan aura of mystery and romance persisted, stimulated no doubt by their elusive literacy, a glyph-based wealth of inscriptions that stubbornly resisted deciphering. In this rarefied atmosphere, historical imaginations overheated, and by the 1950s Classic Mayan civilization had become a sort of theocratic theme park—a benevolent society run by scholar-priests enlisting industrious slash-and-burn peasant farmers scattered across the countryside to build great ceremonial centers, where the clerics might then pursue their esoteric calendrical and celestial calculations in splendid isolation.[13] According to this interpretation, the likenesses carved in Mayan stone monuments must have been gods or astronomers, and the glyphic texts solely concerned with time and religion. Peace was assumed. For in such an environment competitive politics and warfare would have been basically irrelevant.

This vision of ancient Mayan society, a reconstruction at odds with the posited dynamics of virtually every other prehistoric civilization, has been recently challenged as highly fanciful.[14] At the core of this reinterpretation is a series of dramatic breakthroughs in deciphering Mayan writing, propelled by the work of Tatiana Proskouriakoff, Linda Schele, and David Stuart, among others. Although much decoding remains to be done and nearly all texts contain elements that cannot now be completely interpreted, it is already clear that the inscriptions are not primarily religious but are mostly concerned with commemorating key events in the lives of the governing elite—political advertising, in other words.[15] Leaders with names like "Jaguar Claw," "Stormy Sky," and "Shield Jaguar" carefully and ostentatiously recorded the dates of their accession, the dynastic marriages they managed, and the names of those whom they captured in battle—chronicles altogether more compatible with the traditional profile of human rulership.

In a parallel fashion, accumulated insights into Mayan settlement patterns have pushed contemporary theories of their demographic and subsistence bases much more into line with those of equivalent ancient civilizations. For increasingly careful studies of the areas immediately surrounding the ceremonial quarters gradually revealed not vacant complexes but true urban centers with high population concentrations—so high, in fact, that they seemed incapable of being supported by slash-and-burn agriculture.[16] The puzzle persisted until the early 1970s, when B. L. Turner, flying over the central Yucatán, noticed the remains of extensive terracing compatible with intensified farming.[17] This discovery was quickly followed by others that have now established that the ancient Mayas practiced a range of sophisticated techniques, including extensive ditching and raised field agriculture, analogous to the superproductive *chinampas* of the Mexican basin.[18] Very suddenly, estimates of Mayaland's carrying capacity zoomed, making it clear that this was a very large society, probably totaling in the millions. But while the scale and intensity of agriculture imply labor inputs and organization comparable to those of hydraulic societies elsewhere, the role of politics and warfare points in another direction entirely.

Politically, the world of the ancient Mayas was and remained fragmented, a patchwork comparable to Old World multinodal balances except that the focus of competition and the degree of participation differed markedly. To explain its operation Joyce Marcus recently proposed a "dynamic model" of regional states organized around a four-tiered hierarchy of primary centers, secondary and tertiary localities, and villages.[19] But if the records carved in stone are to be believed, principal centers during the Classic period never waged war against each other. Instead, the source of conflict was rebellious secondary centers seeking independence or predominance. Critical here were matters of prestige and the political gravity necessary to forge new alliances or perpetuate old ones. This was plainly a matter of personalities, for few would dispute historian Ross Hassig's judgment that Mayan warfare was normally restricted solely to elites.[20] A wealth of inscriptional evidence indicates that the essential object was neither territory nor conquest but the capture of political rivals.

Typically, the unfortunate prisoner was delivered back to the abode of the captor, where, after a period of public humiliation, he was ceremoniously beheaded.[21] This process, elaborated to the point of psychodrama, not only was the culmination of Mayan martial endeavor and sure to be commemorated in stone but was symptomatic of a general Mesoamerican preoccupation with seizure and human sacrifice, which eventually reached a bloodcurdling crescendo among the Aztecs.[22] But the Classic Mayas limited it mostly to the leadership and did not resort to eating their rivals.

Besides, there were alternatives. Inscriptions indicated that dynastic marriages were an important Mayan political instrument, as was the so-called ball game. It was exactly that, a sort of full-contact combination of basketball and soccer, played by rivals, including rulers and their opponents, on elaborate courts that were among the centerpieces of municipal complexes throughout Mayaland. While first played by the Olmecs and perpetuated by later Mesoamerican societies, the game was a passion for the ancient Mayas, arguably representative of their penchant for ritualizing combat and politics.[23] For if the rules generally produced a tie, a lucky shot could leave the loser without his head. Frustration punctuated by sacrifice at the hands of the victor—metaphorically speaking, this was the essence of Mayan politics.

For the system that emerged leaves the impression of a kind of perpetually immature balance of power—not focused on conflict among primary competitors but preoccupied with the consolidation of the individual components and unable to achieve it. This serves to explain the frequently asked question of why no hegemonic state emerged from the welter.[24] Competition at this level did not and really could not exist. Each time a primary center moved to expand, its jealous, weakly united constituents would take the opportunity to revolt.

Just why this was so remains unclear. However, the roots of frustration possibly can be traced to several factors. This was a highly fragile environment supporting a large population, and it may be that the ecological dynamics simply would not permit the kind of mass participation frequently found in balance-of-power systems.[25] It has also been suggested that the likelihood that large centers had equal access to the same set of resources--a condition not true in the Valley of Mexico— held down competition at the primary level.[26] Similarly, the difficulties involved in transporting all but precious items probably hindered the kind of differential accumulation of wealth that might have helped cement the dominance of primary centers.

This is not meant to imply that Mayan polities lacked central direction. As we have seen, contemporary scholarship has shown them to be more, rather than less, like other large societies based on intensified agriculture. This implies the organization and ability to enlist labor on a scale sufficient to build and maintain the ditches and other field works associated with this type of farming. Still more obvious are the multiplicity of temple complexes, festooned with that ubiquitous manifestation of public industriousness, the pyramid. As with ancient Egypt—the evolution of which was significantly less influenced by war than other Eurasian societies—these Mayan piles of masonry and rubble can be viewed as energy sumps, outlets for the excess vivacity liberated by such a large social concentration. But the dynamics of recruitment remain obscure, since, outside of the inscribed comings and goings of the elite, very little still is known about the lives of the Mayan masses. Yet the physical evidence points to polities subject to societal dynamics found elsewhere, heating up sufficiently—even in the absence of pastoral pressure—to require significant palliative measures.

To a limited extent this potential is reflected in the military record. For throughout ancient Mayan history there were isolated outbursts of fortification, including, for example, the very important site at Tikal.[27] Plainly, this indicates that occasionally Mayan warfare reached a stage of intensity that threatened entire populations. Recently, however, archaeologist Arthur Demarest uncovered several fortified sites in the Petexbatun region, most notably at Dos Pilas, which have led him to hypothesize that the whole pattern of Mayan warfare changed toward the end of the seventh century A.D. He argues that the rulers of Dos Pilas began a campaign of true territorial conquest, not only carving out an empire for themselves but setting off a chain reaction that intensified war through the entire region—radically nucleating settlement patterns, disrupting the fragile ecology, and ultimately contributing significantly to the collapse of Classic Mayan civilization.[28] But as appealing and logical as Demarest's argument undeniably is, it remains unconvincing. For one thing, the so-called collapse can be adequately explained by a congruence of factors, including elevated populations living in a fragile environment subject to severe drought and disastrous tropical storms.[29] In such circumstances the bottom might have dropped out of the subsistence base before war or anything else could have had much of an impact; or alternately, these conditions simply could have led to a long period of enervating decline. Meanwhile, the absence of extensive destruction at abandoned sites and the rare and sporadic pattern of fortification prior to and long after this point also argue against Demarest's thesis.[30] Finally, if we shift our focus to a much more competitive environment, that of the Mexican Basin, the equivalent absence of circumvallation seems to drive the point home.

III

The Aztec imperium toppled by Hernando Cortés was the culmination of more than eighteen hundred years of societal evolution in the Valley of Mexico.[31] Nevertheless, when judged by all but the most permissive standards of human behavior, it must be viewed as an aberration, a culture driven by war to a point most comparable to that reached by ancient Assyria. And by tracing the Aztecs' steps on their path to empire, the psychological and behavioral impact of incessant aggression is revealed in the starkest possible terms. Yet there is another dimension to be considered. For as compulsively warlike, hated, and feared as the Aztecs unquestionably were, their potential victims did not normally find it necessary to build walls around their homes and families. This remarkable fact highlights the impact of the differential evolution of warfare in the New World and the Old, pointing to how in the former case the absence of pastoral aggression served to mitigate its effect on noncombatants, while nonetheless allowing the behavior of those doing the fighting to reach a predatory pitch, which we today have great difficulty fathoming. Yet it is these bizarre incongruities that make the Aztecs so interesting, opening for us a window on war unlike any other.

The Aztecs were newcomers, having wandered unwelcome into the basin sometime during the twelfth century A.D. to begin painfully carving out a place for themselves as the last of the hegemonic Amerindian societies that flowered here.[32] But even in their imperial splendor they remained marked as parvenus, ceaselessly looking over their shoulders, searching like the Assyrians for legitimacy amid the dominant cultures that had preceded them. If we are to understand what the Aztecs grew into, we must have a sense of what they grew out of—the political legacy they inherited from their predecessors. But there will be no inscriptions here to guide us, for the peoples of the basin were never truly literate. There remain only the faint voices of tradition and the silent ruins of the places they once lived.

The first and perhaps most interesting examples are centered at Teotihuacán, twenty-five miles northeast of Mexico City. Here the impressive remains can still be seen, a broad central avenue flanked by a multitude of public structures including the Pyramid of the Sun and the Pyramid of the Moon, the former being the largest single-stage structure ever built by an indigenous New World society.[33] And it is entirely representative of the overall scale of things here. For archaeologist Renée Millon has carefully mapped a surrounding city divided into planned quarters and districts, the whole of which may have accommodated as many as two hundred thousand residents.[34]

This extraordinary size was explained when a team made up of William Sanders, Jeffrey Parsons, and Robert Santley established that, between A.D. 100 and 700, the rulers at Teotihuacán had radically nucleated settlement patterns by basically moving everyone in the valley into this one huge urban complex.[35] Evidently this was not a matter of defense, since the city never would be fortified.[36] Rather, it seems to have been a drastic approach to consolidation in an environment that lacked wheeled and animal transport—the same issue that led to Mayan frustration. And it worked for a long time, ensuring political control and providing the labor to hydraulically intensify agriculture sufficiently close to the capital to allow food products to be effectively delivered, even in the huge quantities required.[37] So effectively, it appears, that more and more of the city's residents (eventually about a third) could be removed from the fields to produce the goods demanded by a burgeoning foreign trade network spreading across Mesoamerica.[38]

Ultimately, however, Teotihuacán could not sustain itself, as an overworked environment, systemic inefficiencies, and interruptions of trade conspired to drag the city into decline and then collapse. There is a heavy concentration of martial imagery and other evidence of increasing warlike activity after A.D. 600.[39] There was also a continuing tradition of military-related human sacrifice and possibly cannibalism.[40] Yet in the end the city suffered little physical damage; some monuments were burned, but not much else was touched.[41] So the Aztecs were left very much aware of this gargantuan ghost town, and may have been referring to it when they spoke reverently of Tollan.[42]

Most, however, believe they meant the Toltec site at Tula, which was the sociopolitical focal point of central Mexico between the tenth and twelfth centuries A.D. Whoever they were, the rulers of Tollan could do no wrong in Aztec eyes. According to Friar Bernadino de Sahagun's remarkable *General History of the Things of New Spain*, the closest thing we have to an Aztec oral history, "The Tolteca were wise. Their works were all good, all perfect, all wonderful, all marvelous."[43]

This is a matter of opinion. Physically, the ruins of Tula point to a locality with as many as 125,000 residents, though less concentrated than Teotihuacán and far less impressive in terms of monumental architecture and urban planning.[44] The city's economy also appears to have been focused on craft specialization and trade, and the city was and would remain similarly unfortified.[45] Nonetheless, there is plenty of evidence of military activity. Ceramics and stone carvings indicate Toltec warriors had access to a wide variety of weapons, and it appears that they were used not just by the elite but by formations numbering in the thousands, divided into projectile specialists and those armed for close combat.[46] As with much else regarding the Toltecs, the role of this force in the culture's expansion out of central Mexico and deep into the Mayan Yucatán remains to be clarified.

But the available evidence indicates that this was not a gentle people. In particular, archaeologists at Tula discovered a rack resembling those used by the Aztecs to display the skulls of their enemies, along with sculptures thought to be receptacles for hearts obtained through human sacrifice. Meanwhile, large numbers of human bone fragments found scattered in the ruins give us ample food for thought regarding cannibalism, especially in light of what happened later.[47] In spite of this, when Tula did collapse—albeit suddenly, late in the twelfth century— it was, like Teotihuacán, spared from more than the destruction of its central ceremonial precinct.[48] The rest remained untouched, simply abandoned.

This, in essence, was the politico-military environment the Aztecs-to-be encountered when they first arrived as displaced hunter-gatherers[49]—the tradition of Teotihuacán and Tula. There had been divergent interludes, times of chaos when certain peoples took to the hills and built fortifications. But in the main the political landscape continued to be dominated by groups living out in the open on the valley floor engaged in a kaleidoscopic competition for hegemony—a world of shifting alliances, endless warfare, prisoner taking, human sacrifice, and, it seems, cannibalism.

Here the newcomers fared poorly at first, proverbial victims driven from place to place, until at last they found a sanctuary on an island in Lake Texcoco, the shallow body of water that once covered a goodly portion of the valley.[50] It was these lacustrian whereabouts that would ultimately provide a key to supporting the Aztec metropolis that grew here under the name Tenochtitlán and is the present site of Mexico City. But for the moment it constituted a point of departure, a haven from which Aztecs could begin to participate in the local power balance. At first they did so as very junior members of the Tepanec alliance, earning a reputation for desperate ferocity as warriors.[51] More years of fighting brought them increasing success, until toward the end of the fourteenth century the Aztecs helped the Tepanecs' tyrant, Tezozomoc, achieve a major victory and were rewarded accordingly with large tracts of land. Meanwhile, the association with Tepanec had also led to a taste for tribute, which began to evolve into a distinct part of the Aztec economy, and one linked directly to the leadership and its martial ambitions.[52]

Subsequently, both the ruling hierarchy and its religious ideology became increasingly militarized. Most important here was the emergence of the imperialist cult of Huitzilopochtli, a god with an insatiable lust for the hearts and blood of prisoners, a hunger that would one day leave Aztec society marooned on an accelerating treadmill of war, capture, and human sacrifice.

But there are indications that this warlike tack initially met with less than unanimous support, and that a crisis was reached early in the fifteenth century around the time the leadership decided to challenge Tepanec. The commoners did not want the fight, and some even suggested appeasement through surrendering the cult image of Huitzilopochtli.[53] Yet the war party's counterproposal was still more drastic. "If we do not achieve what we intend, we will place ourselves in your hands, so that our flesh becomes your nourishment. . . . You can eat us in the dirtiest of cracked dishes so that we and our flesh are totally degraded."[54] Judging by other evidence, this was no metaphorical offer. The gamble worked, however. Tepanec fell in 1428, leaving the militarists permanently in charge and cementing the Triple Alliance, a partnership with Texcoco and Tacuba that would prove the instrument for further Aztec aggrandizement. The die was cast.

Historian Inga Clendinnen calls Tenochtitlán "a beautiful parasite," and the captive host that grew to feed it was very much a cash cow, a tribute empire most resembling in its gross functions, at least, that of the Assyrians.[55] There was the same extortionary lust for payment—everything from scorpions to gold, bulk foodstuffs to the most delicate of featherwork—all administered through a similarly jerry-built structure of control, suspiciously configured in a manner that increased the possibility of revolt, reconquest, and still higher exactions.[56] Like the Assyrians, as military obligations grew, so did imperial levies of tributary manpower, a key Aztec military advantage being the sheer size of its fielded forces.[57] Even their tactical deployments were not dissimilar, being based on massed archers and projectile specialists backing a core of elite warriors armed for close combat.

Yet a good look at an actual battle would have revealed some startling differences. For the link between political and military intent was rendered entirely ambiguous by the behavior of the warriors, particularly elite combatants. Their aim was not primarily to destroy the enemy by killing as many as possible; quite the contrary, it was to capture opponents in one-on-one combat—bringing them down with a blow to the hamstring or knee, and then grappling them into submission so that minions could bind and remove them to the rear.[58] Battles did include other types of fighting (projectiles, particularly arrows, must have taken a considerable toll), and there was clearly a corporate pursuit of victory through breaking opposing formations.[59] Hostilities also proceeded in a familiar enough sequence, beginning at long range and intensifying as armies drew together. But the core and essence of everything was still capture—this is why elite weapons were designed to bleed and weaken, not to kill, and why rules of engagement were so carefully defined to promote individualized fighting and unambiguous outcomes in terms of who captured whom.[60] For a man's prisoner was his future, the key to rank and privileges at home.[61]

This, in turn, appears to have given rise to the phenomenon of "Flowery Wars"—ritual contests for the elite, and arranged between the Triple Alliance and local opponents strictly with the intent of producing prisoners. While some insist that the motivation remained primarily political, these fights, which took place at a designated time on sacred ground, are some of the purest examples of intra-specific aggression applied to human warfare ever recorded, comparable only to the staged combats of the ancient Greeks and medieval tournaments.[62] Yet the aura of ceremonialism—the stress on individual combat, the careful matching of weapons and opponents on what amounts to neutral territory—was bizarrely con-trasted by the fate of the prisoners, which was literally a matter of predation. For life in Tenochtitlán had come to revolve around their consumption.

The ritual killing of war prisoners was clearly part of the regional military tradition and practiced throughout central Mexico,[63] but among the Aztecs it had accelerated steadily, reaching astonishing proportions by the late fifteenth cen-tury.[64] While the process went on daily throughout the year, single ceremonies on occasion accounted for literally thousands of victims. Most notorious was the dedication of the temple of Huitzilopochtli in 1487, an extravaganza conservatively estimated to have cost twenty thousand lives. But numbers pale beside the image of four lines of prisoners each two miles long, all moving slowly toward the central pyramid where teams of priestly executioners awaited them, zealously ripping out hearts literally day and night until the necessary bloodshed was finally accom-plished.[65]

If for no other reason, the sheer magnitude of the slaughter cries out for an explanation, and historians have been grappling with the problem for what amounts to centuries. But recently the debate has become polarized between rel-ativists inclined to downplay the significance of the killing and those who have gone so far as to call the imperium of Tenochtitlán a "cannibal kingdom." While exaggerated claims have been made on both sides, it borders on the absurd to maintain, as does one authority, that "Aztec warriors went to war 'to feed the sun' with the same conviction that many twentieth century soldiers battled to 'make the world safe for democracy'."[66] For any reasoned defense of this particular native American civilization must ultimately come to grips with the obsessive quality of Aztec war making and human sacrifice, not to mention the matter of anthropophagy.

For more than anything else the debate was sensationalized by Michael Harner's claim that cannibalism was at the heart of Aztec aggression, that their military motivation could be best explained as a quest for protein in an environment lacking any large meat-producing domesticated animals, besides humans.[67] Quite possibly Harner's thesis might have provoked less outrage had there not been what even critics concede to be "an overwhelming corpus of evidence" showing that Aztecs did in fact eat those they sacrificed.[68] The real questions boil down to issues of quantity and quality and their roles in driving the Aztec war machine.

At this point the Mexican basin was a crowded place. The extraordinary productivity of *chinampas* agriculture—artificial islandlike gardens built up in freshwater lacustrian shallows—combined with the ability to deliver large quantities of produce by boat to and from communities clustered around the shores of Lake Texcoco had allowed populations to climb to levels conservatively estimated at over a million and quite possible a great deal higher.[69] Given these numbers, the idea that the meat from even fifteen thousand captives a year could have significantly upgraded the protein and fat content of the average diet was quickly judged implausible and used to discredit the whole thesis.

This has allowed apologists to maintain that cannibalism at Tenochtitlán was simply a matter of ritual and symbolism—warriors and their dependents daintily nibbling on anthropophagic hors d'oeuvres while lamenting the possibility of such a fate for themselves.[70] The problem is that this version is wildly out of phase with evidence reflecting the obvious gusto surrounding the custom—the care with which corpses were divided, the recitation of recipes, even the pirating and eating of slaves destined for merchants' tables by roguish warriors.[71] We are dealing with the most massive example of cannibalism in human history, and there is every sign that these were enthusiastic diners, partaking in what amounted to prestige protein. As much as he has been criticized for hyperbolic phraseology ("cannibal kingdom" was his), Marvin Harris's ultimate point that the meat derived from capture was very much a part of the Aztec warrior's motivational package deserves careful consideration in light of the evidence.[72] Clearly, it is not the whole story, but neither can its importance and that of human sacrifice be minimized through a selective reading of the record. For in the deepest sense, the Aztecs were what they ate.

Despite the productivity of the *chinampas*, the Aztecs portrayed themselves as living on the ragged edge. Every hand was encouraged to plant, and much marginal land was placed under cultivation.[73] Even so, peasant diets were mostly a matter of maize and a few vegetables. Severe overkills had obliterated the game supply, so instead the masses received their animal protein from things like tamales stuffed with water flies and worms.[74] Pond scum was also a favorite.

Yet even meager fare was not assured. For the basin's relatively high elevation and attendant frosts were constant threats to crops in the field. Everyone remembered the disastrous famine of the year One Rabbit, but there are also indications, such as outbreaks of rioting at feasts, that dietary stress was not an infrequent occurrence. This was certainly nothing unusual among large agricultural societies, but in a place where people customarily ate other people, it would have taken on a whole different cast. For both myth and reality point to cannibalism as not just a reward but a threat.[75] By the time the Spaniards arrived, it was not only war prisoners who were being sacrificed. A growing class of slave merchants were supplying victims from the ranks of malefactors and the improvident.[76] The message was clear enough: bad conduct could lead to the sacrificial stone and the stewpot.

Even the sympathetic Inga Clendinnen calls Tenochtitlán "a startlingly violent place," a place where thousands upon thousand of skulls were racked and proudly displayed, where men ran through the streets wearing the flayed skins of others, and even children observed fellow humans pinioned to altars and butchered by the clergy.[77] Nevertheless, Sahagun's informants sought to present a picture of propriety and control, a society of hard work, social cohesion, and good conduct. For other than eating it, the pleasures of the flesh were frowned upon. Unauthorized consumption of alcoholic pulque or hallucinogenic mushrooms was strictly forbidden, and habitual drunks ran the risk of having their heads publicly bashed in by club-wielding executioners.[78]

Sexuality was similarly bridled. Modesty was encouraged, and nudity equated with the humiliation of defeated foes.[79] Homosexuality was described as "a defilement, a corruption, filth," and its practitioners were subject to enslavement.[80] Prostitutes were also pointedly compared to sacrificial victims, and unscrupulous women physicians were denounced as having "a friction loving vulva."[81] This unease with female sexuality also was reflected in persistent tales of young men left drained, their "well dried up" by repeated acts of intercourse with insatiable partners.[82]

Most decidedly, this was a society where women were expected to know their place, and that was in the home, as "a skilled weaver . . . a preparer of good food."[83] Presumably, this lesson was reinforced during a day-long festival in the month of Tititl, when young boys roughed up any girls or women they found in the street, ritually beating them with bags made especially for the occasion.[84]

One group, however, was allowed much more latitude—the freedom to wear distinctive clothes, eat special food, even take concubines (discreetly)—and this was made up of successful warriors. But even they were tied to a quota system, the greater the number, and ultimately the quality of their captures, the more extensive their perquisites. There could hardly have been a more accurate reflection of the social structure and its intent. For at birth each male child was greeted by a midwife chanting: "My precious son, my youngest one. . . . Thou hast been sent into warfare. War is thy desert, thy task. . . . Perhaps thou will receive the gift, perhaps thou will merit death by the obsidian knife."[85]

This was the lot of those born to a society locked into a pattern of endless war. Geoffrey Conrad and Arthur Demarest describe the inner workings of a competitive strategy in which the acquisition of tribute, mercantile activities, and even intensified agriculture were orchestrated to serve the ends of continuous aggression and imperialism. Yet driving it all, they argue, was a religious ideology based on the ravenous hunger of the gods for captives, particularly Huitzilopochtli. For only human blood could keep them alive, and the more the better. Thus, the entire cosmos depended on the supply of victims—lots of captives meant a healthy, well-nourished pantheon; the reverse implied starvation and the end of the universe.[86] So staring entropy in the face, the sons of Tenochtitlán felt compelled to keep divine tummies full.

While this thesis certainly does address the obsessive quality of Aztec aggression, it is possible to see beneath it the open-ended, self-amplifying nature of warfare itself. For if Aztec gods fed on humans, war fed on societies. And in the case of Tenochtitlán, the destruction of previous traditions and the reordering of governmental and religious institutions during the struggle with Tepanec left the Aztecs basically without institutional curbs on aggression. The result was a society dedicated to war—a place of pervasive brutality where women were subordinate, warriors were given free rein, priests fought in the streets and killed for a living, and even the merchants did what they could to act like soldiers.

But it was not just a matter of roles. The Spanish found a society in the grips of a thanatotic pathology, an environment where obsessive death dealing had triumphed over the life forces, a hive of bloodthirsty prudishness. The similarities to Assyria, that other thrall of war, are too obvious to miss, and given the gaps in time and distance, they say a great deal about the ultimate psychological and behavioral impact of warfare given free rein.

Yet there was one crucial, even startling, difference. Whereas the victims of Assyria invariably hid behind battlements, the Basin of Mexico remained a world largely without walls. Neither Tenochtitlán nor any of her traditional enemies found it necessary to resort to circumvallation. This remarkable fact, the significance of which has largely been overlooked by archaeologists, testifies better than any other to the essentially ceremonial roots of warfare in the New World. Tradition dictated that rival cities not be sacked; instead, conquest was signified by firing only the main temple, a practice very much in line with evidence pertaining to the end of Teotihuacán and Tula.[87] It also appears that the great majority of the sacrificial victims remained male combatants, taken from the battlefield, not from their homes.[88]

Yet the passing of time seemed to be undermining the consistency of this profile. For example, the victims used to consecrate the temple of Huitzilopochtli were made up of rebellious Huaxtecs, including men, women, and children.[89] Nor was this likely to have been simply an isolated occurrence. The accelerating demand for captives would have dictated fewer scruples in their taking, a trend paralleled by the growing domestic trade in sacrificial slaves. Problems with famine, diet, and the culinary fate of the victims all point toward growing demographic pressures having assumed an important role in motivating Aztec aggression. Even Conrad and Demarest, who favor ideological causes, concede that overpopulation in the basin was accelerating the pace and scale of warfare.[90] As this happened there were signs that its quality was also changing. In particular, the rise of the Tarascans, who did sack opposing towns and relied heavily on their own fortifications, spelled something new. For when the Aztecs invaded in 1478, they proved willing to fight with absolute desperation, driving the intruders back with over twenty thousand casualties.[91] The Tarascans were outlanders, but a defeat of this magnitude would not have gone unnoticed, particularly if potential victims feared that capture and cannibalism might well be extended to kith and kin. Indeed, had Cortés arrived a century later, he might have found the landscape dotted with walled towns. For even without the impetus of epidemic disease and a pastoral heritage, it was clear which way the winds of change were blowing. Nevertheless, the fact that he actually found so little fortification also says a good deal.

IV

And that message would echo across the peaks and valleys of imperial Peru. For here again the Spaniards would find an empire without walls,[92] a civilization that had sprung from a tradition which allowed war to evolve slowly and in a manner which normally—though not always—kept it from peoples' doorsteps. This is particularly important for the major premise of this book, and not simply because it took place here independently and fundamentally free from outside influence. For Tawantinsuyu, the realm of those we call Incas, was the beneficiary of a kind of pastoralism, but one based largely on a single and very limited camelid, the llama. So, faced with an environment unique among civilization's incubators for both its rigors and its diversity, the growers of plants and tenders of animals remained united, working out a way of life that made the most out of the possibilities of sharing and cooperation. War would find a place here, but the societies that emerged do not give the impression of being obsessed with organized violence. Clearly, it was a social lever, but it does not seem to have been at the fulcrum.

The history of Peru was about getting high and coming down, about altitude and how it could be used to survive and even to thrive.[93] The two traditional sources of Peruvian civilization, one in the north near sea level and the other in the south at altitude, would both expand by learning how to exploit an environment literally stacked in the favor of those adept at vertical strategies.

In the former case, sedentary living was very old, based initially on villages exploiting the extraordinarily rich coastal fisheries as early as nine thousand years ago.[94] Yet these groups settled several miles inland to take advantage of the natural *lomas*, or fog meadow vegetation, in an otherwise extraordinarily dry environment. And geography would dictate further penetration, for the Peruvian coast was divided by a series of parallel river valleys running from the Andes cordillera to the sea through a fifty-mile band of the driest desert—fingers of water pointing inland.

So as population increased (an estimated thirty-fold during the Preceramic period) and the scale of fishing grew, the requirements for more fibers to weave more nets led to the cultivation of cotton along the various watercourses. This, in turn, would graduate to intensified agriculture—irrigation, terracing, and eventually the exploitation of resources as high as six thousand feet but still close to the coast—a cooperative economy made up of specialists adept at getting the most from each productive horizon.[95]

At some point, however, these discrete river valley complexes would begin to undergo consolidation—a process that led anthropologist Robert Carneiro to theorize that in an environment so hemmed in by desert and thus "circumscribed," only hegemonic warfare and attendant state formation could have been the outcome.[96] While Carneiro may have been right in an ultimate sense, on the whole the process took a great deal longer and was, initially at least, more peaceful than hitherto believed.

Most notable in this regard was the appearance of truly monumental architecture as far back as the third millennium B.C. While there are at least nine major sites along the coast, one known as El Paraíso contained some one hundred thousand tons of quarried stone and is estimated to have consumed more than a million days' worth of labor to build.[97] Despite its scale, El Paraíso did not necessarily imply tight political organization, for the shrine was associated with no particular settlement and apparently was constructed through the cooperative endeavors of several localities.

A notably precocious example of consolidation did appear to take place as far back as 1800 B.C. in the Casma valley, where Shelia and Thomas Pozorski have found evidence indicating that the whole area may have been united into a single polity, perhaps through elite warfare.[98] Moreover, at Pampas de las Llamas–Moxeke they have uncovered what appear to be a series of massive warehouses where goods and foodstuffs were differentially distributed by a bureaucracy. However, even here the Pozorskis note that "residential architecture occupies a surprisingly small area," possibly indicating that much of the population remained disbursed.[99] And while the Casma centers do seem to have provided a prototype of future Peruvian secular polities, the Pozorskis concede they may have been "ahead of their time," disappearing around 900 B.C. without apparent successors.[100]

Meanwhile, the tradition of gradualism generally persisted elsewhere all the way up to the so-called Chavin horizon that flourished between 800 and 200 B.C. Some debate continues as to the role of the magnificent complex at Chavin de Huantar, but it is generally understood as a kind of religious and cultural energy center, deriving its influence through the peaceful radiation of stylistic and conceptual messages. Thus, even though its impact can be documented far along the coast and well inland, the themes represented are conspicuously lacking in political content, scenes of conquest and submission, or divine affirmation of royal authority.[101]

To be sure, societal consolidation did ultimately accelerate, culminating in the fifteenth-century A.D. Chimu state centered in the Moche River valley—a veritable exemplar of a tightly knit, bureaucracy-bound, hydraulic despotism. But despite an obvious military tradition and the forging of an empire encompassing two-thirds of the irrigated land and population of the north coast, the seaside metropolis and capital of Chan Chan remained without encircling walls, with only the various palace complexes being fortified.[102]

During the process of transvalley unification there had been certain episodes of circumvallation, the castles of the Santa valley being the most notable.[103] But centralized control always spelled an end to bastion building and resulted in the movement of populations away from fortifications. This was the norm, the central vector of Peruvian politics at sea level. And while social evolution proceeded somewhat differently inland, this factor at least remained largely constant.

If vertical strategies were advisable in the river valleys along the littoral, they became critical at altitude. For this was an environment that literally towered above civilization's other cradles, a realm anchored on lands over two miles high but using a horizon ranging from forty-five hundred to over fourteen thousand feet above sea level. This was a world of elevated solar radiation, erratic rainfall, howling wind, numbing cold, anoxia, and constant physiological strain, a place of ecological extremes where resources were distributed not horizontally across the landscape but up and down. Survival here, where only one harvest in three was bountiful, demanded that bets be hedged—that niches literally from top to bottom be exploited, that special agricultural techniques and hardy crops such as tubers, which accumulated their carbohydrates underground, be emphasized, that food always be stored against bad times, and that people work together. These were the adaptations that made the world of the Incas possible, adjustments at the core of everything they stood for.

But this took time. Evidence indicates that people had been living in the Andean highlands roughly as long as on the coast.[104] However, it appears that the rigors and vertical possibilities of the environment encouraged very early plant domestication, for a broad and well-preserved botanical assemblage dating back almost ten thousand years has been recovered from the Guitarrero Cave at 7,750 feet, including the oldest cultigens yet recovered in the New World.[105] This constituted the beginnings of a very long and basically gradual march toward societal complexity, but one also marked by the rise of several truly innovative cultures.

One of these was Tiwanaku, a metropolis that grew into an empire between A.D. 375 and 700 on the broad altiplano surrounding Lake Titicaca, the highest substantial body of water in the world. One key to Tiwanaku's success was its sophisticated agriculture based on the construction of ridged and irrigated fields, which not only watered but thermally mitigated the effects of frost so effectively that yields are estimated to have vastly exceeded contemporary harvests, in some cases by a factor of seven.[106] But as archaeologist Michael Mosely adds, Tiwanaku was best termed an agropastoral state, since its fortunes were linked to agriculture and camelid herding.[107]

Llamas and alpacas had been domesticated as early as 3800 B.C., yet their tending had not led to a separate pastoral existence. For although they were a ready source of fibers and useful in carrying light loads on steep ground, they could not be milked or hitched to wagons, their meat was not particularly tasty, and, perhaps most important, there were no other available animals to make up for these shortcomings.[108] So instead of a pastoral assemblage, there were just camelids—too slender a base to support true independence for any significant human population, particularly in this rarefied environment. Instead mutualism prevailed, diversified agriculture coupled with camelid herding in the very high puna grasslands, survival here being facilitated by the ability to freeze-dry potatoes to produce *chuno*.[109] In this context, then, pastoralism, rather than acting as a goad to war, became just one more stratum in the layered lifestyle of the Andes.

And Tiwanaku made the most of it, building up vast herds in the zones far above the altiplano and then turning them to trade as pack animals in great caravans that snaked through the mountains as far as northern Chile, and gradually knit the entire south central cordillera into an economic network, remarkable for both its extent and its longevity.[110] There are signs—evidence of male human sacrifice and the expropriation of monuments—that this process did include some militarism. However, the expansion of the culture associated with Tiwanaku generally appears to have been peaceful, being based primarily on colonization followed by the nonviolent assimilation of local peoples.[111]

This profile was basically reaffirmed in the face of a radical new competitor, the dynamic Huari of the Ayacucho region. Quite probably, this culture rose in response to a severe drought, which ice cores indicate began abruptly in A.D 562. Appropriately enough, Huari success rested on an ability to exploit still another underutilized layer in the Andean sierra—in this case the vast *quichua* zone between the altiplano flats and the high grasslands. The key was irrigation and terracing on a massive scale, quite literally carving out productive niches in the hillsides and then directing water to them from highland streams. Yet this demanded an investment in labor much greater than other agricultural communities were making at the time.

The Huari solution, Mosely argues, was analogous to marketing a major invention, a new technology wrapped in a conceptual package that in this case was based on hierarchical organization and reciprocity.[112] In essence, participants were encouraged to work together and to party together, compulsory labor exactions being carefully disguised with the trappings of communal living and feasting.[113] It was not only a brilliant innovation—apparently the foundation for what Marxists refer to as the "Inca mode of production"[114]—but also, by and large, a peaceful one.

For the Huari appear to have coexisted with their neighbors without extensive violence. In one case, a deep penetration into Tiwanaku territory at Cerro Baul, the Huari inhabitants found it necessary to fortify. Yet the colony was eventually withdrawn, and a buffer zone established between the two states. Meanwhile, the site associated with this episode was in marked contrast to the nature of other Huari centers, which were typically situated on easily accessible flatlands and left without encircling walls or even adjacent forts.[115] This was a highly dynamic culture whose influence can be documented over a broad span of territory. Some even maintain that the Huari invaded and conquered the coast. However, signs of the Huari here are more compatible with the pure attractiveness of their ideas;[116] for the most part this appears true of upland cultural diffusion in general.

Indeed, it is plausible that the evolution of warfare in the Andean highlands was moderated by the very nature of the subsistence strategies that developed here. For verticality promotes a unique and conditional approach toward territory, one requiring access to a range of noncontiguous holdings at several separate altitudes, each of which might or might not be valuable, depending on essentially capricious microclimatic variables. It was a world of productive bits and pieces, "vertical archipelagoes" where cooperative utilization, more than exclusive possession, made sense.[117] For the most part this was a central dynamic of cultural development here, and one that argues against forceful expropriation and violence in general. It is true that during times of political breakdown there was a tendency to head for the hills and build fortifications. There are even indications that lopsided agropastoralism (differential emphasis on herding or farming) might have exacerbated these trends. Yet, inevitably, political reintegration brought a quick end to circumvallation and reestablished the peaceful equilibrium of subsistence patterns.

This, in essence, was the context in which Tawantinsuyu arose. The collapse of Tiwanaku and the Huari had led to some scattering of peoples and an increase in fortified hilltop sites. By dint of war and politics the Incas did manage to reverse these trends, reintegrating on a scale never before approached, but the nature and degree of the prior dispersion, along with the amount of force required to change its course, remain open to question. Empire builders the Incas unquestionably were, yet inveterate conquerors and warlords? That is debatable.

The tradition of the "conquering Incas" that held sway until very recently paints a picture of a rather typical hegemonic transnational tyranny—ever aggressive, ever expansive, relying on what amounted to a professional army, superior logistics, a web of carefully engineered highways, and heavy-handed government to create what was probably the largest empire on earth at the time of the Spanish incursion—a sort of high-altitude, native American version of Rome.[118] This is the problem. While portions of the evidence remain compatible with such a profile, the model as a whole is simply too pat and parochial, not taking sufficient account of the enormous weight of environmental factors and the long tradition of special adaptations that made mass society here possible.

For example, there is the issue of origins. The traditional interpretation based on verbal accounts (the Incas had no written language) contends that the state's formation was catalyzed by a decisive and unexpected military victory achieved by Cuzco over its traditional rival, the Chanca. The hero of this campaign, Pachacuti ("Transformer of the Earth"), after uniting the local region, began to expand his rule through a series of military victories, which set the wheels of empire in motion.[119] Moreover, the reforms associated with Pachacuti are believed to have been so profound that they not only led to the establishment of the Inca state but immediately transformed the Andean social and political order.[120]

The problem is that this explanation does not fit well with archaeological findings. For example, while there is contemporaneous evidence for fortified settlements in the Titicaca area and elsewhere, one glaring exception is the supposedly warlike Cuzco region, where archaeologist Edward Dwyer found no indication of circumvallation or the location of sites on defensible ridges.[121] And this was not just a matter of local politics. Radiocarbon dating does support the very rapid expansion of the Inca state beyond the Cuzco region.[122] However, the nature of that expansion has recently been called into question by evidence uncovered by archaeologist Brian Bauer. In a survey of eighty-five sites in the nearby province of Paruro, he found no evidence of fortification prior to the coming of the Incas.[123] Not only was there no indication of conquest on their part, but subsequent Inca rule appears to have brought little change in settlement patterns or social organization.[124]

Yet the implications of these findings should not be carried too far. They do tend to show that the Incas inherited and built upon earlier traditions, and that one of these traditions was a relatively subdued concept of warfare. Nevertheless, there is ample evidence that, once having established themselves, the Incas did pursue a policy of calculated aggression, and that they were considerably more warlike than the cultures that preceded them—particularly as time went by. Thus, upland resistance was methodically stamped out by storming hilltop bastions, and the two poles of Andean civilization were amalgamated through the brutal dismemberment of the Chimu state.[125] Yet it also appears that many provinces were invaded so as to cause as little damage and as few casualties as possible, since "they will soon be our people, as much as the others."[126] Indeed, the very rapidity of the Incas' expansion across the cordillera in the space of but a few generations argues against sustained and intransigent opposition. The Incas were definitely on the march, but apparently many others were quick to fall into step.

Inca militarism has been explained in several ways. Of particular interest to Conrad and Demarest was the impact of the royal ancestor cult and the system of split inheritance practiced here and elsewhere in pre-Hispanic Peru. For although the death of the Sapa Inca, or emperor, caused the rights of governance, taxation, and waging war to devolve to his principal heir, the new sovereign received no material legacy. Instead, his predecessor—now mummified—along with the remaining descendants of the male line continued to enjoy full rights to all previously accumulated property. Thus, each new emperor was forced to acquire his own possessions virtually from scratch.[127] This was a matter of some urgency, for the emperor's ability to requisition labor was based on a reciprocal ability to feast that labor and otherwise take care of it—the Huari innovation that was the essence of the economic system. And the quickest way out of this conundrum was conquest, the rapid acquisition of productive agricultural lands, which Conrad and Demarest argue wed the Incan Empire to constant and expansive warfare.[128] Yet such new territory inevitably brought more mouths to feed, which complicates things considerably.

Even by the elevated standards of other agricultural tyrannies Tawantinsuyu grew into a very populous place, embracing some ten million souls.[129] Given the capriciousness of the environment, this was an extraordinary figure and bound to have had an impact on the system's evolution. Particularly suggestive in this regard was the almost compulsive construction of food warehouses, or *qollqas*. Finely wrought of masonry and deliberately placed on hillsides where they would be visible at great distances, *qollqas* were obviously intended for display and built in numbers that could not fail to impress—literally thousands surrounded Cuzco and equivalent numbers overlooked regional Incan centers.[130] While such facilities would prove highly useful to armies on the move, their implications plainly went well beyond military logistics. At enormous cost the Incas had erected a tangible and psychological hedge against the uncertainties of mountain agropastoralism, a sort of corporate billboard stating in unmistakable terms: "Come what may, we can feed you."

Part of that ability would have had to do with expansiveness—but less as a matter of territory than of consolidation. For faced with the discontinuous nature of vertical archipelagoes and the elevated populations that had ensued, massive and rapid integration of the kind only warfare could produce was plainly in the larger interest. There was certainly resistance at the time, along with residual bitterness, particularly among those who had been singled out for rough treatment. But ultimately environmental and demographic expediency best explain both the aggressiveness of Tawantinsuyu and the substantial acquiescence of those it conquered. The subsequent stewardship of the Incas—the swift construction of a huge net of roads and bridges, the repetitious resettlement of entire communities—produced familiar signs of a society living right at the edge of technological possibility. Coercion and warfare necessarily served an important organizational function in the societal equation, but factors of environment and tradition ensured it would be a limited one—at least until the Spanish arrived.

V

The very magnitude of the odds still raises questions. In essence, how could so few overcome so many—literally hundreds against millions? But, in fact, they were not alone. It was something akin to an all-court press—an invasion not simply of conquering humans but of conquering cows, horses, and sheep, conquering weeds, and, most lethally, conquering bacteria and viruses. All were representative of an entirely more competitive environment, toughened survivors fearfully adapted to sweep aside the defenses of a world long insulated from the harshness that prevailed elsewhere.

The results spoke for themselves. Within a century herds of feral horses and cattle hundreds of thousands strong roamed the Argentine pampas, and would soon do the same on the Great Plains of North America. Pigs were everywhere. And so were weeds. For once all these animals got loose, explains historian Alfred Crosby, they gobbled their way through species after species of native American plants that had evolved in the absence of such herbivores, only to be replaced by ever-opportunistic Old World weeds, which long before had developed evolutionary strategies to protect them.[131]

The fate of native American humans was even harder. For the gamut of Old World epidemic diseases—smallpox, measles, mumps, cholera, gonorrhea, influenza, malaria, and yellow fever—descended on the immunologically defenseless population to create one of the greatest pandemics in history, a tempest of pestilence that rapidly reduced populations in many places to barely a tenth of their pre-Hispanic levels.[132]

The cultural onslaught was no less devastating. For the politics, warfare, and weapons of the Old World cut through native American civilizations like red-hot knives through butter, leaving their populations not just conquered but so psychologically debilitated that the way was similarly opened for the rapid advance of European language, religion, and institutions in general.[133]

Militarily the campaigns in Mexico and Peru were twins, if not identical then at least fraternal, characterized by purposeful voracity on one side and virtual paralysis on the other. For the Mexicans and Peruvians proved barely more capable of defending themselves against the tactics and weaponry of the Spaniards than they were of repelling their diseases. In particular, they were awed by the Europeans' firearms and cavalry. Sahagun's informants and Bernal Díaz repeatedly refer to Aztecs stupefied by the reports of guns, while their auditory effect appears to have been similarly debilitating among the Andeans.[134] Meanwhile, the capacity of men on horseback to terrify was reaffirmed. For just as centaurs endured in the imagination of the Old World, native Americans tended to perceive mount and rider as one gigantic invincible behemoth, and reacted accordingly.[135] Add to this the cumulative impact of crossbows, steel blades, and effective armor, and the full magnitude of the tactical imbalance becomes apparent.

Yet the martial handicapping of the Aztecs and the Peruvians went far beyond tangibles. Above all there was a uniform incomprehension of what they were facing, and what war meant to these bearded strangers. As befit their own military tradition, both Moctezuma and the Inca Atahualpa did basically nothing to prevent the Spanish from establishing themselves in key urban centers, lapses that led directly to the capture of each. Of course there were no walls, and even if there had been the Europeans carried the most portable and lethal of siege engines—epidemic. Yet tactical and strategic hopelessness should not be confused with real cultural barriers to understanding.

The Spaniards, on the other hand, knew exactly what they wanted (gold), and what it meant to control cities, to dominate the political and economic nervous system. They represented a political culture recently honed by the thinking of Niccolò Machiavelli, and they practiced divide and rule—particularly against the Aztecs—to shred alliances and activate old hatreds. When this failed, they were instinctively ready to use indiscriminate violence to bring entire populations to their knees.

This was less a matter of cruelty, military technology, or even expediency than of tradition—the mainstream of a martial heritage that spread across the face of the globe during the age of imperialism and has come to be known as the Western way of war. The civilizations of the New World had represented a distinctive version of war's origins among large agricultural societies, one that demonstrates how organized violence might have evolved had there been no split between agriculturalists and pastoralists. Yet the future would no longer tolerate the requisite isolation, and in the new context the most violent practitioners held all the trumps.

Transcendence involved much irony, for the ascendant martial mainstream was rooted in ecological and historical accident, and tended to destroy its most adept and enthusiastic practitioners. So if the Aztecs and Incas were doomed, so was Spain destined to fight itself into debility. As we should have seen by now, war is an institution capable of considerable variability. Yet it is also prone to amplification and inherently difficult to moderate. For among most forms of traditional societies, besides submission the only option to force was more force. This is why the standards of violence tended to rise until finally all-out war became literally suicidal—a veritable blind alley.

Bundled in the Western skein of war, however, was a subordinate thread representing a unique and alternative approach to organized violence. Unlike war's other original sources, it stemmed from societies primarily dependent on neither plants nor animals—nor land, for that matter. Indeed, that they occupied territory at all was simply a matter of necessity. For they took to the sea for a living, and they looked to the sea for protection.

{13}

THALASSOCRACY

I

The priestess's bare breasts heaved as she watched the long boats gathered in the harbor below depositing warriors, horses, and chariots on the beach. "So it has come to this. The two-faced Achaeans are back, this time to conquer us. How have we come to be so cursed? Why, Mother, have you forsaken us?" her sobbing voice drowned in the wailing crowd of on-lookers.

Six times the seasons had danced past since the Bull had pulverized Thera and thundered across the sea on the great wave that had buried Knossos' fleet. Darkening the sky with the cloud kicked up by his hooves, the great beast had left the island of dreams helpless—her labyrinths shattered, her land poisoned by dust, her sons and daughters slowly starving. When the men of Tiryns and Athens and Mycenae came and saw her condition, they cried false tears and oozed sympathy: "Let us carry your oil and wine in our ships; for we are partners as always." But nothing had come in return, and soon the Achaeans were back demanding more goods: "You have no choice. Your fleet is gone; it is our black ships which still sail the foam-flecked seas." This from a race that had quaked at the sight of our swift ships, fawning when we cast them the crumbs from our trade.

Now their ingratitude was our greatest malediction. But none—none—was a bigger hypocrite than the *Lawegetas*[1] Theseus. His voice was loudest in accusing us of killing Athens's children, of dragging them to Crete to be devoured by our bulls. Certainly, some had died; our own sons and daughters had died also. Leaping the Bull would always be dangerous. Yet no others had reveled in it more than the Achaeans. Now they use it as a pretext for invasion, a shroud of lies to cloak their greed and ambition. We begged the Egyptians for protection.[2] But they could do nothing, and now we are alone, without friends and without hope.

Crete's time had passed. The shining culture, which for so long had managed to avoid power's brutal necessities, would at last succumb—a victim more of bad luck and plate tectonics than of any societal shortcoming. For the civilization that developed here was a remarkable combination of old and new—at once a logical outgrowth of the original Neolithic social dynamic and a mirror of the future, nurtured and protected by a novel slant on human subsistence. The island of Crete originally had been settled at the beginning of the sixth millennium B.C.[3] well before pastoralist depredations began transforming inland agricultural settlements into walled social crucibles. Instead, Crete was left to evolve peacefully and gracefully in insular safety until reaching a takeoff point of an entirely different order. For the omnipresence of the surrounding waters combined with a fortuitous mix of resources suggested a very promising way of making a living.

Trade is ancient, clearly predating human sedentism. Yet among the landlocked, transportation always posed significant difficulties, normally limiting the range of transferable products to highly portable luxury goods.[4] With sufficient organization and determination, obstacles to large-scale terrestrial shipment might be overcome, but it was not easy. Meanwhile, what generally passed for trade in the Bronze Age—perhaps due to its focus on the precious—was dominated by coercive agents and frequently lacked any measure of reciprocity, amounting to little more than tribute—the Assyrian brand of mercantilism.

The sea raised other, more expansive possibilities—inherently neutral space, with the potential for transporting far heavier loads, and even contributing an inexhaustible motive power, the wind. Like trade, maritime travel was very old— witness the arrival of the Cretans. But during the third millennium several societies located in and around the eastern Mediterranean were learning to construct seaworthy craft, propelled by oar and sail, capable of carrying cargoes measured by the ton.[5] With this development the spectrum of appropriate products was broadened in a most significant fashion. Meanwhile, their growing transit rather quickly revealed the rough outlines of supply and demand—that goods plentiful in one area could bring handsome profits in places where they were scarce.[6] From here it was but a short step to the notion of value added—that products might be processed or created in ways that made them uniquely valuable, generating even broader demand and still larger profits.

This had revolutionary potential. Unlike prior means of subsistence, it was an open-ended proposition, theoretically without limits as to the wealth it might generate. Free exchange, manufacture, and innovation were self-reinforcing processes, which played upon the uniquely human capacities for reciprocity, complex communications, and invention to raise the prospect of a future vastly exceeding anything nature had as yet rendered possible—and also one more in tune with our heritage. For it promised to restore a measure of the mobility that agriculture had stolen from our species, not just to those actively involved in long-distance trade but eventually to large numbers of people by creating the discretionary funds and transregional economies that would one day make travel and changes of venue truly feasible for the masses.

But this would require time. Meanwhile, the prototype cultures that began evolving in this direction remained subject to many of the same constraints that stymied other types of societies existing in the same time frame. There is evidence that overpopulation continued to be a problem. A rudimentary understanding of natural processes, an immature technology base, and the lack of a premeditated developmental philosophy combined to set fairly rigid bounds on the productivity and sophistication of the developing material culture.

Then there were issues of defense and the use of force. Initially, archaeological remains from both Crete and Harappa led some to conclude that these were peaceful cultures that had somehow managed to avoid the relentless spread of warfare.[7] More recent evidence, however, appears to contradict this original view, leading to counterclaims that in their own way they were likely to have been every bit as warlike as other kinds of concurrent societies.[8]

A more balanced view seems in order. Almost by definition these tended to be relatively small, rich entities, virtual magnets for potential aggression. Moreover, the desirability of establishing and maintaining far-flung trade routes and entrepôts logically called for some kind of protection. And in all cases but Harappa, the primary means of this protection would have been navies, the perishability of which naturally left less evidence than an equivalent string of stone forts. So there was both the need for defense and the likelihood that efforts devoted to it would be underestimated by posterity.

Nonetheless, available evidence still reveals a notable lack of martial themes and a recurrent though not uniform absence of fortifications. These cultures appear to have resorted to force selectively and pragmatically—a sort of continuation of trade by other means. Although cases exist of mercantile states driven into poverty by the loss of access to a particular product or market, there frequently were alternatives. For ships could go elsewhere and trade for other things, so in this context war seems to have been a matter of circumstance and geography, not a compulsive reaction to basic societal disequilibriums. This is most significant, for it supports the thesis that warfare among humans is purely a cultural institution and largely a function of subsistence patterns. Therefore, if the facts fail to show that early mercantile societies were not warlike, it is still important that, compared with agricultural despotisms or mounted pastoralists, they were considerably less warlike.

II

Greek mythology was always immensely entertaining, but virtually no one suspected it of containing a core of fact until Homer's tales led Heinrich Schliemann, the Attila of archaeology, to the site of ancient Troy and the grave circles of the kings of Mycenae.[9] Arthur Evans was similarly influenced in his life's work, the excavation of the ruins of Knossos. In this case it was a series of stories surrounding King Minos, who reputedly united Crete, established history's first fleet, and then built a naval empire sufficiently far-flung to enable him to exact a blood tribute from distant Athens. Each year the city was obliged to deliver seven youths and seven maidens, who were then taken to Crete and devoured by the hideous bull-headed Minotaur in a huge maze known as the Labyrinth—a practice that continued until Theseus, with the help of Minos' daughter Ariadne, finally slew the beast.[10]

With these legends in mind and several peculiarly engraved sealstones in hand, Arthur Evans arrived in Crete determined to find evidence of the island's ancient kingdom and proof that it was literate. Almost immediately his excavations were rewarded with success. Barely a week passed before Evans found a baked clay bar inscribed with the ancient script he had hoped to find. Very soon after, he came upon "an entire hoard of these clay documents, many of them perfect."[11] Then on April 13, 1900, barely three weeks into the dig, Evans's trench hit upon a spectacular find, a chamber whose well-preserved frescoed walls contained both stone benches and what appeared to be a throne—"the Council Chamber of Minos," he would excitedly label it, calling the civilization itself Minoan, a designation that persists to this day.[12] But perhaps most impressive to Evans was the sheer age of everything he was finding, "nothing Greek—nothing Roman . . . nay, its great period goes well back to the pre-Mycenaean period."[13]

Soon other rooms saw the light of day, a number with vivid and remarkable frescoes. Gradually it became clear that they were part of a vast six-acre complex, irregular and filled with passages—a veritable labyrinth. Everywhere there were reminders of bulls—bull-leaping frescoes, a bull-horn altar, and various other bullish accoutrements—along with repeated representations of a stylized double ax, the Anatolian word for which was *labrys*.[14]

Nor was the connection with the Greek mainland overlooked. Evans was very much aware that the relics recovered from Mycenae and elsewhere seemed deeply influenced by the style and culture of Crete, and he drew the logical conclusion that Knossos had ruled there also.[15] While this judgment would be spectacularly overturned, it remains true that Evans was always more impressed with the art and sophistication of his Minoans than with their power. For the most part his perspective would be supported by evidence gathered by other teams at similar but smaller sites scattered about the eastern half of the island at Phaistos, Hagia Triada, Gornia, and Mallia.

Even Edwardian eyes could not miss the joy and sensuality of a culture that H. E. L. Mellersh would aptly characterize as "hedonistic with an active undercurrent of religiosity."[16] Frescoes revealed a trim, athletic, and nature-loving people, consumed with the pleasures of life. The human form was an object of exuberant embellishment, with female attire typically plunging sufficiently low to fully reveal the breasts and male garb consisting simply of a loin-cloth, often incorporating a codpiece. Youth and beauty were celebrated, and from a decidedly co-ed perspective. And nowhere was this more evident than in the depiction of bull leaping, which showed its obvious dangers being shared by male and female alike.[17] To archaeologists of the early twentieth century this must have been a startling revelation of equality between the sexes, not only in the context of their own backgrounds but also in terms of the other ancient societies their colleagues were digging up and presenting to the world.

Nearly as surprising must have been the lack of emphasis on things military. A few weapons were found, but there were no depictions of fighting, nor, apparently, any fortifications. It was clear that the labyrinths' exterior irregularity was due to their being built outward from a ceremonial courtyard rather than inward from surrounding walls, seemingly without regard for protection.[18] There was not even an effort to separate the complexes from surrounding houses, which occasionally occupied higher and more defensible ground.[19]

Indeed, there is every sign that these edifices, which Evans and his fellow researchers were quick to label palaces, played a continuous role in community life. It was equally apparent that much of what went on inside had to do with religion. Evans realized that a number of areas within the Knossos labyrinth were likely shrines and recognized the religious symbolism of the double ax and the bull. Plainly aware that the central deity would have been an offshoot of the very ancient Great Mother goddess, he wrote that the place "teems with religious suggestion" and reminded him of Anatolian initiatory sanctuaries.[20] This was an important insight, reflecting the clear differences between religion here and the grandiose tabernacles and sky-god imagery of Indo-European imperial despotisms. Room dimensions, artifacts, frescoes—everything at Knossos was on a human scale. If these were palaces or temples, they were homey workaday renditions.

And this is meant literally, because it was also obvious that the complexes were centers of industry, crammed with workrooms and storage capacity—in the case of Knossos, between 60,000 and 120,000 liters.[21] Here skilled craftsmen—potters, lapidaries, metalworkers, and faience makers—would have produced luxury goods for palace use and export, their labor commissioned by Evans's hypothetical priest-kings and recorded by their scribes. The archaeologist would find evidence of three separate scripts at Knossos—a hieroglyphic scheme and two more abstract systems, which he labeled Linear A and Linear B. Though Evans would succeed in deciphering none of them, on the basis of context and recognizable ideograms he assigned Linear A (which remains basically untranslated) a "sacral use" and

Linear B the role of keeping "business records, such as accounts and inventories."[22] Substantively, these designations may have been plausible, but in assuming all scripts represented the same language he made, as we shall see, a pivotal mistake.

On the whole, though, Evans performed a great service. While his restorations of Knossos remain controversial, he was a meticulous collector who captured the essence of Minoan culture and grasped that this was a society heavily dependent on commodity production and overseas trade. Evans was weakest in explaining the florescence of ancient Crete—he leaned heavily on Egyptian influence—and its demise.[23] Smoke-blackened stone led him to conclude that one spring day around 1400 B.C. his beloved palace had been consumed by fire, but the chain of events leading up to the conflagration escaped him.[24]

Nearly a century has passed since Arthur Evans began digging at Knossos, and in the interim much has been learned. Gaps and controversies certainly still remain, but it is now possible to address Crete's rise and fall in terms of causal chains drawing on a much broader band of evidence than was available at the turn of the century. Although less is known about the society's beginnings than its end, the nature of Minoan religion, the probability that its people spoke a non-Indo-European tongue, the prominence of the women, and the generally pacific cast of the material culture—all argue for direct and uninterrupted development from Neolithic roots.[25]

Because the island's useful land was limited and its water supply too meager for irrigation, a sharp cleavage between agriculture and pastoralism never materialized, nor did intensification proceed in separate directions.[26] Instead, a mixed economy evolved which encouraged the generation of secondary products—woven and dyed woolens, along with olive oil, wine, and a host of ceramic vessels used both as containers and purely for decorative purposes. Such value-added commodities were natural objects of trade, and their production and transfer would be further stimulated by the availability of another prime staple, timber—stands of oak, maple, and, in particular, extensive cyprus forests covering the now-barren hills and mountainsides of Crete.[27] It can be assumed that the Cretans had been building boats out of wood since their arrival, if for no other reason than to gain access to the profusion of seafood swimming in local waters.[28] But gradually during the third millennium B.C., the ready sources of timber, combined with surpluses of exportable commodities and a notable scarcity of metal resources of all kinds, would have led to more and longer voyages by a growing fleet of larger and increasingly seaworthy craft.[29]

This period, known as the Early Minoan, (c.2800–1990 B.C.), also witnessed the appearance of protolabyrinths, such as the ones at Palaikastro and Vasiliki, which were plainly paced by commercial development and likely sites of production themselves. This was very much part of the accelerating process of development. But unlike the social compression stemming from the circumvallation of agricultural settlements elsewhere, consolidation around these centers would not necessarily have been its primary impetus. Rather, a case can be made that it was the very nature of maritime trade—the continual stimulation from contact with far-off peoples, unfamiliar environments, and novel customs and techniques—that acted as the key factor invigorating the Cretans and pushing them toward a truly novel version of social complexity.

At any rate, shortly after the year 2000 B.C. the process of boat building, commodity production, and labyrinth construction passed a critical juncture, and Minoan society began maturing very rapidly. There was a marked increase in the number of occupied sites, and this continued until virtually all areas suitable for habitation were being used.[30] Population densities were also plainly on the rise, particularly around the major centers such as Knossos, which expanded to an estimated forty thousand by around 1500 B.C. The infrastructure apparently grew apace, with evidence remaining of harbor improvements, aqueducts, bridges, and roads linking the key centers.[31] Meanwhile, the introduction of wheeled transport would have further rationalized internal distribution and the delivery of products to ports of departure.

The result of all this development was a dramatic gain in prosperity, and while some clearly became wealthy, the appearance of general well-being is more suggestive. Skeletal evidence, although limited, exhibits few signs of malnutrition or early death.[32] Specialized roles clearly proliferated, but this did not necessarily lead to much stratification. Instead, as archaeologist Peter Warren points out, Minoan society seems to have remained unusually free of class or territorial divisions.[33] Little more is known about the specifics of governance than when Evans hypothesized priest-kings; it remains clear, however, that whoever was in charge successfully preserved internal peace.

This brings us to the matter of external defense. It is now apparent that by the end of the third millennium a rich though insular culture such as the Minoans' was subject to raiding and even invasion. For archaeologists Stella Chryssoulaki and Maria Agouli have recently uncovered at Karoumes Bay on Crete's eastern end a complex of fortifications dating from approximately 1900 B.C. and obviously designed to provide early warning of attack. These discoveries bring to mind the legend of Talos, a kind of Bronze Age Robocop, who reputedly patrolled the island, casting rocks upon intruding vessels. More significant perhaps are indications that after around two centuries the forts were no longer in use.[34] Since threats of this kind rarely disappeared of their own accord, it is logical to conclude that the Cretans had come to rely on a more proactive means of securing their shores—in effect, a navy. And the evidence indicates that they did not confine themselves to home waters.

Vessels at this time were reasonably seaworthy, but they were basically day sailers. Therefore, prudence and convenience dictated that trade routes be interspersed with bases spaced at intervals of approximately one day's travel time.

Indeed, archaeologists have been able to piece together two lines of Minoanized island settlements, one pointing toward Asia Minor and the other aimed at the Greek mainland—an "Eastern String" and a "Western String," to use the terminology of archaeologist Jack L. Davis.[35] Together these two strings encompassed much of the lower Aegean Sea, an area that, if it was to remain pacified and kept open to trade, must have been guaranteed by what most authorities now concede amounted to a Minoan thalassocracy after around 1700 B.C.[36]

Its exact operation, however, remains elusive. The image of a mighty flotilla—an ancient analogue of the British fleet—vigilantly patrolling the high seas and, if necessary, engaging in epic naval battles probably overstates the case by a considerable margin. The possibility that the Minoans employed specialized warships with rams has been examined and rejected, though it probably deserves to be reconsidered on the basis of recurrent and suggestive evidence.[37] Meanwhile, however, the consensus remains that Crete's sea power would have been based on the unrivaled possession of enough generalized vessels to deliver troops to any point in the Aegean, a supposition bolstered by a fresco discovered on the island of Thera apparently depicting an amphibious landing.[38] While there are physical signs that the Minoans did suppress what are presumed to have been pirates in exactly this manner at Ayia Irini and Phylakopi, their use of force should not be overemphasized.[39]

For the most part the Minoan expansion in the Aegean seems to have been peaceful, a judgment based on the relatively few signs of violence in the archaeological remains. Frequently, the Minoans managed to live amid or alongside indigenous islanders for long periods.[40] Indeed, it is quite possible that the phenomenon some call "Pax Minoica" was based as much on cultural prestige, mutal self-interest, and diplomacy as it was on the Cretans' ability to enforce it.[41] Iconographic evidence indicates that the Minoans did employ mercenaries, African troops and, in particular, Mycenaeans.[42] These aggressive Achaean Greeks, more warlike and skilled sailors in their own right, appear to have been co-opted by the Minoans. Captivated by their culture and enriched by their trade, there are signs that the Mycenaeans came to assume the role of enforcers, the muscle behind the Minoan thalassocracy.[43] But this, too, should not be overemphasized. For the central impression that has emerged from the archaeology is that of a commercial nexus, more commonwealth than empire. Protection and expansion were clearly factors, but as long as the Minoans remained dominant, the watchwords seem to have been peace and prosperity.

But ancient Crete was no Utopia. Recent scholarship has tended to reveal another, darker side to Minoan life—not just signs of endemic social problems but also evidence that the culture was founded, quite literally, on shaky ground. First there was the matter of population pressure. Because the earliest commercial societies were characteristically based on limited territory, overpopulation was a recurring problem. One productive outlet, plainly serving the ends of long-range trade, was colonization; another was expatriation, the sale abroad of the same skills fostered by commercialism at home. In Crete's case the Eastern and Western Strings would have accommodated a certain number of excess people, while itinerant Minoan artisans may have wandered as far as Kahun in Egypt.[44] Characteristically, however, commercial colonies of this sort were small, as were the number of spaces for the ancient equivalent of "guest workers." Faced with similar conditions, the Phoenicians as well as the Hellenic Greeks resorted to infanticide. Until recently, such a practice among the life-loving Minoans would have been considered unthinkable. In 1979, however, the excavation of a shrine near Archanes produced clear evidence that a seventeen-year-old boy had been sacrificed there.[45] Around the same time, archaeologist Peter Warren uncovered in a house, about one hundred yards from the Bull's Head Sanctuary at Knossos, a mass of children's bones from which the flesh had been removed—bringing to mind the legend of the youth-devouring Minotaur. While Warren maintains that human sacrifice was rare in Minoan culture, to some extent the evidence speaks for itself.[46]

Whatever the ultimate impact of overpopulation upon Minoan culture, it would have been mild in comparison to what nature had in store. For it was Crete's misfortune to be located in an area of extreme geological ferment. And repeatedly during the first half of the second millennium B.C. its residents found the stability of their lives being pushed off the Richter scale. Around 1700 B.C. one or a series of earthquakes shook Crete, causing major damage to all the labyrinths on the island.[47] The resilient Minoans, however, responded with a rebuilding program dedicated to producing new complexes even more elaborate than the originals. At Phaistos, for instance, the eager renovators knocked down everything, leveling the site with a thick layer of rubble and cement in an apparent effort to make the successor structure earthquake-proof.[48] Elsewhere, the architecture took on a new richness and delicacy that exemplified Minoan civilization and ushered in the so-called Neopalatial period, during which the culture reached unprecedented levels of refinement and prosperity.

But it, too, was destined to be shattered by disaster, in this case a veritable Bronze Age big bang. For around 1470 B.C. the thriving volcanic isle of Thera, seventy miles to the north, suffered a catastrophic caldera eruption that totally blew out the center of the once-circular island and left a crater about the same size as Krakatoa, which perished in a similarly spectacular fashion in A.D. 1883.[49] The blast—equivalent to perhaps one hundred megatons—would have delivered almost instantly a shock wave sufficient to crack and even collapse many of the exposed mud brick structures in northern Crete.

Yet the secondary effects must have been far worse. A tsunami traveling in excess of a hundred miles an hour in deep water, and then building to a height of between 90 and 150 feet, would have crashed into the coast, shattering ports, along with any ships they held.[50] As likely as not, this marked the end of the Minoan fleet and with it its control of the thalassocracy. Soon after, a giant cloud would blot out the sun and begin depositing a choking layer of white sterilizing ash—smothering crops and rendering the farmland in eastern Crete useless for several years.[51] All told, it was a blow from which no society could recover quickly. Indeed, of all the labyrinths on the island, only the one at Knossos would subsequently be fully restored and occupied.[52]

Researchers at Knossos, including Evans, had long been aware of an increase in things military in and around the complex during the late period of inhabitation. Rather suddenly, graves began to include large amounts of arms and accoutrements. Certain frescoes depicted what could be interpreted as shields. Finally, clear pictograms of horses and chariots were found on Linear B tablets, the first time either had been seen on anything Minoan.[53] Further questions were raised in 1939 when a cache of Linear B tablets were found in Greece at the Mycenaean site of Pylos, but this could also be interpreted as further evidence of Minoan penetration.[54] Evans, at least, never doubted that Minoans remained dominant.

Then, in 1952, eleven years after his death, came a stunning revelation. A young English architect and part-time classicist, Michael Ventris, using techniques similar to cryptoanalysis, succeeded in deciphering Linear B. It was written in Greek—an archaic form, but still recognizably Greek.[55]

Very quickly the signs of militarism at Knossos, the unprecedented appearance of horses and chariots, the very use of Linear B all snapped into focus—Achaeans had come to rule at Knossos. It now appears that the explosion of Thera had left the Minoans fatally vulnerable, and the opportunistic Greeks, having been more shielded from the tsunami, seized the moment. Half starved, with neither fleet nor fortifications, the Minoans must have found resistance hopeless, explaining why there were no signs of violence. Indeed, the coming of the Greeks may have been more akin to a leveraged buyout than an invasion—the new Mycenaean management team arriving to remove the old Cretan board of directors and taking the reins of Thalassocracy Incorporated.

In some respects things would continue much as before. Knossos, at least, gave the impression of renewed prosperity.[56] Goods were produced, carefully cataloged, and shipped to many of the same customers. Thus, in Egypt it was only necessary to register the change in tomb paintings of Cretans delivering goods by painting over their codpieces and replacing them with the kilts favored by the Mycenaeans.[57] In other ways, however, much changed under the new management.

Minoan culture had evolved from what amounted to pristine Neolithic beginnings—a world predating the pastoralist depredations that had given war such a predatory cast over much of Eurasia. Spared this looming menace, life at home had developed gracefully and with a degree of equiponderance, its pleasures and responsibilities shared more equally between men and women. Granted, there is evidence of overpopulation, disease, and fears of the outside world. Yet the impression left by Minoan culture remains sunny and relatively peaceful. When its merchant-directors did resort to force, it was apparently applied on a limited and pragmatic basis, likely with the help of mercenaries. This was appropriate and characteristic of their calling. Sustained success in commerce was essentially about profits, not preponderance or territory. Access to markets and self-preservation were fighting matters, but very little else qualified. So, paradoxically, the Minoans' ancient Neolithic perspective—amounting to a prewar view of war—served them well in an occupation that promised the future.

The Mycenaeans were a different breed entirely. They knew war long before they knew trade, and they never forgot it. Indo-Europeans to the core, they arrived in Greece around 2000 B.C. carrying all the baggage wrought by the agropastoral split—sky gods, male dominance, horses, chariots, and an irrepressible zest for combat.[58] The society they established here vividly reflected this heritage—aristocratic, built around fortified strong points, and regulated ultimately by the brutal use of force. Yet geography and the Minoans showed them another way. For the sea drew them to it, bidding them to build boats and expand their horizons. And the Minoans shared with them their culture and with it very possibly the lucrative notion that goods could be made and exchanged far and wide.

But with the Achaeans it was never the same. Traders they might be, but they were also freebooters, slavers, and mercenaries—full-service Bronze Age entrepreneurs. What's more, they gave every appearance of having been proud of it. H. E. L. Mellersh makes the point that the *Iliad* and the *Odyssey*, those indelible testimonies to Mycenaean existence, only pay homage to the swashbuckling antics of the leadership, never mentioning the heart of the mercantile establishment, the bureaucracy that kept meticulous account of inventories and transactions in Linear B.[59] This could be simply a reflection of the simpler Hellenic society that produced the poems four centuries after the fact, but it also may have been how the Mycenaeans themselves wanted to be remembered, what they chose to tell others about themselves, the core of the oral tradition. And on these grounds fighting was what mattered.

There can be no doubt, however, as to their prowess as traders. For upwards of two centuries after their arrival in Crete, Achaean Greeks were vending their wares virtually everywhere in the eastern Mediterranean—in Egypt, along the Canaanite coast, across the Aegean, and in and around Asia Minor.

Yet they were troublemakers, by their own admission raiding and pirating but also meddling in local politics, acting as mercenaries, and carving out territorial spheres at the expense of more established powers. This was a period of increasing turmoil in the region, and as things grew worse the Achaeans apparently grew more active—playing power politics with the Hittites, who referred to them as "Ahhiyawa," and perhaps even joining the "Sea People's" assault on Egypt, this time being remembered as "Akwasha."[60] Their most intemperate move was the decision to eliminate by force what was probably their biggest rival and impediment to their trade. The desultory siege of Troy, which fell around 1240 B.C. may have given us the *Iliad*, but it appears to have exhausted the Mycenaeans. From this point they encountered nothing but trouble, until less than a century later another band of Greeks, the Dorians, moved in behind their overextended commercial empire and swept over the Greek mainland. All told, this had been no way to run a thalassocracy.

III

The Phoenicians managed better. And they did so after long exposure to war. Unlike the Minoans and the Mycenaeans, whose early development took place in relative isolation and safety, the Phoenicians were surrounded by danger from the beginning. Yet their response remained characteristically oblique, falling back on the traditional strengths of the mercantile—wealth, flexibility, and the capacity to go elsewhere. And if aggressors had to be bought off, at least it was with the expectation that the bribes would be returned with interest by future trade. It wasn't heroic, but it kept the Phoenicians long afloat in a sea of troubles.

Nonetheless, they continue to suffer from what amounts to a bad historical press. In large part this is because most of what we do know about them derives from the writings of others, Phoenician literature having been almost entirely lost (more than a little ironic for the inventors of the alphabet).[61] And to a world ruled by war, the shrewd passivity of the mercantile was bound to be misunderstood. Thus Plutarch, a Greek, was typical in calling them "a people full of bitterness and surly, submissive to rulers, tyrannical to those they rule, abject in fear."[62] In fact, they do appear to have been preternaturally gloomy. Yet this may merely reflect the somberness of a small and wealthy race, forced to navigate a passage littered with imperial behemoths.

Like the Minoans, the Phoenicians' true calling took time to evolve. They began as Canaanites, descendants of wandering Semites who started trickling into the three-hundred-mile coast of what is now Syria, Lebanon, and Israel as early as the fifth millennium.[63] Here they found a band of rich farmland, bounded just a few miles inland by a cedar-covered chain of mountains, rising in places to a height of nine thousand feet. As they filled in this strip, establishing towns along the coast, population began to grow beyond the productivity of the fertile but limited acreage—the onset of what would become a chronic problem.[64] Yet the proximity of timber bid them to build boats and engage in local trade, while promising bigger things in the future. Down through most of the second millennium B.C., however, traffic in the fabled cedars of Lebanon as well as domination of the coast itself remained largely the province of outsiders—the treeless river valley societies of Egypt to the south and Mesopotamia to the east, as well as Hittites to the north. For the Levant constituted the essential land bridge between all three, and as such became the focus of their imperial dreams and well as those of their successors. The resulting turmoil provided the political context in which the coastal ports grew into separate little merchant kingdoms, each struggling to survive amid a grinding train of invasions, sieges, and imperial exactions.[65] To say the least, it was an environment that encouraged flexibility. Yet the obvious source of alternate opportunity, the sea, was blocked for a long time by the thalassocracy, particularly after it fell under Mycenaean management.

Then around 1200 B.C. came dramatic change. In short order the Hittites collapsed, the marauding "Sea People" arrived, Egypt slipped into penultimate decline, and the Dorians cut the heart out of the Achaeans' trading empire.[66] Amid the smoldering ruins, the coastal Canaanites were left to discover that in confusion there is profit, and in the process became Phoenicians.[67]

The transformation of the six principle Phoenician cities—Aradus, Berytus (Beirut), Byblos, Sarepta, Sidon, and Tyre—into commercial dynamos took several centuries, and just as with the Minoans it was based as much upon ingenuity and industry as on pure transport. So if the famous biblical passage in Ezekiel 27 does compare Tyre to a great merchant vessel, it is notable that the goods enumerated here are imports, many of them raw materials.[68] For in actual fact Tyre had become a giant workshop, and the concept of value added proved central to its operation. Thus, stacks of lumber became ornate furniture in the hands of expert cabinetmakers, while giant shell piles accumulated as testimonies to the Phoenicians' skill and persistence in extracting from the murex snail the precious dye that was the basis of their signature product, royal purple. At sixty thousand snails per pound of dye, this meant a lot of work, but at an estimated selling price equivalent to twenty-eight thousand dollars for a single pound of fabric tinted to the highest standards, it was well worth the effort.[69] Meanwhile, Phoenician know-how itself proved to be a valuable export commodity, enabling King Hiram of Tyre to earn a yearly supply of oil and grain from the agricultural Israelites by sending his workmen to design and construct the Temple of Solomon, and later having his mariners sail the Jewish king's merchant fleet, for which they received a substantial sum in gold.[70] Projects this remunerative were plainly the exception, but they do serve to illustrate the demand for Phoenician skills.

Yet there is another side to all of this industry. Archaeologists and art historians have failed to define any really distinctive Phoenician aesthetic canons.[71] For this was a people who borrowed from nearly everyone but created very little that was truly unique, and this is clearly reflected in much of what they produced. Their pottery was dull and poorly made. They did some fine work in gold, but the vast majority of their jewelery was simply trinkets. Other metal castings were frequently crude and careless. Glasswork focused on beads and gewgaws. Behind this shoddiness loomed mass production. For the Phoenicians were among the first to produce and trade manufactured goods in truly huge quantities. At this they were highly skilled, targeting their products to the dictates of the market, but in doing so they apparently succumbed to one of the eternal shortcomings of mass merchandising, the temptation to make money rather than things of lasting value.

Of course, there was one key exception, and that was the alphabet. The Phoenicians may have been aesthetic mimics, but when it came to their language and their script they clung to them fiercely.[72] And well they might have, since their system of writing constituted a truly revolutionary improvement in communications. It is probably not accidental that a trading people was involved, since commerce placed a premium on superior means of record keeping and long-range communications. At any rate, the process was completed by the thirteenth to twelfth century B.C., and—as might be expected with the Phoenicians—was based heavily on borrowing.[73] But in this case they were highly selective and inventive, apparently combining a limited number of Egyptian hieroglyphs with the notion that these, in turn, could be stripped of all graphic meaning to represent single consonants.[74] While the inclusion of vowels would have to await the Hellenic Greeks, the net effect was an almost endlessly flexible system capable of representing virtually any idea or information—an applicability that would be startlingly reaffirmed by the recent cracking of the genetic code, showing, in essence, that life itself is based on a kind of alphabet.

In the case of the Phoenicians it is probably safe to conclude that their major literary outlet was recording financial transactions. It is also apparent that they accumulated a substantial body of historical, religious, and philosophical writings, but since all that remains are restricted and monotonous inscriptions, it is impossible to know what it contained.[75] Direct knowledge of Phoenician existence is further compromised by the fact that in the Levant the key sites are overlaid by modern towns and cities. Nonetheless, certain facts of life remain clear.

Without exception, urban sites were founded either on rocky promontories or on small coastal islands, situated so they provided sheltered anchorages and maximized protection from land attack.[76] All urban centers were surrounded by massive curtain walls, which in combination with their placement severely restricted the room inside. Typically, substantial industrial quarters would have further limited living space. Thus, an Assyrian relief of a Phoenician city shows that residents built up instead of out, with private dwellings reaching several stories high. On the other hand, truly monumental architecture was conspicuously absent. In part this can be explained by the close quarters, but it also seems to reflect the capacity of this type of society to fully channel the energies of its citizenry into productive endeavor, without recourse to the architectural energy sumps that were so much a part of states based on large-scale intensive agriculture.

Population levels, however, must have remained a problem. Little is known about the demographic impact of communicable disease. However, logic indicates that a mercantile society's continual contact with alien pathologies and infected transients not only would have subjected its residents to a somewhat greater incidence of contagion but also gradually would have built up immunities and created a disease-experienced population. But if this sword potentially cut both ways, the spatial limits placed on population growth were less ambiguous.

This brings us to the controversial topic of Phoenician infanticide. Contemporaries—both classical and biblical sources--record numerous examples of child sacrifice, stigmatizing the practice as barbarous and outrageous.[77] These accusations have received archaeological support with the discovery, first in the Punic west but later in the Levant, of sanctuaries, or tophets, containing the remains of children, sometimes in the thousands.[78] It continues to be argued that in both eastern and western Phoenicia this was essentially a rite of propitiation reserved for emergencies and exceptional circumstances.[79] Others, while conceding the religious context, cite evidence that the sacrifice of children, as opposed to animal substitutes, increased as populations grew.[80] It is also worth noting that the skeletal structure of infant humans is largely cartilaginous and does not preserve well. Therefore, the practice could have been more widespread than the remaining evidence indicates, especially among the classes who could not afford to protect the remains in burial urns. It continues to be clear, however, that space imposed severe and inflexible limits on Phoenician numbers. As with the Minoans, itinerancy and colonization provided some safety valve, but also a limited one. This left infanticide.

For those who reached maturity, life within the tight confines of a Phoenician city was probably quite tolerable. Although much wealth was concentrated in the hands of a few great families, the high skill levels demanded by this type of economy, along with limited archaeological evidence, point to a fairly broad and relatively prosperous working class.[81] Society does appear to have had a lower rung, and, although little is written about domestic slaves, they were clearly an important item of trade and almost certainly existed—though perhaps in small numbers in the Levant.[82] For the most part, however, the majority of inhabitants seem to have occupied some sort of economic middle ground.

Nor was government particularly oppressive. Characteristically, cities in eastern Phoenicia were ruled by monarchs, but sovereigns hemmed in by a series of functionaries and representative bodies.[83] It is now believed that in the east general city assemblies not only existed but apparently wielded real influence.[84] Meanwhile, in the west and especially Carthage, where the evidence is better, potentates disappeared entirely, replaced by a system of republican magistracies. In reality, however, power everywhere in Phoenicia probably remained basically in the hands of the great merchant families—essentially an oligarchy of wealth. Yet they could not afford to ignore the general welfare, for it is important to realize that in the Levant, Phoenicia as a cohesive entity never existed. It remained to the end a string of independent-minded city-states, possessing some cultural affinity but little in the way of political loyalty. Whatever fidelity existed took place primarily within the confines of the city walls. Rich and divided, it was not a formula calculated to deter aggression.

Phoenicia's respite from big-power intrusion lasted only several centuries, followed by an even worse scourge. It is hardly an exaggeration to say Assyria tormented Phoenicia, threatening, invading, extorting, and generally terrorizing its inhabitants for nearly two and a half centuries. It began early in the ninth century B.C. when Ashurnasirpal II conducted several military forays into the region, imposing tribute and forcing many towns, including Byblos, Sidon, and Tyre, into submission.[85] It was a pattern that continued basically unabated until Assyria fell in 612 B.C. Indeed, some of the best epigraphic evidence of what Phoenician towns actually looked like depicts Assyrians gleefully looting them. With their customary tyrannical flair, Assyrian rulers gloried in detailing just how much tribute they expropriated and exactly what indignities they inflicted upon Levantine rulers. Esarhaddon, for example, is shown holding the king of Tyre on a leash![86]

Yet in the face of all this domination, the enduring wealth of Phoenician city-states and their continuing penchant to throw off the Assyrian yoke raise some questions as to the exact nature of the relationship. For between the lines of Assyrian bluster it is possible to detect hints of another reality entirely. It is notable, for instance, that in contrast to most of the rest of Syria and Palestine, Tyre, Byblos, and Aradus were able to preserve their independence in the face of continuing Assyrian aggression.[87] Further, it has been maintained by a number of scholars that overland trade within the Assyrian Empire was actually conducted by Phoenicians and Aramaeans—a proposition that, if correct, not only says a good deal about roles and power relationships in this case but, as we shall see with the Harappans, potentially expands the terrestrial possibilities of early mercantile societies in general.[88] In any case, it is apparent that the Assyrians could neither acquire directly nor control the flow of goods brought in by sea. Meanwhile, as historian Moshe Elat points out, the Assyrian appetite for their products could be satisfied only with "the cooperation of the Phoenicians themselves, since many branches of the Phoenician economy remained beyond Assyrian control."[89] So the Assyrians were forced to make concessions. Certainly, the Levantine cities paid substantial tribute—all of which and more was likely returned by Assyrian customers—and endured periodic sieges—destined either to fail or to destroy a valuable source of goods. In any event, it was the Phoenicians' very way of life that provided a degree of immunity from the raw military power of imperial agriculture, which Assyria represented. And the essence of this standoff was inadvertently captured by the Assyrians themselves in a telling inscription that depicts King Luli of Tyre slipping a five-year siege, escaping literally out his city's back door, to joint the fleet and the sea, which remained Phoenicia's last refuge.[90]

It is now believed that the Phoenicians had been sailing as far as Spain in search of metals at least since the twelfth century B.C. but that they did so without establishing permanent way stations, so there is little in the way of an archaeological record.[91] This began to change late in the ninth century B.C. Perhaps partly to escape Assyrian pressure and also in anticipation of the Hellenic Greeks, who also were starting to move into these waters, Phoenicians began to plant colonies dotting the shores of the western Mediterranean, the most famous being Carthage. Unlike the Greeks, however, the Phoenicians had no interest in controlling the hinterlands, confining themselves instead to enclaves that served as trading posts and havens for shipping, set at intervals of around one day's sailing time. Characteristically, the sites sought to approximate the isolated and protected settings of Levantine cities.[92] Here they appear to have established basically pacific relations with the surrounding indigenous peoples. Indeed, it has been suggested that one key to Phoenician commercial success was the ability to enlist aboriginals as agents to market their goods locally.[93] But although few doubt that their primary motivation was profit, not conquest, the Phoenicians were clearly less passive out here than they were at home. For on the sea lanes linking their bases they were a match for anyone.

The Phoenicians were renowned throughout antiquity for seamanship. Their voyages of discovery were the stuff of legend; they apparently succeeded in circumnavigating the African continent at the end of the seventh century B.C. and then in the fifth century sailing past the Pillars of Hercules to explore the coast as far south as the Gulf of Guinea and possibly reaching Ireland and England in the north while searching for tin.[94] There is even evidence that the Carthaginians headed out into open ocean, penetrating the Atlantic as far as the Azores.[95]

As might be expected, the Phoenicians were excellent shipwrights, exploiting the most advanced construction techniques available.[96] And this was true of their warships as well as their merchant vessels. Thus, one ancient tradition has the Phoenicians, not the Greeks, inventing the trireme, the standard fighting vessel in the Mediterranean from the seventh to the fourth century B.C. Similarly, the principle behind its replacement, the quinquereme—the idea of placing more than one man at an oar—appears to have originated in Carthage.[97] Yet it is also interesting that the evolution of Phoenician warships, and those of their chief maritime rival, the Greeks, apparently differed in a fundamental way. As far back as the Geometric period, perhaps even to Mycenaean times, Greek warships were clearly differentiated from merchant bottoms by their fine lines--long, narrow hulls optimized for bursts of speed useful in battle.[98] Early Phoenician naval construction, on the other hand, tended toward more of a compromise—a vessel, in the words of Lucien Basch, "equally suited for commerce and for war."[99] And while specialized Phoenician military craft did emerge later, they characteristically retained a more beamy aspect than their Greek equivalents.

This alternate developmental path can be interpreted as speaking to intent. Like the Minoans, the use of force for the Phoenicians appears to have been a matter of expedience, a necessary part of doing business in an increasingly competitive environment. Phoenicians certainly engaged in a number of massed battles at sea— most notably Salamis—and were known as capable tacticians and fighters. But formalized naval warfare can hardly be considered a major preoccupation, a fundamental mechanism of their society. Keeping trade lanes open was essential, but this would have been accomplished primarily through relentless coastal patrols aimed at suppressing piracy—more of a policing role than a military function.[100]

This circumscribed approach to war seems to have been very much a reflection of the type of society Phoenicia represented, for organized conflict in this context could not be made to serve in the same capacity it had assumed with imperial agriculture. At sea the occupation of territory was irrelevant. Colonial expansion could absorb some surplus population, but the existence of tophets argued that other means were necessary. Central to everything here was wealth generation; this, rather than mass combat or monumental architecture, was the primary outlet for inhabitants' excess energy and a key means by which a modicum of stability was maintained in an intensified urban environment. Unlike the residents of Crete, the Phoenicians were a people who had grown up literally in war's cradle, and the fact that they did not resort to intensified militarism seems to point directly to the limited possibilities of warfare in this type of society.

The great exception was Carthage. It alone of Phoenician foundations created an empire and came to rely heavily on war, not only at sea but also on land. To explain this phenomenon, ancient historian C. R. Whittaker argues brilliantly that it was the result of a unique set of conditions that transformed the nature of Carthaginian society. Among Phoenician colonies, he notes, only Carthage was formally restricted territorially, bound by a pact with neighboring Libyans to confine itself to a narrow neck of land.[101] As Carthage's population began to swell, this initially encouraged the formation of its own set of colonies and a growing reliance upon long-distance trade based on the Phoenician model. While the Carthaginians did respond forcefully to earlier Greek pressure, down to the fourth century B.C. their policy remained fundamentally peaceful and imperial control mechanisms nonexistent.[102] From this point, however, Carthage became more aggressive, gradually being caught up in a cycle of war and imperialism that would ultimately lead to its destruction at the hands of the Romans.

Meanwhile, at home in Carthage, the legalistic territorial restrictions had given land possession and agrarian wealth a very un-Phoenician prestige. This resulted in the treaty being breached, and by the fourth century B.C. Carthaginian aristocrats began a rapid agricultural expansion into native lands.[103] As this occurred, Whittaker maintains, the values of the leadership changed; their notions of trade and overseas commitments came to be conditioned by new ideas of territorial possession, which in turn led to increased social inequality, fears of foreign incursions, and the predisposition to physically exclude rivals from spheres of influence.[104] Even so, unlike the citizen-based navy, the land armies they raised inevitably were composed of mercenaries, procured in what were essentially commercial transactions. Nonetheless, it is still reasonable to conclude that the Carthaginians came to rely on war precisely because they themselves came to be more like the societies that had always relied on war. Yet this may not have been inevitable.

IV

Of all the venues where pristine civilizations bloomed, one remains truly enigmatic. "I have often thought how singular the Indus valley civilization is in this respect," writes the distinguished archaeologist Colin Renfrew.

> For it possesses very large urban centers . . . worthy of comparison with Teotihuacan. The centers have "citadels" with large granaries which were clearly the nub of a complex redistribution exchange system. A range of traded material is seen. Yet nowhere . . . is there the superabundant personal wealth so characteristic of the early civilizations of Egypt, Mesopotamia, and China. Nor has there been found the exceedingly complex and monumental religious symbolism characteristic of the Mesoamerican early state modules. Nor yet, despite the existence of a script, is there the vainglorious assertion of personal power.[105]

Nor, we might add, is there much in the way of fortifications, or even very many signs of war.

That the mystery persists is not for lack of evidence. New and important sites are continually being uncovered, adding to what already is conceded to have been the most extensive Old World Bronze Age civilization, sprawling over an area presently estimated at around a half million miles, or roughly ten times the size of other ancient river valley societies.[106] But as the artifacts pile up, so too do the contradictions, leaving us a culture that continues to resist the familiar categories into which other early civilizations have been satisfactorily pigeonholed.

At first, however, it did not appear to be so. After John Marshall undertook systematic excavation of the two major sites at Mohenjo-Daro and Harappa along the Indus River in the 1920s, the archaeological mainstream interpreted the findings as characteristic of a river-based despotism, perhaps even one heavily influenced by Mesopotamia. For here was an urban literate culture reflecting a very high degree of planning and standardization, which in turn implied a centralized and probably authoritarian government typical of those in the Middle East.[107] This diffusionist paradigm was further buttressed by the apparent abruptness of Harappan civilization, having burst out of what was thought to have been primitive agricultural roots and then matured and disappeared in the space of only around five centuries—now dated between 2300 and 1750 B.C. Gradually, however, this thesis has been undermined by the accumulation of archaeological data on pre-Harappan farming communities, showing them to have been both densely distributed and quite sophisticated. If Harappan civilization came from elsewhere, it was not for lack of an indigenous cultural base.[108] Meanwhile, the artifacts and configuration of Harappan sites gradually came to be seen as products of a highly original and independent developmental pattern, and one that really matched none of the other river valley societies very well.

Almost entirely missing were the hallmarks of imperial despotism. If city walls had existed at the major sites, neither Marshall nor later excavators could find them.[109] While there were signs of weaponry, they were singularly undeveloped—mostly slingballs and maces but not effective arrow-or spearheads.[110] There was relatively little in the way of religious imagery, and what was recovered depicted female deities and bulls rather than male sky gods.[111] More surprising, perhaps, was the conspicuous absence of monumental architecture, no outsized edifices that obviously filled the role of palaces or temples, no pyramids or ziggurats. "In the Indus valley," explained Marshall, "the picture is reversed and the finest structures are those erected for the convenience of the citizens."[112] Nor was this simply a matter of housing. Skeletal evidence later indicated that the entire population, and not just an elite, was well nourished and notably free from dietary stress.[113] Vast disparities of wealth, if they existed, were not obvious. Indeed, there were few signs of any great measure of social inequality.

Instead, there was a remarkable sameness about Harappan sites, not just within them but across the entire geographic spectrum. Marshall himself remarked on the striking uniformity of the culture, the amazing similarity of localities and artifacts found hundreds of miles apart.[114] And while some maintain that this is overstated, the basic point still stands.[115] What remains of Harappan culture is notably standardized and utilitarian. Some art has been found, but there are few signs of decoration for its own sake. Houses and public buildings were apparently without ornamentation. Tools and household items are well made but plain and lacking any vivid style that might provide insights into those who crafted them.[116] So we are left with a mask of sameness, compounded by an undeciphered script that continues to deprive us of even the shadows of history—peoples' names, transactions, events they thought important enough to record. But if much remains opaque, there is still the possibility of resolving some basic contradictions.

In the early 1980s J. G. Shaffer proposed a fundamental reconsideration of Harappan culture, suggesting that in the Indus valley a technologically progressive, urban, and literate society was achieved without recourse to hereditary elites, centralized politics, and war.[117] As an alternative Shaffer suggested "that the similarity in style and manufacture among objects of Mature Harappan material culture reflects the existence of an extensive and intensive internal redistribution and communication system . . . throughout the vast region."[118] The Harappans represented, in other words, an inland version of the mercantile state—an intricate net of entrepôts distributing mass-produced goods in return for produce from an agricultural substrate without primary reliance on territorial control or coercion. To support the hypothesis, Shaffer points to excavations at Allahdino, which showed that even very small and presumably rural sites had all the basic features of the larger urban society, including literacy and virtually every artifact category of Harappan culture, yet lacked any capacity to manufacture them locally.[119] Further attesting to the effectiveness of the distribution system, Shaffer points to the widespread availability of metal tools in Harappan society at a time when such objects "were the playthings of the elite further west."[120]

The thesis plainly has merit. Not only does it match the quantity and quality of the evidence, but, by substituting the concept of coexistence for control, it addresses the central enigma of how a society so apparently lacking in martial skills could dominate an area so vast. What seemed an empire very likely was simply two societies living side by side.

How exactly it worked remains to be explained. For example, ships are depicted in Harappan seals, and it can be assumed that there was considerable traffic along the Indus and the Persian Gulf coast.[121] However, a great many sites were farther inland, leaving unanswered the crucial question of how goods, in what must have been very large quantities, were transported throughout the interior with such efficiency. This, in turn, raises issues of administration. Centralized granaries and possibly public buildings were located on what appear to have been citadels, implying some governmental control over redistribution. Beyond this, however, very little can be said about who ran the operation and on what basis. But whatever model is applied, the remains of the Harappans still give off an aura of tight and meticulous organization that begs for explanation as to how it was imposed.

This brings us back to the subject of force. Applying available evidence and considering the behavior of mercantile entities in general, it is logical to assume that the Harappans did not expand primarily through military means. On the other hand, the types and quantities of weapons that have been recovered are probably consistent with the preservation of internal order, especially in what appears to have been a tranquil society. And there is still defense, potentially no small matter for a culture this rich and dependent upon peaceful interchange. Certain outlying Harappan sites, such as those in the Kutch-Saurashtra region, were fortified and appear to have served as military outposts.[122] As noted, however, settlements in the heartland did not seem to find circumvallation necessary. Since all signs point to the Harappans having been a non-Indo-European culture,[123] this profile is appropriate for a people still unacquainted with the predatory style of war emblematic of those with pastoral origins. And there are some authorities still inclined to equate the Harappan collapse with the coming of the Indo-European Aryans, a horse chariot–based warrior culture, typical of those that spilled off the steppe elsewhere.[124] But because the chronology seems to indicate that the Aryans arrived centuries later, most attribute the Harappan fall to disease, natural disaster, ecological causes, or some combination of factors.[125]

It almost doesn't matter. Even if the Harappans had lasted longer, their fate was sealed. For they also would have been utterly vulnerable to the kind of military power the Aryans represented. That Alexander found India thickly sown with forts and strongholds when he arrived in the fourth century B.C. was simply inevitable.[126] For there could have been no other response to this virulent strain of war, born on the steppe and destined to infect the entire Eurasian land mass. The Harappans represented an earlier, more gentle path, and through geographic good fortune had been given time to evolve creatively along the Indus in relative safety and isolation. But their days were numbered from the beginning.

There is much irony here. For they represented a window on the future. Unlike the Carthaginians, the Harappans appear to have demonstrated that the mercantile model, even at this level of technology, could be superimposed over the country-side without reverting to agricultural hegemony. As we saw, not a great deal is known about how they actually managed it. Nor is it possible to say that the kinds of products they generated or the quantities they exchanged in any way approximated the productivity of a modern economy. But in terms of wholesale reliance on occupational specialization, the Harappans and other mercantile enti-ties marked a new phase in human subsistence, and one destined to eventually rule the earth. But their emergence was premature. Like infants cast in the wild, they were full of promise but unable to defend themselves in the cruel heartland. So war would marginalize them geographically, pushing mercantile states to the very edge of the littoral, where they might survive through strategic retreats to the sea and safety. Here they would remain like mammals in a world of dinosaurs until the time came when the style of life they represented was sufficiently powerful and dynamic to dominate on its own terms—and, in the process, unseat the Second Horseman.

{14}

CONCLUSION

THE HORSEMAN'S FALL

I

More than three thousand years after the Harappan collapse, an unprecedented struggle known as the cold war took a dangerous turn. During the early 1980s Vladimir Kryuchkov, then head of the Soviet state intelligence bureau (KGB) and future leader of the failed coup against Gorbachev, became convinced that the Americans were planning a surprise attack—presumably with their new Pershing II missiles, whose earth-penetrator warheads and extremely short flight times seemed tailor-made to execute a decapitating first strike against deeply buried Soviet command bunkers.[1] On his advice the Soviet Union's leadership mobilized their intelligence assets in a program of global vigilance. KGB stations in all key Western capitals were pressured to report immediately on anything that might indicate preparations for war.[2]

These fears were without substance. The Pershing IIs were yet to be deployed and had never been tested at anything like the ranges necessary to hit Moscow.[3] Yet the antique leaders of the Kremlin were spent men presiding over a crumbling empire—particularly their chairman, Yuri Andropov, scion of the KGB and a sick man. Conditioned by Russia's traumas in the twentieth century, they responded to their system's mounting troubles with growing fury and unbridled suspicion, much of it focused on the man who rhetorically relegated them to "the scrap heap of history," Ronald Reagan.

It was in this context that U.S.-Soviet relations continued their downward spiral. By June 1983 Andropov was describing the situation as "marked by confrontation, unprecedented in the entire post war period."[4] Less than two months later, when a Russian interceptor deliberately shot down a Korean 747 air transport, killing 269 people, the leadership sought to justify the act as an appropriate response to a "spy mission." To drive the point home, on September 29 the by now very ill Andropov wrote the Supreme Soviet that recent events had dispelled the possibility of better relations with the present American administration "once and for all."[5] On November 7 Politburo member and fellow hardliner Grigoriy Romanov called the international situation "white hot, thoroughly white hot."[6] From his perspective this was no exaggeration. For the leadership was all but convinced that the Americans were about to launch the attack.

The pretext was supposed to be Able Archer, a NATO command-post exercise scheduled to run from November 2 through 11 to test nuclear release procedures.[7] But to the other side this might be no test. American and British monitors were surprised to note a sharp increase in the volume and urgency of Eastern bloc communications. Incredible as it must have seemed to the listeners, the Warsaw Pact showed signs of expecting an imminent nuclear attack.[8] Yet it was true. They were on a strategic intelligence alert, the visible portion of an iceberg of Communist fear.[9]

Back in the Kremlin, this fear very likely was compounded by confusion over who was in charge. Andropov was dying. As far back as July he needed support walking, and he was known to suffer from both diabetes and a heart condition as well as a severe kidney ailment. Reportedly, his condition grew worse through the fall, and by November, his mental state and degree of control remained open to question. Probably it devolved upon his septuagenarian colleagues on the Politburo either to carry out his will or to improvise. By November 9 they were apparently near the jagged edge of panic. Moscow sent all KGB residencies in the unsuspecting capitals of the West an urgent call for any scrap of data that might pertain to an attack.[10]

Disaster beckoned. Deterrence theorists' worst nightmares seemed about to materialize. For the first time since the Cuban missile crisis we were "eyeball to eyeball"—only now one side was acting crazy and in the midst of a leadership crisis, while the other had little idea what was going on. The prospects could hardly have been worse.

Days passed, but nothing happened. Able Archer wound down, and still nothing happened. One by one the Eastern bloc units stood down from their alert and returned quietly to their daily routine. Gradually, it must have dawned on the Soviet leaders that they would live to see 1984. Their system remained doomed, but it would collapse peacefully, without benefit of mushroom clouds. This was a crisis made for miscalculation and catastrophic overreaction; if history had been willing to give World War III a chance, this would have been the moment. But it had come to nothing.

The previous scenario may well strike some readers as an exercise in techno-fantasy, the likes of which have made authors such as Tom Clancy household words. The difference is that, given the limitations of the historical record, this chain of events actually appears to have taken place. And while this may be of some surprise to the uninitiated, the outcomes will strike most as thoroughly predictable. From the perspective of the mid-1990s, it makes sense that even desperately misled men should have abstained from responding to their fears when the probable outcome was suicide on a global scale.

Consider now what would have been the reaction of Ashurbanipal the Assyrian had his soothsayers presented him with such a tale; or how would Moctezuma have coped if the priests of Huitzilopochtli had begun philosophizing on the finer points of deterrence theory? Would these despots even have recognized the conditions and possibilities involved as pertinent to war as they knew it? Could they even have conceived of a crisis this grave, and perceived provocation this monumental not leading to immediate warfare? Would any of this make the slightest degree of sense to them? I doubt it.

The fact is that we live in an era utterly different in its fundamental dynamics than the one that brought large-scale warfare to fruition. Granted, the human material remains much the same, but the societal and technological factors that were once war's lifeblood either have gradually ceased to have much impact or have found other less destructive outlets. In short, the Second Horseman's prognosis is not good. But to come to grips with his humbled condition, we would do well first to summarize what we have learned about his origins and then track his erratic path through modern times. It is likely we will never know the exact moment when humans climbed aboard the grim charger war, nor can we say when the ride will absolutely reach its end. Nonetheless, what we have found does appear to have important implications for the future of our kind, and how we might deal with our abiding capacity to inflict violence on each other.

II

War is not simply armed violence. Rather, it is a specific institution—premeditated and directed by some form of governmental structure; concerned with societal, not individual, issues; featuring the willing (though perhaps not enthusiastic) participation of the combatants; and intended to achieve lasting, not ephemeral, results. These characteristics point to a requisite level of human social evolution and imply that warfare is a mechanism intended to perform certain functions, which logically varied in range and intensity as communities developed in different directions. Given such a definition, it becomes possible to differentiate simple blood feuds and extended acts of revenge from what is meant by war in historical terms. This is important since it makes it apparent that humankind was not born to war but came to it late in our existence and as the result of fundamental shifts in subsistence patterns.

Yet this definitional filter, when applied in another direction, raises some profound questions. In the world of nature, although it becomes possible to dismiss virtually all forms of hostility among animals as being not truly warlike, there remains one critical exception—ants, besides ourselves the most social and well organized of creatures. In this case we are driven to the startling conclusion that a number of species within the vast family *Formicidae* qualify as real war makers. Here, not among ourselves, can we find the origins of true warfare. And the fact that ants wage war flows logically from their way of life, much as it eventually would with human warriors. There is one crucial difference, however. Individually these ants are genetically predestined for a martial existence, for they are haplodiploid reproducers. Such creatures sacrifice for the group because it is their best chance to perpetuate their own genes, death in battle being a trivial matter compared with the success of an army made up of genetic near replicates.

It is eminently apparent that we humans are not haplodiploid but fully sexual. Nor is it accidental that—with the exception of Africa's naked mole rats, who share some of the breeding characteristics of social insects—all mammals besides ourselves limit their sociality and self-sacrifice to close relatives. So, very suddenly, it is no longer most pertinent to ask whether humans are inherently warlike but instead to question why it is possible for us to wage war at all, or even to cooperate sufficiently for large societies of essentially unrelated individuals to function.

Although the matter is far from resolved, support has begun to emerge for a dual-inheritance model of human development, positing not only separate mechanisms for genetic and cultural evolution but also a subtle interplay of both to produce the most efficient behavioral patterns. Thus, it seems at some point we became smart enough and our cultural interactions vital and advantageous enough, under certain circumstances, to override our immediate reproductive advantage. It was this profoundly liberating evolutionary innovation, allowing us to rely upon the much more rapid and communicable medium of ideas rather than flesh, that made us flexible and opportunistic in a way never before possible with living organisms. Eventually, it would permit us to wage war, not because our genes compelled us but as a premeditated response to external conditions.

Meanwhile, we still had to eat. Humans evolved as hunter-gatherers, living for 99 percent of our line's history in pack-sized bands dictated by the availability of food sources and genetic affinity. Probably, the lives of our distant ancestors roughly mirrored the patterns revealed by ethnographic studies of recent hunting and gathering societies—a relatively low-key existence emphasizing personal independence, general equality among group members, including women, consensus-based decision making achieved through open and protracted discussion, and freedom of movement, particularly as a means of conflict resolution. Weapons possession would have been virtually universal among males, but employed for hunting and not dominating the group, or in conflict with other groups. Meanwhile, the status of females would have been reinforced by their role as gatherers—the steadiest providers of the most calories.

It is unlikely that we were any less aggressive or less inclined to commit violence against each other than during later times. Indeed, hunting likely provided us with much of the behavioral raw material necessary one day to construct armies. Nevertheless war, as we have defined it, would have been basically irrelevant in a context where personal property had to be limited to what could be carried, seasonal diversity more than territory dictated the availability of food, and the necessity of outbreeding made it advisable to avoid alienating other local gene pools.

It was in such an environment that our big brains evolved, along with speech and our advanced sociality. And it stands to reason that we were deeply affected by this heritage. Thus, in many respects it appears that we retain the souls of hunter-gatherers, preferring a way of life that permits a measure of equality, freedom of speech, freedom of movement, and even an absence of war. While there may be some genetic involvement, its role in behavioral predispositions this complex cannot presently be demonstrated. Nonetheless, it does seem logical that our most primary and intimate social contacts would have continued to reflect the values inherent in our earlier existence and that these would have been passed from generation to generation, regardless of the circumstances in which we came to live. But the explosive growth of our cerebral cortex and our capacity for culture also left us, if not infinitely plastic, then at least capable of an unprecedented level of adaptation—able to undergo a transformation that in relatively short order would find us living in vast despotic societies that were, in their gross outlines at least, analogous to those of the social insects.

It belabors the obvious to state that domestication was critical here. Yet contemporary students view the process as far more mutual and subtle than previously thought, a symbiosis that has the affected species domesticating us virtually to the same degree that we domesticated them. The first great change began to occur around 10,500 B.C. in the Middle East with plants, opportunistic cereals that lured humans into harvesting and eventually cultivating them through evolutionary plasticity in the direction of producing more and tastier food product. Before we knew it, we had become farmers, our ancient mobility compromised and our population swelled to the point there was no going back to hunting and gathering—a veritable plant trap.

Shortly after, domestication struck again, uniting us with the beasts of the field. Yet in this case the mechanisms were somewhat different, as were—more importantly for our purposes—the outcomes. For in the latter instance the accumulation of pastoral animals naturally led to them being pushed toward the outskirts of farming communities, where they could not eat or step on growing crops. In the Middle East a fairly broad assemblage of useful ruminants encouraged their human keepers to strike out independently, leading to two separate ways of life forming around domesticated plants and then animals. While both types of societies would continue to interact and display certain dependencies, there was also a considerable basis for antipathy.

More specifically, it has been argued here that, beginning around 5500 B.C., the phenomenon of increasing numbers of Middle Eastern farming communities building walls around their domiciles can best be accounted for by raids on the part of pedestrian pastoral nomads. Because of the need to prevent pursuit, along with the basic discordance between the two ways of life, it stands to reason that these attacks would have assumed a fairly brutal aspect—terrifying enough to result in the gradual abandonment of open village sites and the concentration of populations in fortified townships. The raiding itself would have been sporadic and geographically irregular, but behind it would have been economic and even ideological motivation sufficient to mark it as the beginning of something approaching true warfare among humans.

Furthermore, recent evidence from the Ukrainian steppe indicating that horse riding extends back to around 4000 B.C., along with remains of domesticated horses in the Middle East from late in the fourth millennium, raises the possibility that by about 3200 B.C. equestrian nomads had penetrated far enough south to be in a position to establish their own pattern of raiding. If this was the case, their inherent military advantage, despite what were almost certainly very small numbers, could have been sufficient to further spark consolidation behind walled enclaves. This is suggestive in that this time frame coincides with the rise of the state and the emergence of true urban societies in Mesopotamia.

It also may be that the ultimate influence of pastoral nomads, either pedestrian or equestrian, was as much indirect as direct. For war tends to spread in a manner analogous to contagion, with hostilities often establishing themselves at higher rather than lower levels of violence. And among those lacking an acceptable means of escape, aggression can be met only by submission or by an equivalent level of violence. So once the precedent of attacking whole settlements was set, it is quite possible that it would have been continued by the original victims—jumping from agricultural community to agricultural community until circumvallation and eventually siege warfare became the norm. In any case, it is still plausible that the origination of warfare as a function of the agropastoral split at the juncture of Europe, Asia, and Africa imparted a significantly more predatory cast to the institution than had it evolved solely out of the interaction of Neolithic farming communities.

But while true warfare may have begun as an act of theft destined to be repeated again and again by future pastoral nomads spilling off the Inner Asian plateau, among the agricultural it would become something considerably different. Social development in the Near East, being focused and compacted behind walled fortifications, was intensified and accelerated. Armed male elites rose quickly to prominence, fostering the erection of governmental structures, differential access to resources, and the coercive organization of labor. The rise of plow agriculture and especially the elevation of force as the central arbiter of human affairs put women at a special disadvantage, turning them into baby machines and forcing them to rely on sexual attraction as the basis of their influence and respect. Population dynamics took on an entirely different aspect. On one hand, sedentism and an almost exclusively carbohydrate diet promoted fertility and long-term population pressure. Yet close quarters also encouraged the spread of epidemic disease, which when combined with increasingly intensified and famine-prone agriculture produced demographies approximating the architecture of roller coasters, threatening serious social dislocation and even collapse at any time.

More than anything else, it was this problem that wars and armies traditionally addressed among the agricultural, not just maintaining social control but acting as stabilizing agents. During periods of overpopulation armies could conquer new lands or, at worst, self-destruct and no longer have to be fed. And when numbers fell, new laborers could be appropriated, a process that serves to explain the traffic in slaves and repeated transfers of entire peoples. It should be emphasized that warfare was never more than a crude equilibrator, simply the most effective available, and one that could be consciously applied.

The external manifestations of this syndrome were first exhibited in Sumeria, initially in a mulinodal balance of power, a geopolitical form that would be repeated through much of human history. In such a system, besides addressing internal causes, war assumed a primary external role as enforcer of the balance among competitors. And because military advantage could be quickly countered through alliance and opportunism, the net result at the higher level was usually rough equilibrium—or general frustration. For as events in Sumer and elsewhere demonstrated, such systems are only metastable, being composed basically of a pack of potential hegemonists. Should one triumph, there is the possibility of a total systemic transformation to an imperial format.

The rise of great hydraulic despotisms was of profound significance, if for no other reason than that the bulk of humanity came to live in them. It was also probably inevitable, since the basic feedback loop of population growth, irrigation, and economies of scale seems to have built upon itself until the carrying capacity of the environment and the limits of available technology caused a developmental plateau to be reached. This is where empires existed, precariously poised on the outer edges of possibility. Here also was war's center of gravity—not because these societies were necessarily the most adept militarily but because they were most in need of warfare's demographic ministrations. This is why imperial armies so frequently met disaster. Soldiers were the mediums of exchange, and battles—orchestrated to produce their death, capture, or return with more labor or land—were the key mechanisms by which energy was transferred from state to state. This was the plant trap's legacy at the highest level of intensification, humans reduced to tokens to be traded on war's version of the market.

But this was also a matter of degree. As we have seen, pristine societies—contingent upon their origins, location, and internal dynamics—varied considerably in their dependence on war. Thus Egypt, sheltered geographically and blessed with environmental factors that moderated the swings of demography, placed primary emphasis on monumental architecture to work off excess societal energy.

On the other hand, Assyria, born surrounded by enemies, came to be driven by war, pursuing it almost for its own sake until finally destroyed by belligerence. This is suggestive, since war alone among the basic factors governing agricultural societies seems capable of something approaching open-ended amplification. Thus, the very feature that made warfare a serviceable equilibrator—that it could be initiated as a matter of choice—also allowed it to become an all-consuming pursuit.

But if this was Assyria's fate, it was not China's. Here the manner in which war first arrived and the patterns of social and political evolution predisposed the Chinese to remain wary of the institution and its possibilities. Given the continuing pastoral threat from the steppe, the necessity of armies and defense could never be ignored. Yet war would be employed gingerly, shackled by all manner of intellectual and governmental restraints. There were costs in terms of both military efficiency and demographic stability, but China would endure when other societies were devoured.

Perhaps most interesting from a theoretical perspective is the independent development of warfare in the New World. Here the absence of a decisive agropastoral split and resulting urge to huddle behind walls caused social and political consolidation to take place in a less abrupt fashion under conditions that reduced demographic instability. War was plainly a part of the process, but its role was more exclusively political—a matter of elites, not peoples. Nonetheless, agricultural intensification and population increases, most notably in the Basin of Mexico, eventually led to an intensification of war and policies more akin to those of Old World hydraulic despotisms. So here, too, humanity remained caught in a giant and increasingly lethal plant trap.

As we saw in the last chapter, there was the slim opportunity to break loose, even using Bronze Age techniques, for a few societies clustered along the littoral, pursuing a mercantile existence and retreating to the sea in the face of aggression. This may have been a prelude to the future, but for most this was no option. Populations had grown too large and technology remained too crude to help them escape from endless drudgery and compulsion. Instead, for the great majority of humans, history was destined basically to repeat itself over the next four thousand or so years, and with it war ground on, the clumsy balance wheel of this rough-hewn clockworks. Automata with a genetic stake in the mechanism might have kept it running indefinitely. But we remained through it all the brightest of animals, with a legacy utterly contrary to the spirit of the great machine. So when at last opportunity beckoned, our most ancient values would prompt us to tear it apart and begin anew.

III

In 1989, just as the cold war dissolved, political scientist John Mueller put forth the audacious proposition not only that was war obsolete but that it was finished simply because most people had come to find it repulsive and uncivilized. "Like dueling and slavery, war does not appear to be one of life's necessities . . . War may be a social affliction, but in important respects it is also a social affectation that can be shrugged off."[11] As we shall see, there is some validity in Mueller's thinking. But by focusing on effects, not causes, he trivialized the subject—the equivalent of "just saying no" to the drug problem. It has been argued here that for a very long time warfare was deeply imbedded in the structure of civilization, and it follows that if war has indeed become defunct, then it must be the result of societal changes of sufficient magnitude to undermine its fundamental utility. In short, this is one Gordian knot that demands unraveling, not cleavage.

Enhanced technology would provide the key to the breakout, but it would require a long period to accumulate the necessary stock of improvements. Appropriately enough, the introduction of vastly more productive "high farming" on the northern European plain in the eighteenth century A.D. was the likely catalyst, freeing (dispossessing, really) significant numbers of peasants and in general allowing nonagricultural populations to grow substantially.[12] The resulting labor pool, the application of new financial methods, and the very rapid evolution of machine technology combined to set off the Industrial Revolution—first in northern Europe and then in an ever-expanding zone around the globe. Thus, for the second time in human history there transpired a fundamental reordering of subsistence patterns, a transformation, in the words of William McNeill, "as radical in its way as the resort to agriculture had been before the dawn of recorded history."[13] The promise of the early mercantile societies was finally being fulfilled.

Although elemental shifts in economic roles and functions were the basis of the transformation, it was the manner in which change cascaded into matters of health, reproduction, and politics that truly metamorphosed the way people lived. And while this process is very clearly still proceeding, the basic implications for human existence have become apparent. Most fundamental has been the stabilization of demographic patterns in industrial societies. Pestilence was the first to fall. By the mid–eighteenth century Europe had become basically disease-saturated, with the general population possessing a high degree of immunity to a number of traditional killers, now largely relegated to childhood maladies.[14] After the middle of the nineteenth century this was compounded by significant improvements in diet, the scientific conquest of several other deadly epidemic diseases, and the introduction of public health procedures featuring new measures in water management and sanitation. The result was a dramatic drop in urban morbidity. For the first time since cities had come to exist, they became able to support and even increase their populations without migration from the countryside.[15]

But these developments did not preordain a hopeless Malthusian spiral. For roughly contemporaneously, effective and convenient means of contraception were becoming widely available. This factor and several others—increases in education, the prolongation of childhood, and the individual emphasis on improved living standards—conspired to lead urban couples to have much smaller families. Thus one by one, beginning with France, the various populations of industrial Europe were able to voluntarily stabilize themselves, bringing under conscious human control matters of life and death that once had been considered the province of fate. Populations in what we have come to call the developed world would continue to grow for a long time, but the jagged Sturm und Drang oscillations driven by famine and pestilence would come to be a thing of the past.

Given its traditional role, it seems inevitable that the fate of the Second Horseman's fellow travelers would have profound implications for his own future. If human societies could and then did find alternate means of stabilizing their own demographics, then the time-honored function of war would be undermined in the most basic way. For instance, in an environment where agriculture was no longer the basic means by which people supported themselves, then the absolute criticality of territory and its forceful defense and acquisition might, and eventually would, diminish. Similarly, the coercive acquisition of labor and particularly slavery had always been a manifestation of agriculture; and to a world whose future was clearly industrial, it would quickly be deemed inappropriate and eradicated. The usefulness of armies as disposable accumulations of pawns would take longer to undermine, but it too would follow almost as inevitably.

For closer to the surface of events there occurred an accompanying political revolution that would dramatically reinterpret the worth of the individual human being. I have argued that the despotic hierarchies dictated by traditional labor-intensive agriculture, while pragmatically necessary for the survival of very large populations at this level of subsistence, were fundamentally dissonant to human nature as it had evolved during the millions of years we spent as wanderers, hunters, and collectors. Cultural plasticity may have enabled us to live like ants, but our most ancient social values would never allow us to stop dreaming of a more congenial existence. In this context it is possible to see documents like the Bill of Rights and the Declaration of the Rights of Man as virtual hunter-gatherer manifestoes, reassertions of what was fundamental to human nature as it had actually developed. Indeed, the tone of these declarations, the use of phrases such as "inalienable rights" and "self-evident" truths, indicates that the authors sensed that what they were composing truly mirrored the soul of humanity, that there was in all of us a predisposition to be a certain way and that this was the basis of all justice.

Quite plainly, industrialism, especially in its earliest stages, would prove to be far from a perfect incubator for what were perceived to be the natural rights of individuals, but there was a growing understanding of how this new adaptive stage might prove better able to accommodate them. Thus, if true consensus decision making was implausible in mass society, then at least the will of the people might be reflected through representatives, whose position was dependent upon regular election by their peers. Similarly, it would be possible to guarantee ancient proclivities such as freedom of speech, movement, and even the right to bear arms. And if people's lives remained far from equal, the principle of equality could nevertheless be viewed as basic to law and government. Inevitably, this would come to be true for women as well as men. For during this period females would gain control of their reproductive functions, see their economic independence progressively restored, and come to exist in a society where physical strength was of only marginal value—changes that would pave the road to true liberation.[16]

The actual manifestations of the great political transformation would vary from society to society, and periodically there were considerable setbacks. Yet as political philosopher Francis Fukuyama has maintained, the general momentum was irrepressible and the direction irrevocable.[17] In part this was because its progress was abetted by particularly powerful examples of what biologist Richard Dawkins calls "memes"—terms such as *freedom* and *equality*, which spoke directly to our most basic values and were liable to produce an inordinate response even from those presumed numb to politics.[18] Yet the driving force at the core of everything was a recognition that with membership in our species came certain inherent rights, not the least of which was the prerogative not to be killed capriciously.

All of this, it should be emphasized, would prove fundamentally disruptive to the conduct of war. Although millions would initially flock to military standards intent on defending their versions of the revolution, eventually the contradictions between the essentials of that revolution and the traditional premises of warfare would become starkly apparent. For at the heart of war was an aristocratic willingness to tyrannize, brutalize, and ultimately squander the lives of common soldiers. This outlook had not changed measurably from the time of Homer to that of the Compte de Saint-Germain, who epitomized the attitude in the phrase "the army must inevitably consist of the scum of the people."[19] And soldiers, by an extension of this logic, were simply objects to be dressed up, marched around, and then cut down. This was about to become intolerable. Armies would be made up of the sons and even the daughters of enfranchised citizens, and their sacrifice would be a matter of the deepest concern. The eventual development of the electronic media would only compound matters. For war had always thrived out of sight; when television brought its brutality to the dinner table, it would prove to be an unwelcome guest indeed.

But if the slaughter would prove impossible either to hide or to rationalize, developments in another sphere were ensuring not simply its perpetuation but its amplification to hitherto undreamed-of levels. The application of machine technology to weaponry began innocently enough, essentially along the lines of other laborsaving strategies. For if the military ranks of the industrialized were to be filled with what amounted to cross sections of the people—short-term conscripts enthusiastic and capable, but with neither the willingness nor the time to master traditional weaponry—then it made eminent sense to modify armaments to make them easier to use. Not uncoincidentally, the lessons derived from the gun's initial proliferation and the resulting firestorm of war that engulfed Europe during the sixteenth and seventeenth centuries had gradually been forgotten. Even if this had not been the case, any intimation that weapons development was inherently dangerous and had been, in effect, fenced in by the ritualization of battle in the eighteenth century would have been lost on the likes of Hiram Maxim, John Browning, and the other inventive souls busy cultivating engines of destruction. Instead, they were concerned with making money and carrying the doctrine of progress to what was considered to be a technological backwater.[20]

Thus, during the last three and a half decades of the nineteenth century—a period of relative peace punctuated only by short and satisfying wars such as those of Prussia—the forces of the newly industrialized accumulated arsenals truly worthy of their respective economies' manufacturing prowess. Indeed, they were seen as the epitome of modernity. And rather than facing up to the paradox of weapons that killed too easily, Westerners found it expedient to use their arms far afield to conquer virtually all of Africa and considerable chunks of Asia, meanwhile reassuring themselves that such devices would never be used at home. The next century had a very different agenda in store for the legions of optimists. But only a few Cassandras like Ivan Bloch (who foresaw the weapons-induced stalemate of the Great War's trenches) and H. G. Wells (who even envisioned the possibility of nuclear weapons) had any idea of what was coming.[21]

This was characteristic of disasters in general—processes in which deceptively powerful forces operate for considerable periods without visible consequence, only to erupt with cataclysmic results. So it was with war on the brink of the twentieth century. Yet for most the notion that an institution so apparently fundamental as warfare could possibly have outlived its purpose and remained merely dangerous would have seemed not just implausible but completely at odds with what was perceived to be happening around them.

For the impact of Darwinian thinking, combined with the widespread feeling that urban-industrial life was basically enervating, led many to view war as a palliative, an equipoise to the tedium and uncertainty of daily existence. Relatively bloodless and decisive conflicts such as the War of 1870 and the Spanish-American War allowed citizens of all classes to convince themselves that warfare held out the prospect of adventure in an overcivilized world. It was idealized as a short interlude during which young males might test themselves against the traditional warrior code, to be delivered back to their former lives purged, hardened, and much the better for the experience.[22] Casualties would remain a regrettable but necessary consequence of the process. But modern weapons, by raising the tempo of military operations, generally were expected to keep the killing to a minimum and confined mainly to the "unfit."[23] While this line of reasoning reached a sort of crescendo in the bellicose pronouncements of peripheral figures such as German General von Bernhardi and Frenchman Charles Maurras, there is every reason to believe that such notions were widely held, not only among mainstream politicians like Winston Churchill and Theodore Roosevelt (both of whom took the "cure") but also by broad segments of people in every corner of the industrialized world.

This time span also marked the genesis of modern pacifism, and events such as the two disarmament conferences at the Hague indicate at least some apprehension in governmental circles as to the potential consequences of future warfare. But pacifists remained very much a minority, and not a particularly insightful one. There was some talk of economic futility, but for the most part the overt rationale for rejecting warfare went no deeper than its perceived immorality and barbarism.[24] This strain of pacifism may have reflected more profound fears and suspicions, but they were not clearly articulated. Meanwhile, judging by the happy throngs gathered in Europe's capitals, awaiting their respective leaders to announce formal belligerency in August 1914, it appears that the vanguard of humanity had voted with its feet.

There had never been a conflict remotely like the Great War, as World War I was invariably called before 1939. Two massive field armies lay siege to each other, locked in a double line of trenches stretching from the Swiss border to the North Sea like parallel serpents. Until March 1918 no assault would succeed in moving the lines even ten miles in either direction, while millions of men would die on the narrow band of territory separating the two snakes of earth.

If anything, the war at sea was even more bizarre, with two huge fleets of dreadnought battleships—themselves a product of an arms race that helped to poison the prewar climate—glowering across the North Sea but virtually helpless to engage in significant combat. In the meantime, that pariah among naval weapons, the German submarine, feasted on Britain's seaborne commerce, curbed in the end more by the inhibitions of its users than by effective countermeasures.

"I don't know what is to be done—this isn't war," bemoaned Lord Kitchener, the personification of England's army and soon to be drowned by a naval mine. Behind it all loomed military technology, enforcing the stalemate with a series of weapons, many of them such newcomers to the battlefield that opposing forces had little idea how to cope with them. Worse still, armaments did not simply dominate, they made a mockery of the warrior ethic. Strength, swiftness, skill, cunning, and bravery were rendered largely irrelevant. Combatants were gassed, torpedoed, mined, bombarded by unseen artillery, or mowed down by machine guns more or less randomly; there was hardly a valiant death to be had in this sanguinary burlesque. Indeed, the conflict would render heroism, that ancient transmitter of warrior values, "wanted," in the words of critic William Pfaff, "only by fanatics, the emotionally crippled, adolescents, and, in knowledge of its nature, by a few military ascetics."[25]

The lesson here was that weapons were turning against war itself, making it increasingly difficult to fight except at a cost far greater than the potential gains. World War I is burned into our collective memories as much for the quality of the slaughter as the quantity—the uselessness of the deaths and the perception that the victims were valued members of society. Little productive land or labor was exchanged. The sheer magnitude of the deaths (about ten million), while utterly appalling to public sensibilities, constituted but a small fraction of the combatants' total populations, and was destined to have only a marginal impact on future numbers. So war as a demographic lever had become merely repellent.

But if the phrase "a war to end all wars" was prophetic in concept, it hardly matched the topography of events after Versailles. In retrospect, it appears that the period between the two world wars was simply an interlude during which the industrialized world girded itself again for renewed combat. Yet the shock induced by the Great War was sufficient to cause a radical disjunction between thought and action, a schism of fundamental significance not just for the future of warfare but also in determining the context for the whole complex of economic, technological, and political changes that were transforming human existence.

At one level—primarily intellectual and most apparent in the liberal democracies—there was a renunciation of war and an acceptance that it was inappropriate to the kind of world that was forming as a result of agriculture's displacement as the basis of subsistence. However, pacifism and attempts at arms control during the twenties and thirties were also disembodied from contemporary political reality—moral and cerebral exercises studiously partitioned from the here and now. Yet they were not necessarily fecklessly utopian. Although the 1925 Geneva Protocol banning poison gas in combat was compromised by reservations, and the 1928 Treaty of Paris outlawed war by fiat and nothing else, both still registered a growing awareness, soon to be ratified by nuclear weapons, of warfare's approaching obsolescence and the concept that certain weapons were simply too horrible to use.

The immediate future, however, would be held hostage to a very different paradigm. As is common in times of great change, refuge was sought in the constructs of the past. And so large segments of the developed world, battered by the impact of the Great War and the cumulative oscillations of the business cycle, sought to resurrect the institutions of hydraulic despotism and apply them to an industrial environment, bringing into existence modern totalitarianism. It was not an entirely illogical syncretism. Given the apparent evolution of machine technology at this point in the direction of economies of scale, centralized management, and the regimentation of huge labor pools, it made a certain sense to conclude that forms analogous to the great pyramidal hierarchies that had once dominated the ancient world's river valleys could be superimposed on an environment dominated by machines. For, if nothing else, tyranny promised the appearance of stability and compatibility with military institutions deemed appropriate during a time of great insecurity. For rather than rejecting war, the totalitarian temper, employing essentially agricultural criteria, drew the opposite conclusion. Aggressive warfare was to be the key mechanism by which such societies fulfilled their destinies. Thus, imperial agriculture's hunger for territory would find an equivalent in the supposed necessity to physically control the natural resources necessary to feed an industrial economy. There was also the same drive to exploit labor through force of arms—to move and eliminate whole populations both within and without. And implicitly there was the willingness to squander large armies in the pursuit of these objectives, a prospect that military technology was eminently ready to accomplish.

Ultimately, totalitarianism would be shown to be an anachronism based on a profound misreading of human nature, technology, and the role of organized violence, but this was not obvious at the time, nor until very recently. The totalitarian disposition was no aberration; it was a logical extrapolation from what many believed was the true course of events. And it should not be forgotten that important strains of this thinking existed in the liberal democracies, just as elements of individualism, capitalism, and pacifism continued to persist in these supposed monoliths.

And so the industrialized world was plunged into another bout of horrific warfare. Only this time the numbers killed—approximately five times the magnitude of the Great War—would be far less an accidental consequence of technologies barely understood than a conscious application of machinery in the cause of bloodletting. For the major combatants of World War II all had worked quietly to perfect and develop doctrine for the weapons that had emerged during the Great War, especially the tank, the airplane, and the submarine;[26] when the time came, all would be employed effectively and ruthlessly.

But it was totalitarianism itself that would set the tone for World War II, and it was on the Russian front that the issue was basically decided. Here, in battles encompassing hundreds of square miles and millions of participants, human blood would be spilled like never before, as the two exemplars of modern tyranny, Nazi Germany and the Soviet Union, waged an all-out fight to the finish, which brought to fruition the possibilities of industrialized warfare. No doubt both assumed that their struggle and its outcome would determine the future of humanity. But this was not to be.

The signal event of World War II took place on July 16, 1945, when the first atomic bomb was tested at Alamogordo, New Mexico, sending, among other things, a mushroom cloud 41,000 feet into the stratosphere and breaking a window 125 miles away. The military-technical developmental sequence that had begun around a century earlier with the first major improvements in small arms had finally produced a weapon so powerful that, not much more than a generation after its first use, it would reduce unlimited war between major powers to a logical absurdity.

IV

The postwar period, however, would be filled with uncertainty. For the cold war, with its Manichaean struggle between totalitarian Communism and democratic capitalism, would mark the climax of our species' initial efforts to come to grips with a new way of living. The stakes were enormous and the outcome by no means assured, since true mismanagement increasingly implied not simply victory or defeat but the probable end of civilization itself. And the magnitude of the danger was primarily a function of the spiraling arms competition, which would prove the pacesetting event of the entire confrontation.

In retrospect, it does seem that weapons acquisition on both sides consistently reflected a commitment to deterrence but that the respective interpretations of the concept were divergent. The Soviets, for their part, appear to have taken Vegetius' classic definition ("Let he who desires peace, prepare for war")[27] literally, stockpiling weapons, which in their quantities and qualities would make them as useful as possible in what were perceived as realistic wartime scenarios. The problem was that such a strategy left their adversaries no clear criteria to differentiate preparing for war from intending war—a very dangerous ambiguity to a generation deeply influenced by the so-called Munich syndrome.

Meanwhile, the approach of the other key contestant, the United States, more clearly reflected what would come to be understood as the essential message of the nuclear age: a major East-West war would be an unrelieved disaster. Consequently, one central thread of U.S. strategic armaments policy—though often pursued in an unspoken and even unconscious fashion—would be intimidation aimed at making such a war appear completely futile. Pursuant to this objective and driven by a profound suspicion of their adversary's intentions, Americans would apply their superior technological and economic resources to generate a succession of radical advances in weapons performance. The Soviets, in turn, would focus a vast scientific and technical espionage network and the heart of their massive neohydraulic industrial base in an unrelenting but hugely inefficient effort to keep up. One side trudging doggedly, the other leaping and bounding—a contest between the proverbial tortoise and hare—it was this dynamic that would propel the arms race to undreamed-of destructive potential and help set the tone for cold war politics.

Nonetheless, caution would be the watchword. And while there is a natural tendency to dwell on the potentially catastrophic consequences of the cold war's several crises—the one cited at the beginning of this chapter being perhaps the most bizarre—the degree of moderation consistently shown by both sides was in fact remarkable and must be attributed to a growing recognition that nuclear weapons had truly changed the rules of international politics. Philosophically and operationally, the Soviets were plainly the more reluctant to accept this new condition. But multiplying nuclear stocks and ever-more-reliable means of delivery invariably defeated their war planners' repeated efforts to circumvent the grim logic of mutually assured destruction.[28] So at the highest level of engagement it gradually became clear that war on the grand scale had truly been thwarted. Accidents were still possible, but as a premeditated act of policy between nuclear-armed members of the developed industrialized community war was truly defunct.

Less appreciated, however, is the stifling effect the nuclear standoff had on the cold war's chief players' attempts to employ force at a lower tier. The American experience was particularly instructive. In both Korea and Vietnam the United States' unwillingness to apply sufficient force to achieve victory can be traced to concerns about provoking general war. In the case of Vietnam, critics of the American strategy of graduated escalation and our refusal to invade the North overlook the probability that both stemmed from fears of drawing China into the conflict and ultimately sparking World War III.[29] So the urge to be perceived as prudent and always allowing our adversaries acceptable alternatives to all-out war came to outweigh the pursuit of victory. At the same time, this halfhearted belligerency would be further undermined by television, which exposed the public to a vision of warfare's brutality never before seen. The transformation of Vietnam from an intervention of secondary strategic import into a national tragedy has been blamed on many factors, but relatively little significance has been attached to the inefficacy of war itself. Yet all the signs are there—the public revulsion at the killing, the excessive destructiveness of the weaponry, and the inability to articulate exactly what was to be gained from the fighting.

The Soviet Union was more successful in steering clear of direct combat, while sponsoring surrogates to wage so-called wars of national liberation. Nevertheless, when the Soviets themselves invaded Afghanistan, their troops were soon caught in a Vietnam-like quagmire, the first in a series of calamities leading directly to the regime's collapse.

Judged by any standard, the downfall of Communism and the fragmentation of the Soviet Empire were events of major historical importance. Yet they were particularly indicative of warfare's debility and the cumulative effects of militarism. In particular, the lumbering Soviet performance in the arms race, by relentlessly prioritizing the military-industrial sector, not only consumed much of what was most productive but choked, through secrecy and segregation, what little technological creativity the system could muster. So while Americans were able to draw upon a very healthy civilian technology base to generate advanced weapons, Russian forays into high technology lacked synergy between civil and military spheres, and led nowhere. Rather, they were fated to meet ever-greater costs during heroic Soviet attempts to approximate Western weapons in sterile defense industries. By the early 1980s, after decades of sacrifice and hard work, the Soviets were truly losing the arms race, and losing it precisely because of what that sacrifice and hard work had built.

It was emblematic of the entire system. For the superimposition of what were essentially the institutions of despotic agriculture on an industrial economy had produced masses of cement and steel but little that was truly useful. The instruments of tyranny were adapted to run simple societies based on relatively few variables; but only systems that left their members free to operate on their own initiative and fully develop their own talents could cope effectively with the manifold complexity of truly developed economies. So Communism and much of what it stood for failed.

Yet the manner in which this took place and the actual outcomes are as important to consider as the nature of the failure. That we are alive at all is no small matter. The very fact that a ruling class—armed to the teeth and wedded to an institutional culture that lionized coercion—would give up without a fight speaks volumes on the inutility of warfare. For even the most hardheaded and skeptical must admit that the peaceful resolution of the cold war is an event basically without historical precedent. That a confrontation this fundamental and dangerous could have been brought to a close without major warfare is plainly indicative of a profound change in the mechanisms that shape events. Most salient has been the appearance of nuclear weapons, a factor which even the strictest of realists now admit has had an extraordinarily chastening effect.[30] But it is also apparent that war has been undermined along a number of other fronts, particularly in the industrialized world. What was once a useful, if costly, institution simply has been supplanted by other expedients more effective in an environment that reflects a fundamentally different form of subsistence.

This is a matter of both internals and externals. For beneath the surface political cacophony that has replaced the cold war run strong currents of accelerating internationalism, especially among the developed, trends that constitute the most significant challenge to the monolithic nation-state since the seventeenth century. This is not essentially a matter of countries voluntarily ceding elements of their sovereignty to supranational bodies, although this constitutes a significant element, particularly in Europe. Globally, however, the astonishing growth of international trade (a tenfold increase between 1970 and 1992)[31] and the rise of multinational corporations are well on the way to creating truly stateless economies. At the same time the consciousness of far-flung citizenries is being knit together by the revolutionary forces of the electronic media, the traditionally universalist claims of organized religion appear to be undergoing something of a renaissance. The state system is unlikely to disappear—indeed, its multiplicity remains perhaps the best guarantee against hegemonic tendencies. Nonetheless, its components will be increasingly composed of peoples and institutions with profoundly divided loyalties. This will not preclude certain kinds of violence, but war as defined here and traditionally understood will likely prove far too blunt an instrument to wield effectively in such a variegated environment. And to what end would it be waged—to seize land or resources basically unusable without the cooperation of international economic institutions inherently wary of violence; to brutalize obnoxious elements only to be captured by the electronic media and subjected to world condemnation, sanctions, or worse; to address population problems far better handled by contraceptive and public health measures; to destroy an enemy only to be destroyed in return? All are still technically possible, but are they useful? Institutions thrive because they are functional. Based on the obverse, war's prognosis must be considered bleak.

V

There is a catch, however. The patient lingers. For we live in a world divided. Perhaps as many as two-thirds of our species continue to subsist in agricultural economies or those based on the direct exploitation of natural resources—an area where warfare not unexpectedly plays an entirely more significant, if not exactly constructive, role. Here conquest and the violent expropriation of wealth and peoples still can be held up as attractive options, often under the guise of anti-colonialism, irredentism, or religious crusades. Frequently, politics is dominated by armed force and overseen by unrepentant despots who seek to use war as a social safety valve and a moneymaking proposition. In this respect a ruler like Saddam Hussein has more in common with his Assyrian forebears than with the Western heads of state who were intent upon thwarting his ambitions to corner the world's oil supply. Meanwhile, the victims of such aggression—especially those without something particularly valuable to the developed world—must still face the ancient martial quandary: submit or fight back. So war retains a modicum of its infectious potential.

What is termed civil war presents special problems, especially in this environment. There is plainly a range of motivations that vary conflict by conflict, but in general it is possible to say that revenge—or as it has been termed here, negative reciprocation—plays a much larger role than in true aggressive warfare. Thus in the Bosnian conflict, while there are certainly subthemes of Serbian aggression and genocidal "ethnic cleansing," the fighting seems primarily motivated by an overwhelming urge to "get even," frequently rationalized in the context of historical misdeeds but more realistically in response to the previous week's massacre. But if the former Yugoslavia remains a basically rural member of the industrialized world, occupying what is essentially border territory, civil war in the heart of the agricultural sphere is much more frankly about killing, with politics playing the role of fig leaf. Thus in contemporary Africa, where population continues to rise and fall largely on the basis of inadequate contraception, famine, and epidemic disease, we see repeated acts of true mass slaughter, the latest being in Rwanda. While aggressive war is hardly unknown here, genocidal campaigns typically are acted out in the context of civil conflict. This is probably not accidental.

To some it may seem like hairsplitting to draw such a fine distinction between these two types of human fighting, but careful taxonomy is necessary to highlight why civil conflicts are so difficult to quell. Because such conflicts are internal, nonaggressive outside intervention remains fraught with ambiguities and dangers. Thus, what is termed "peacekeeping" is nearly always accompanied by agonizing debates over who should go and under what circumstances, while actual operations must tred a very fine line between passivity to the point of ineffectiveness and being drawn into belligerent status. It is perfectly apparent that present institutions are not up to the task. Military commanders are the first to admit that conventional armies are not trained to perform such a role, and the difficulties encountered by the United Nations in this sphere also indicate that political mechanisms have a long way to go.

The situation is far from hopeless, however. Because such conflicts are primarily driven by chains of revenge, leverage can be best applied at the very beginning or at the end, when the combatants are truly exhausted. This is obviously no universal prescription, since it implies intervention in practically every nascent dispute or, alternately, waiting until the worst has happened. But at least it points to when it might be most efficacious to become involved, and lays down some basic guidelines for success. Conflict resolution is still very new, and it can be assumed that over time appropriate tools and institutions will be developed— so-called nonlethal means and activities lumped under the term "peace studies" are just two examples. Finally, it should be remembered that civil conflicts hark back to the kind of fighting that might be expected to have occurred prior to the invention of the state, and do in fact take place during varying degrees of anarchy. While this is not reassuring in terms of Africa and parts of central Asia where the institutional structure shows signs of true collapse, the rest of the globe remains fairly well regulated, with the mechanisms of social control gradually adapting to change in ways more in tune with the essentials of human nature. So in this context civil conflict can be expected to be the exception rather than the rule.

Meanwhile, even the least developed areas remain subject to many of the same phenomena that have marginalized true warfare elsewhere. To a degree sometimes unappreciated, atrocities and mass death are greeted here with the same concern and disapprobation as elsewhere. Moreover, when war does come to such areas, it is destined to be fought with the same modern weapons, whose excessive destructiveness gave it such a bad name among the industrialized. And if the legacy of the recent past—Vietnam, Afghanistan, and Iraq—is any indicator, winning or losing will make little difference; war waged with such implements will continue to be almost a certain prescription for national disaster and subsequent poverty. Meanwhile, those who argue that war's future lies in fights over ecological issues—access to water, rain forests, and the like—do so by ignoring war's present wantonness. For nothing we do is potentially more polluting.

But perhaps more important than these specifics is the overall direction of history. Development is spreading, not fast enough for those living outside its cocoon of relative wealth and safety, but spreading nonetheless, and with it will come the factors undermining war's vitality. The expansion of democratic governments is demonstrably a potent factor in this regard. While it may not be instantaneously toxic to war—witness postrevolutionary Iran—over the longer haul governments that depend on truly participatory politics have been shown to be decidedly more peaceful among one another. Thus it is reasonable to hope that as the sphere of democracy grows, the province of war will shrink.

It can certainly be argued that the Falkland Islands campaign and Desert Storm constitute examples of the democratic industrialized world not only waging war but doing it successfully. But this misses a larger point. Force was applied primarily to correct naked acts of aggression, as much a matter of principle as anything else. Especially in the latter case, extraordinary moderation was shown. Indeed, the campaign resembled a group of white blood cells surrounding and then eliminating an infection. But, significantly, no concerted effort was made to destroy the source of the infection. Rather, allied forces wielded their power in the Gulf War in a manner that confounds conventional military logic and demonstrates instead just how suspect the use of force has become among the stewards of megadeath. For even unconstrained by the fetters of the cold war, precedent still continued to overshadow victory. And as we have seen in the case of civil conflict, the pressure grows to apply military power to simply stop wars—wherever they are and whatever their motivation. Clearly, this will not happen overnight. It will be an onerous and in many respects a thankless task. And until it is finished it can be expected that war's black heart will flutter on. But one day in the not-too-distant future, it can truly be hoped that humankind will look back on warfare as truly a corpse in armor. Yet this is not guaranteed.

We learned to wage war because it made sense in terms of the kinds of societies we lived in. That is far less true today. Nevertheless, we fought because we were able to fight. We crafted weapons. We demonstrated aggression, both among ourselves and against other species. Our loyalty to our culture could be manipulated through mechanisms such as religion and our urge to reciprocate, which also compelled us to seek revenge. We may live differently, but in these regards we have not changed. So a recourse to war will remain a possibility. We can only hope that the path of history will continue to lead us away from its carnage and despair.

NOTES

1 INTRODUCTION: IN SEARCH OF A BEGINNING

1. Bar-Yosef, "The Walls of Jericho: An Alternative Interpretation," *Current Anthropology* 27/2 (April 1986): 157.

2. W. T. Sanders, J. R. Parsons, and R. S. Santley, *The Basin of Mexico: Ecological Processes in the Evolution of Civilization* (New York, 1979), pp. 94–105; W. Conklin and M. E. Mosely, "Patterns of Art and Power in the Early Intermediate Period," in R. W. Keatinge, ed., *Peruvian Prehistory: An Overview of Pre-Inca and Inca Society* (Cambridge, 1988), pp. 149–53.

3. See, for example, K. R. Otterbein, "Internal War: A Cross Cultural Study," in R. A. Falk and S. S. Kim, eds., *The War System* (Boulder, 1980), pp. 204–5; R. B. Ferguson, "Explaining War," in J. Haas, ed., *The Anthropology of War* (Cambridge, Eng., 1990), p. 26; M. Ember and C. R. Ember, "Fear of Disasters as an Engine of History: Resource Crises, Warfare, and Interpersonal Aggression," paper delivered at Texas A & M University Conference "What Is the Engine of History?" 27–28 October 1988.

4. This series of characteristics is based on a host of sources, but in particular drew heavily on H. H. Turney-High, *Primitive War: Its Practice and Concepts* (Columbia, S.C., 1971), p. 30.

5. W. D. Hamilton, "The Genetical Theory of Social Behavior, I, II," *Journal of Theoretical Biology* 7/1 (1964): 1–52.

6. G. Clark, *World Prehistory in New Perspective* (Cambridge, Eng., 1977); M. Harris, *Cannibals and Kings: The Origins of Culture* (New York, 1977), pp. 27–28.

7. S. Budiansky, *The Covenant of the Wild: Why Animals Choose Domestication* (New York, 1992), p. 3.

8. R. L. O'Connell, *Of Arms and Men: A History of War, Weapons and Aggression* (New York, 1989).

9. G. Daniel, *The Origins and Growth of Archaeology* (New York, 1967), p. 5.

10. Sir M. Wheeler, *Archaeology from the Earth* (Oxford, 1954); Daniel, pp. 247–48.

11. R. J. Wenke, *Patterns in Prehistory: Humankind's First Three Million Years* (New York, 1984), ch. 2.

12. R. Cribb, *Nomads in Archaeology* (Cambridge, Eng., 1991), pp. 66–68.

13. M. N. Cohen, *The Food Crisis in Prehistory* (New Haven, 1978), pp. vii–viii.

14. Wenke, pp. 10–24.

15. Budiansky, *Covenant of the Wild*, p. 25.

16. J. Klein, et al., "Calibration of Radiocarbon Dates," *Radiocarbon* 24/2 (1982): 103–50.

17. N. L. Whitehead, "The Snake Warriors—Sons of the Tiger's Teeth: A Descriptive Analysis of Carib Warfare, ca. 1500–1820" in Haas, pp. 147, 156, 167; R. B. Ferguson, "Tribal Warfare," in ibid., pp. 109–110; R. B. Ferguson, "Tribal Warfare," *Scientific American* 2661 (January 1992): 108–13.

18. J. Mellaart, *The Neolithic of the Near East* (London, 1975), p. 11.

19. J. D. Bernal, *Science in History* (Cambridge, Mass., 1971), p. 662; D. Rindos, *The Origins of Agriculture* (Orlando, Fla., 1984), pp. 8–9.

20. Wenke, p. 19.

21. R. Fox, *The Search for Society: The Quest for a Biosocial Science and Morality* (New Brunswick, N.J., 1989), pp. 23, 45; I. Eibl-Eibesfeldt, *Human Ethology* (New York, 1989), pp. 86–89.

22. C. McCauley, "Conference Overview," in Haas, p. 2.

23. G. W. Conrad and A. A. Demarest, *Religion and Empire: The Dynamism of Aztec and Inca Expansionism* (Cambridge, Eng., 1984), p. 192.

24. V. D. Hanson, *The Western Way of War: Infantry Battle in Classical Greece* (New York, 1989).

25. W. Meacham, "Origins and Development of the Yueh Coastal Neolithic: A Microcosm of Cultural Change on the Mainland of East Asia," in D. N. Keightley, ed., *The Origins of Chinese Civilization* (Berkeley, 1983), p. 153.

2 THUNDER BENEATH OUR FEET

1. M. Talbot, "Slave Raids of the Ant *Polyergus lucidus*," *Psyche* 74/4 (1986): 299–313.

2. W. M. Wheeler, *Ants: Their Structure, Development, and Behavior* (New York, 1910), pp. 472–73.

3. E. O. Wilson, personal correspondence, 15, March 1990.

4. B. Hölldobler and E. O. Wilson, *The Ants* (Cambridge, Mass., 1990), pp. 23–29.

5. Ibid., pp. 398, 573, 584; A. A. Mabelis, "Wood Ant Wars: The Relationship Between Aggression and Predation in the Red Wood Ant," *Netherlands Journal of Zoology* 29/4 ( ): 451–620.

6. Hölldobler and Wilson, *The Ants*, p. 183. It is true that the colonies of certain species harbor multiple queens, and that in some cases several of these are active egg layers. Generally, however, queens form dominance hierarchies, and it is thought that mature colonies of the majority of species have only a single egg-laying queen. Ibid., p. 210.

7. W. D. Hamilton, "The Genetic Evolution of Social Behavior," *Journal of Theoretical Biology* 7/1 (1964): 1–52.

8. B. Hölldobler and E. O. Wilson, *Journey to the Ants: A Story of Scientific Exploration* (Cambridge, Mass.: 1994), pp. 3, 7, 63; H. C. McCook, "Combats and Nidification of the Pavement Ant, *Tetramorium caespitum*," *Proceedings of the American Academy of*

Sciences of Philadelphia (1879), 156–61; B. Hölldobler and E. O. Wilson, "The Multiple Recruitment Patterns of the African Weaver Ant *Oecophylla longinoda*," *Behavioral Ecology and Sociobiology* 3/1 (1978): 19–60; G. J. J. Driessen, A. T. Van Raalte, and G. J. De Bruyn, "Cannibalism in the Red Wood Ant, *Formica polyctena*," *Oecologia* 63/1 (1984): 13–22.

9. E. O. Wilson and J. H. Eads, "A Report on the Imported Fire Ant . . . in Alabama" (Montgomery, Ala., 1949).

10. E. J. Fittkau and H. Klinge, "On Biomass and Trophic Structure of the Central Amazonian Rain Forest Ecosystem," *Biotropica* 5/1 (1973): 1–14.

11. Hölldobler and Wilson, *The Ants*, p. 566.

12. Ibid., p. 330; W. M. Wheeler, "The Physiognomy of Insects," *Quarterly Review of Biology* 2/1 (1927): 1–36.

13. Hölldobler and Wilson, *The Ants*, p. 398.

14. E. O. Wilson, *The Insect Societies* (Cambridge, Mass., 1971), p. 42; J. D. Carthy, "Odour Trails of Acanthomyops fuliginossus," *Nature* 166 (1950): 154.

15. McCook, p. 159; Hölldobler and Wilson, *The Ants*, p. 398.

16. B. Hölldobler, "Tournaments and Slavery in a Desert Ant," *Science* 210 (1976): 912–14; B. Hölldobler, "Foraging and Spatiotemporal Territories in the Honey Ant *Myrmecocystus mimicus*," *Behavioral Ecology and Sociology* 9/4 (1981): 301–14.

17. Hölldobler, "Foraging and Spatiotemporal Territories," p. 301–14.

18. Hölldobler and Wilson, *Journey to the Ants*, p. 69; B. Hölldobler, "Food Robbing in Ants, a Form of Interference Competition," *Oecologia* 69/1 (1986): 12–15.

19. E. O. Wilson, "The Organization of Colony Defense in the Ant *Pheidole dentata*," *Behavioral Ecology and Sociobiology* 1/1 (1976): 63–81; Hölldobler and Wilson, *The Ants*, p. 429.

20. J. L. Brown, "The Evolution of Diversity in Avian Territorial Systems," *Wilson Bulletin* 76/2 (1964): 160–69; J. L. Brown and G. H. Orians, "Spacing Patterns in Mobile Animals," *Annual Review of Ecology and Systematics* 1 (1970): 139–262.

21. Hölldobler and Wilson, *The Ants*, p. 400.

22. Ibid., pp. 396, 401–13.

23. M. J. West-Eberhard, "Sexual Selection, Social Competition, and Evolution," *Proceedings of the American Philosophical Society* 123/4 (1979): 222–34; M. J. West-Eberhard, "Intragroup Selection, Social Competition, and Evolution," in R. D. Alexander and D. W. Tinkle, eds., *Natural Selection and Social Behavior* (Concord, Mass., 1981), pp. 3–17; G. F. Oster and E. O. Wilson, *Caste and Ecology in the Social Insects: Monographs in Population Biology* (Princeton, N.J., 1978); S. D. Porter and C. D. Jorgensen, "Foragers of the Harvester Ant, *Pogonomyrimex owyheei*: A Disposable Caste?" *Behavioral Ecology and Sociobiology* 9/4 (1981): 323–36.

24. S. Higashi and K. Yamasuchi, "influence of a Supercolonial Ant *Formica yessensis* on the Distribution of Other Ants in Ishikari Coast," *Japanese Journal of Ecology* 29/3 (1979): 257–64; D. B. Clark, C. Guayasamin, O. Pazmino, C. Donoss, and Y. Paez de Villacis, "The Tramp Ant *Wasmannia auropunctata*: Autecology and Effects on Ant Diversity and Distribution on Santa Cruz Island, Galapagos," *Biotropica* 14/3 (1982): 196–207; Hölldobler and Wilson, *The Ants*, p. 398–401.

25. Hölldobler and Wilson, *The Ants*, p. 464.

26. Hölldobler, "Foraging and Spatiotemporal Territories," p. 308; Hölldobler and Wilson, "Multiple Recruitment Patterns of the African Ant," pp. 19-60.

27. Hölldobler and Wilson, *The Ants*, p. 462.

28. E. O. Wilson, "*Leptothorax duloticus* and the Beginnings of Slavery in Ants," *Evolution* 29/1 (1975): 108–19.

29. K. Vepsalainen and V. Pisarski, "The Taxonomy of the *Formica rufa* Group: Chaos Before Order," in P. E. Howse and J. L. Clement, eds., *Biosystematics of Social Insects*, Systematics Association Special Volume no. 19 (New York, 1981), pp. 27-35; Hölldobler, "Foraging and Spatiotemporal Territories," p. 313.

30. Wheeler, *Ants*, pp. 442, 444.

31. Ibid., p. 246.

32. E. O. Wilson, "The Beginnings of Nomadic and Group-predatory Behavior in the Ponerine Ants," *Evolution* 12/1 (1958): 24–31; Hölldobler and Wilson, *The Ants*, pp. 576, 588, 594.

33. T. C. Scheirla, *Army Ants: A Study in Social Organization* (San Francisco, 1971), p. 43.

34. E. O. Wilson, "The Beginnings of Nomadic and Group-predatory Behavior" pp. 24-31; Hölldobler and Wilson, *The Ants*, 594.

35. Roger Cribb, *Nomads in Archaeology* (Cambridge, Eng., 1991), p. 29.

36. Denis Sinor, ed., *The Cambridge History of Early Inner Asia* (Cambridge, Eng., 1990), pp. 7-10, 183; Cribb, pp. 22, 39, 59.

37. Hölldobler and Wilson, *The Ants*, pp. 358-59; E. O. Wilson, *Sociobiology: The New Synthesis* (Cambridge, Mass., 1975), pp. 399-412; J. Meyer, "Essai d'application de certains modeles cybernetiques a la coordination chez les insects sociaux," *Insects Sociaux* 13/2 (1966): 127–138; H. Markl, "Vibrational Communication" in F.Huber and H. Markl, eds., *Neuroethology and Behavioral Physiology* (Heidelburg, 1983).

38. E. O. Wilson, *Sociobiology*, p. 430; Hölldobler and Wilson, *Journey to the Ants*; pp. 191-193; R. E. Hutchins, *The Ant Realm* (New York, 1967), pp. 194–97.

39. Hölldobler and Wilson, *The Ants*, p. 359.

40. W. M. Wheeler, "The Ant Colony as an Organism," (1911), cited in Hölldobler and Wilson, *The Ants*, p. 358; W. Wheeler, *Ants*, p. 4.

41. R. J. Wenke, *Patterns in Prehistory: Humankind's First Three Million Years* (New York, 1984), p. 192.

42. D. Rindos, "Symbiosis, Instability, and the Origins and Spread of Agriculture: A New Model," *Current Anthropology* 21 (1980): 754.

43. A. C. Cole, "The Rebuilding of Mounds of the Ant, *Pogonomyrmex occidentalis*," *Ohio Journal of Science* 32/3 (1932); Holldobler and Wilson, *The Ants*, p. 373.

44. Wenke, p. 251.

45. Holldobler and Wilson, *The Ants*, p. 391.

3 THE SOUL OF A HUNTER

1. R. L. Bettinger, *Hunter-Gatherers: Archaeological and Evolutionary Theory* (New York, 1991), p. v.

2. T. Hobbes, *Leviathan* (New York, 1962), p. 100.

3. S. J. Mithen, *Thoughtful Foragers* (Cambridge, 1990), pp. 1–89; K. V. Flannery, "Archaeological System Theory and Early Mesoamerica" in B. Meggars, ed., *Anthropological Archaeology in the Americas* (Washington, D.C., 1968), p. 67.

4. P. S. Martin and J. E. Guilday, "A Bestiary for Pleistocene Biologists," in P. S. Martin and H. E. Wright Jr., eds., *Pleistocene Extinctions: The Search for a Cause* (New Haven, 1967).

5. P. A. Janssens, *Paleopathology: Diseases and Injuries of Prehistoric Man* (London, 1970), pp. 60–63; H. Vallois, "The Social Life of Early Man: The Evidence of the Skeletons," in S. Washburn, ed., *The Social Life of Early Man* (Chicago, 1961).

6. M. Harris, *Our Kind: Who We Are, Where We Came From, and Where We Are Going* (New York, 1989), pp. 147–48.

7. J. L. Angel, "Paleoecology, Paleodemography and Health," in S. Polgar, ed., *Population, Ecology, and Social Evolution* (The Hague, 1975); W. H. McNeill, *Plagues and Peoples* (Garden City, N. Y., 1976), p. 29.

8. H. de Lumley, "A Paleolithic Camp Site at Nice," *Scientific American* 220/5 (1969): 42–50.

9. P. J. Wilson, *The Domestication of the Human Species* (New Haven, 1988), p. 59.

10. I. Eibl-Eibesfeldt, *Human Ethology* (New York, 1989), p. 292; P. J. Wilson, *Domestication of the Human Species*, pp. 32–33.

11. M. Ehrenberg, *Women in Prehistory* (London, 1989), p. 38.

12. Eibl-Eibesfeldt, *Human Ethology*, p. 266; E. D. Wilson, *On Human Nature* (Cambridge, Mass., 1978), p. 140.

13. Ehrenberg, p. 37; Eibl-Eibesfeldt, *Human Ethology*, p. 253.

14. R. D. Masters, *The Nature of Politics* (New Haven, 1989), p. 19; Edward O. Wilson, *Sociobiology: The New Synthesis* (Cambridge, Mass., 1975), p. 559.

15. D. F. Lancy, "Cross-Cultural Studies in Cognition and Mathematics," in *The Coevolution of Culture, Cognition and Schooling*, vol. 8 (New York, 1983), pp. 185–211.

16. P. Weissner, "Hxaro: A Regional System of Reciprocity for Reducing Risk Among the !Kung" (Ph.D. diss., University of Michigan, Ann Arbor, 1977).

17. M. Harris, *Cannibals and Kings: The Origins of Cultures* (New York, 1977), p. 13; P. M. Dolukhanov, *Ecology and Economy in Neolithic Eastern Europe* (London, 1979), p. 23.

18. J. P. Birdsell, "Some Predictions for the Pleistocene Based on Equilibrium Systems Among Recent Hunter-Gathers," in R. B. Lee and I. DeVore, eds., *Man the Hunter* (Chicago, 1968), pp. 229–40; Harris, *Cannibals and Kings*, p. 13.

19. John Riddle (*Contraception and Abortion from the Ancient World to the Renaissance* [Cambridge, Mass., 1992], ch. 7) finds some evidence of potentially effective contraceptives and abortion-inducing substances as early as 1850 B.C. in Egypt but does not suggest any use of birth control in the Paleolithic context. Riddle (pp. 11–12) also notes that, while the infanticide hypothesis has been challenged by Donald Engels, most historians interested in the subject continue to support it.

20. Harris, *Cannibals and Kings*, p. 15.

21. Eibl-Eibesfeldt, *Human Ethology*, p. 711.

22. M. H. Fried, *The Evolution of Political Society: An Essay in Political Anthropology*

(New York, 1967), pp. 101–2; C. McCalley, "Conference Overview," in J. Haas, ed., *The Anthropology of War* (Cambridge, Eng., 1990), p. 11.

23. R. Fox, *The Search for Society: Quest for a Biosocial Science and Morality* (New Brunswick, N. J., 1989), p. 18; Eibl-Eibesfeldt, *Human Ethology*, p. 270; Lionel Tiger, *Men in Groups* (New York, 1969).

24. R. L. O'Connell, *Of Arms and Men: A History of War, Weapons and Aggression* (New York, 1989), ch. 2.

25. Harris, *Our Kind*, p. 279.

26. M. N. Cohen, *The Food Crisis in Prehistory* (New Haven, 1978), pp. 93, 116.

27. L. A. Sroufe, "Wariness of Strangers and the Study of Infant Development," *Child Development* 48 (1977): 731–46; K. Kaltenbach, M. Weintraub, and W. Fullard, "Infant Wariness Toward Strangers Reconsidered: Infants' and Mothers' Reactions to Unfamiliar Persons," *Child Development* 51 (1980): 1197–202.

28. R. J. Wenke, *Patterns of Prehistory: Humankind's First Three Million Years* (New York, 1984), p. 130.

29. Eibl-Eibesfeldt, *Human Ethology*, p. 405.

30. E. H. Erikson, "Ontogeny of Ritualization in Man," *Philosophical Transactions of the Royal Society, London* B251 (1966): 337–49.

31. Mithen, chs. 7–8.

32. J. Ortega y Gasset, *Medications on Hunting* (New York, 1985), p. 67.

33. Edward O. Wilson, *Sociobiology: The New Synthesis* (Cambridge, Mass., 1975) pp. 37–41.

34. Harris, *Cannibals and Kings*, pp. 22–23.

35. M. N. Cohen, *Food Crisis in Prehistory*, pp. 159–60, 181–82, 187.

36. Ibid., pp. 53–54; J. M. Beaton, "The Importance of Past Population for Prehistory," in B. Meehan and Neville White, eds., *Hunter-Gatherer Demography: Past and Present* (Sydney, 1990), pp. 31, 36.

37. P. J. Wilson, *Domestication of the Human Species*, p. 29.

38. E. O. Wilson, *Sociobiology*, p. 547; Harris, *Our Kind*, p. 96.

39. Ortega y Gasset, p. 3.

40. S. L. Washburn, "Tools and Human Evolution," *Scientific American* 203 (1960): 63–75.

41. Richard Potts, quoted in *U.S. News & World Report*, 27 Feb. 1989, p. 54.

42. E. O. Wilson. *Sociobiology*, p. 565.

43. D. Falk, *Braindance* (New York, 1992), p. 2; R. L. Ciochon, "Hominoid Cladistics and the Ancestry of Modern Apes and Humans," in R. L. Ciochon and J. G. Fleagle, eds., *Primate Evolution and Human Origins* (Menlo Park, Calif., 1985), pp. 345–62; Harris, *Our Kind*, p. 18.

44. C. Jolly, "The Seed Eaters: A New Model of Hominid Differentiation Based on a Baboon Analogy," *Man* 5/1 (1970): 5–26.

45. L. S. B. Leakey, "Development of Aggression as a Factor in Early Human and Pre-Human Evolution," in C. D. Clemente and D. B. Lindsley, eds., *Aggression and Defense: Neural Mechanisms and Social Patterns*, vol 5, (Berkeley, 1967), pp. 201–19; Harris, *Our Kind*, p. 40.

46. J. van Lawick-Goodall, "Tool-using in Primates and Other Vertebrates," *Advances in the Study of Behavior* 3 (1970): 195–249; J. Sabater-Pi, "An Elementary Industry of the Chimpanzees in the Okorobiko Mountains, Rio Muni (Republic of Equatorial Africa), West Africa," *Primates* 15/4 (1974): 351–64.

47. An experiment using rock flakes on an elephant carcass yielded a hundred pounds of meat in an hour.

48. Harris, *Our Kind*, p. 41.

49. R. A. Dart, "The Predatory Transition from Ape to Man," *International Anthropological and Linguistics Review* 1 (1953): 201–19; see, for example, Harris, *Our Kind*, p. 41.

50. G. Clark, *World Prehistory: In New Perspective* (Cambridge, Eng., 1977), p. 19.

51. Fox, p. 40.

52. K. Lorenz, *On Aggression* (New York, 1966), pp. 228–65; E. O. Wilson, *On Human Nature*, pp. 102–3; F. Huntingford and A. Turner, "Aggression: A Biological Imperative?" *New Scientist*, 4 Aug. 1988, p. 44–47.

53. Eibl-Eibesfeldt, *Human Ethology*, p. 396.

54. S. S. Kim, "The Lorenzian Theory of Aggression and Peace Research," in R. A. Falk and S. S. Kim, eds., *The War System* (Boulder, 1980), p. 94; R. P. Shaw and U. Wong (*Genetic Seeds of Warfare: Evolution, Nationalism and Patriotism* [Boston, 1989], p. 8) point to a chart of aggression noting sixteen different brain-related triggers for aggression, five different categories of aggression, and four separate biochemical complexes.

55. E. O. Wilson, *Sociobiology*, p. 552.

56. Eibl-Eibesfeldt, *Human Ethology*, p. 361.

57. E. O. Wilson, *Sociobiology*, p. 94.

58. O'Connell, *Of Arms and Men*, p. 14.

59. E. O. Wilson, *Sociobiology*, pp. 242–43.

60. R. N. Johnson, *Aggression in Man and Animals* (Philadelphia, 1972), pp. 14–15.

61. D. McFarland, ed., *The Oxford Companion to Animal Behavior* (Oxford, 1981), p. 7. The antlers of deer and moose have evolved for locking and pushing rather than penetration, while the horns of bighorn sheep are specialized for butting. In cases of predators possessing notably lethal instruments, their use is often avoided or mitigated during intraspecific combat. Thus, rattlesnakes do not bite each other but determine dominance through elaborate wrestling matches (I. Eibl-Eibesfeldt, "The Fighting Behavior of Animals," *Scientific American* 205 (1961): 112–22; C. D. Shaw, "The Male Combat Dance of Croatalid Snakes," *Herpetologia* 4 (1948): 137–45). Similarly, the oryx defends against the lion by attempting to gore it, while conspecifics are fought in a tournament fashion according to specific rules that have the effect of reducing injuries.

62. E. O. Wilson, *Sociobiology*, pp. 247–48; Lorenz, pp. 105–12.

63. A cat is silent while stalking prey, and its attack is specifically aimed at inflicting a killing bite. In intraspecific combat, however, the same cat's autonomic nervous system would be highly activated as rivals make loud and piercing calls and assume maximum threat posture. Sometimes a fight ensues, but the encounter frequently ends at the display stage (Eibl-Eibesfeldt, *Human Ethology*, p. 361). The horn of the Hercules beetle actually exceeds the length of its body! (E. O. Wilson, *Sociobiology*, pp. 322–23).

64. E. O. Wilson, *Sociobiology*, pp. 331-33.

65. Eibl-Eibesfeldt, *Human Ethology*, p. 384; Fox, p. 158; M. Mead, "Alternatives to War," in M. Fried, M. Harris, and R. Murphey, eds., *War: The Anthropology of Armed Conflict and Aggression* (Garden City, N.Y., 1968), p. 220.

66. D. Zillman, *Hostility and Aggression* (Hillsdale, N.J., 1979), p. 99.

67. B. Rensberger, "Skull Fragment Is Oldest Yet Found: Origins of Humans Traced Back an Additional 500,000 Years," "Washington Post," 20 Feb. 1992, p. 1; D. Pilbeam, *The Ascent of Man: An Introduction to Human Evolution* (New York, 1972); E. O. Wilson, *Sociobiology*, p. 548; Falk, pp. 144, 175.

68. Falk, pp. 151-73.

69. P. Lieberman, E. S. Crelin, and D. H. Klatt, "Phonetic Ability and Related Anatomy of the Newborn and Adult Human, Neanderthal Man, and the Chimpanzee," *American Anthropologist* 74/3 (1972): 287-307; "The Archaeology of Perception: Traces of Depiction and Language," *Current Anthropology* 30/2 (1989): 125-38.

70. Falk, p. 4.

71. Ibid., pp. 50, 70-71, 144-45.

72. Masters, pp. 79, 99.

73. C. F. Hockett, *A Course in Modern Linguistics* (New York, 1959); T. A. Sebeok, "Aspects of Animal Communications: The Bees and Porpoises," *ETC: A Review of General Semantics* 24 (1967): 59-83; G. Stebbins and F. Ayala, "The Evolution of Darwinism," *Scientific American* 253 (1985): 72-82.

74. E. O. Wilson, *Sociobiology*, pp. 75-134.

75. Eibl-Eibesfeldt, *Human Ethology*, p. 92.

76. E. O. Wilson, *Sociobiology*, pp. 37-47.

77. "Mole-Rats and Darwin," *Washington Post*, 24 Sept. 1990, p. A3; H. K. Reeve, D. F. Westneat, P. W. Sherman, and C. F. Aquadro, "DNA 'Fingerprinting' Reveals High Levels of Inbreeding in Colonies of Eusocial Naked Mole-Rat," *Proceedings of the National Academy of Sciences, USA* 87 (1990): 2496-500.

78. R. Axelrod, *The Evolution of Cooperation* (Cambridge, Mass., 1983); R. Axelrod and W. D. Hamilton, "The Evolution of Cooperation," *Science* 211 (1981): 1390-96.

79. A. C. Wynne-Edwards, *Animal Dispersion in Relation to Social Behavior* (Edinburgh, 1962); Fox, p. 203; Eibl-Eibesfeldt, *Human Ethology*, p. 93.

80. Fox, p. 23.

81. R. Boyd and P. J. Richerson, *Culture and the Evolutionary Process* (Chicago, 1985), pp. 2-5.

82. Ibid., p. 8.

83. Ibid., p. 116.

84. E. O. Wilson, *Sociobiology*, p. 565.

85. A. C. Wilson and R. L. Cann, "The Recent African Genesis of Humans," *Scientific American* 266/4 (April 1992): 68-73.

86. A. G. Thorne and M. H. Wolpoff, "The Multiregional Evolution of Humans," *Scientific American* 266/4 (April 1992): 76-83; L. Tianyuan and D. A. Etler, "New Middle Pleistocene Hominid Crania from Yunxian in China," *Nature* 357 (June 1992): 404-7; "Pleistocene Population Explosion," *Science* 262 (1993): 27-28.

87. A. C. Wilson and Cann, p. 73.

4 FALSE ALARM

1. M. N. Cohen, *The Food Crisis in Prehistory* (New Haven, 1978), p. 123; P. J. Wilson, *The Domestication of the Human Species* (New Haven, 1988), p. 59.

2. M. N. Cohen, *Food Crisis in Prehistory*, pp. 97, 132, 157, 219.

3. P. Dolukhanov, *Ecology and Economy in Neolithic Eastern Europe* (London, 1979), fig. 4. According to the model presented here, the melt-based sea level rise would have been very high between 8300 and 7000 B.P. amounting to around thirty-six feet during this period.

4. D. R. Yessner, "Maritime Hunter-Gatherers: Ecology and Prehistory," *Current Anthropology* 21 (1980): 727-50.

5. S. K. Kozlowski, "Introduction to the History of Europe in the Early Holocene," in S. K. Kozlowski, ed., *Mesolithic Europe* (Warsaw, 1973), pp. 332-33.

6. P. C. Woodman, "Mobility in the Mesolithic of Northwest Europe: An Alternative View," in T. D. Price and J. A. Brown, eds., *Prehistoric Hunter-Gatherers: The Emergence of Cultural Complexity* (Orlando, Fla., 1985), p. 326.

7. Yessner, p. 6; T. D. Price, "Affluent Foragers of Mesolithic Southern Scandinavia," in Price and Brown, 341-63; M. N. Cohen, "Prehistoric Hunter-Gatherers: The Meaning of Social Complexity," in Price and Brown, p. 105.

8. J. Maringer and H. G. Bandi, *Art in the Ice Age—Spanish Levant Art—Arctic Art* (New York, 1953); I. Eible-Eibesfeldt, *The Biology of Peace and War: Man, Animals and Aggression* (Munich, 1975); P. Taçon and C. Chippindale, "Australia's Ancient Warriors: Changing Depictions of Fighting in the Rock Art of Arnhem Land, N.T.," *Cambridge Archaeological Journal* 4 (1994): 211-48.

9. T. D. Price and J. A. Brown, "Aspects of Hunter-Gatherer Complexity," in Price and Brown, p. 13; Price, "Affluent Foragers of Mesolithic Southern Scandinavia,", p. 351.

10. B. Hayden, M. Eldridge, A. Eldridge, and A. Cannon, "Complex Hunter-Gatherers in Interior British Columbia," in Price and Brown, p. 195.

11. G. Clark, *Mesolithic Prelude: The Paleolithic-Neolithic Transition in Old World Prehistory* (Edinburgh, 1980), p. 5.

12. G. W. Sheehan, "Whaling as an Organizing Focus in Northwestern Alaskan Eskimo Society," in Price and Brown, p. 123.

13. O. Soffer, "Patterns of Intensification as Seen from the Upper Paleolithic of the Central Russian Plain," in Price and Brown, pp. 235-70; P. A. Mellars, "The Ecological Basis of Social Complexity in the Upper Paleolithic of Southwestern France," in Price and Brown, pp. 271-97.

14. O. Soffer, telephone conversation, 10 Dec. 1989.

15. M. N. Cohen, "Prehistoric Hunter-Gatherers," pp. 105-12.

16. See, for example, M. Harris, *Cannibals and Kings* (New York, 1977), ch. 4; A. Ferrill, *The Origins of War: From the Stone Age to Alexander the Great* (London, 1985), pp. 19-22; T. Gregor, "Uneasy Peace: Intertribal Relations in Brazil's Upper Xingu," in J. Haas, ed., *The Anthropology of War* (Cambridge, Eng., 1990), pp. 105-24; R. L. Carniero, "Chiefdom-Level Warfare as Exemplified in Fiji and the Cauca Valley," in Haas, pp. 191-210.

17. N. A. Chagnon, *Yanomamo: The Fierce People* (New York, 1968); N. A. Chagnon,

"Yanomamo Social Organization and Warfare" in M. Fried, M. Harris, and R. Murphy, eds., *The Anthropology of Armed Conflict and Aggression* (Garden City, N.Y., 1968), pp. 109–59; N. A. Chagnon, "Reproductive and Somatic Conflicts of Interest in the Genesis of Violence and Warfare Among Tribesmen, in Haas, pp. 78–100; Sheehan, p. 127; K. M. Ames, "Hierarchies, Stress, and Logistical Strategies Among Hunter-Gatherers in Northwestern North America," in Price and Brown, pp. 155–80; Hayden et al., pp. 181–200.

18. R. B. Ferguson, "Tribal Warfare," *Scientific American* (January 1992): 108–13; N. L. Whitehead, "The Snake Warriors—Sons of the Tiger's Teeth: A Descriptive Analysis of Carib Warfare, ca 1500–1820," in Haas, pp. 147–70.

19. M. N. Cohen, *Food Crisis in Prehistory*, pp. 12–14; Price and Brown, "Aspects of Hunter-Gatherer Complexity," p. 14.

20. G. Clark, *Mesolithic Prelude*, p. 53.

21. R. Carneiro, "A Theory of the Origin of the State," *Science* 169 (1970): 733–38.

22. G. N. Bailey, "Shell Middens as Indicators of Postglacial Economies: A Territorial Perspective," in Paul Mellars, ed., *The Early Postglacial Settlement of Northern Europe: An Ecological Perspective* (Pittsburgh, 1978), pp. 37–63.

23. Maringer Bandi, pp. 132–42.

24. See, for example, Ferrill, pp. 20–22, and Mary Settegast, *Plato Prehistorian: 10,000 to 5,000 in Myth and Archaeology* (Cambridge, Mass., 1987), pp. 58–62.

25. Maringer and Bandi, pp. 114–42.

26. A. A. Formozov, "The Petroglyphs of Kobystan and Their Chronology," *Revista di Scienze Preistorichi* 18 (1963): 91–114; J. Mellaart, *The Neolithic of the Near East* (London, 1975); Settegast, p. 61.

27. P. Taçon and C. Chippindale, pp. 217, 225.

28. Maringer and Bandi, p. 130; Settegast, p. 58; Ferrill, p. 21.

29. Ferrill, p. 22.

5 PLANT TRAP

1. This term was coined by archaeologist V. Gordon Childe as far back as 1925. In his book *Man Makes Himself* (New York, 1952) Childe proposed what was to become a highly influential model as to how agriculture might have begun. Calling agriculture a revolution— "the greatest in human history after the mastery of fire" (pp. 23–25)—he proposed that the progressive desiccation of the environment drove humans, along with certain key animals and food plants, together in an oasis environment where propinquity encouraged the domestication of the latter by the former.

2. In addition to Rindos, others include M. Blumler, R. Byrne, and B. Ladizinsky.

3. D. Rindos, *The Origins of Agriculture: An Evolutionary Perspective* (Orlando, Fla., 1984), p. 110.

4. M. N. Cohen, *The Food Crisis in Prehistory* (New Haven, 1978), p. 5; A. B. Cohen, "Comments," *Current Anthropology* 12/1 (1991): 35.

5. Rindos, *Origins of Agriculture*, p. 284; National Academy of Sciences, *Underexploited Tropical Plants with Promising Economic Value* (Washington, D. C., 1975).

6. M.A. Blumler and R. Byrne, "The Ecological Genetics of Domestication and the Origins of Agriculture," *Current Anthropology* 12/1 (1991): 24.

7. D. O. Henry, "Preagricultural Sedentism: The Natufian Example," in T. D. Price and J. A. Brown, eds., *Prehistoric Hunter-Gatherers: The Emergence of Cultural Complexity* (Orlando, Fla., 1985), p. 370.

8. J. R. Harlan, "A Wild Wheat Harvest in Turkey," *Archaeology* 20 (1967): 197–201.

9. Rindos, *Origins of Agriculture*, pp. 87, 89.

10. Ibid.

11. Ibid, pp. 175–177; T. F. Lynch, "Harvest Timing, Transhumance, and the Process of Domestication," *American Anthropologist* 75 (1973): 1254–59.

12. Blumler and Byrne, p. 33; Rindos, *Origins of Agriculture*, pp. 113–14.

13. Blumler and Byrne, p. 25; Rindos, *Origins of Agriculture*, p. 178.

14. M. A. Hoffman, *Egypt Before the Pharaohs: The Prehistoric Foundations of Egyptian Civilization* (New York, 1979), pp. 88–90; Henry, "Preagricultural Sedentism," pp. 371–72; J. Mellaart, *The Neolithic of the Near East* (London, 1975), p. 30; K. W. Butzer, "Agricultural Origins in the Near East as a Geographical Problem," in S. Struever, ed., *Prehistoric Agriculture* (Garden City, N.Y., 1971), p. 226.

15. D. O. Henry, "The Fauna in Near Eastern Archaeological Deposits," in F. Wendorf and A. E. Marks, eds., *Problems in Prehistory: North Africa and the Levant* (Dallas, 1975), pp. 379–85.

16. O. Bar-Yosef and E. Tchernov, "Archaeological Finds and Fossil Faunas of the Natufian and Microlithic Industries at Hayonim Cave (Western Galilee, Israel," *Israel Journal of Zoology* 15 (1966): 104–40.

17. Henry, "Preagricultural Sedentism," p. 371.

18. Ibid., p. 376.

19. Rindos, *Origins of Agriculture*, p. 31, Henry, "Preagricultural Sedentism," p. 380.

20. Mellaart, *The Neolithic of the Near East*, p. 46.

21. M. N. Cohen, *Food Crisis in Prehistory*, pp. 22–23.

22. Blumler and Byrne, p. 24; Rindos, *Origins of Agriculture*, pp. 184–85.

23. Rindos, *Origins of Agriculture*, p. 183.

24. Blumler and Byrne, p. 25.

25. Rindos, *Origins of Agriculture*, pp. 121–27.

26. Ibid., pp. 186–87.

27. J. L. Angel, "Paleoecology, Paleodemography and Health," in S. Polgar, ed., *Population, Ecology and Social Evolution* (The Hague, 1975).

28. M. Harris, *Cannibals and Kings: The Origins of Cultures* (New York, 1977), p. 29.

29. Rindos, *Origins of Agriculture*, p. 188.

30. O. Bar-Yosef, "The Walls of Jericho: An Alternate Interpretation," *Current Anthropology* 27/2 (1986): 157; K. M. Kenyon, *Digging Up Jericho: The Results of the Jericho Excavations 1952–1956* (New York, 1957), pp. 74–76; K. Kenyon, *Excavations at Jericho*, vol. 3, ed. T. A. Holland (London, 1981).

31. Mellaart, *The Neolithic of the Near East*, p. 49; P. Dorell, "The Uniqueness of Jericho," in R. Moorey and P. Parr, eds., *Archaeology in the Levant: Essays for Kathleen Kenyon* (Warminster, Eng., 1978), pp. 15, 17.

32. Mellaart, *The Neolithic of the Near East*, p. 51.

33. Dorrell, p. 15.

34. Bar-Yosef, p. 157.

35. Ibid., pp. 157–62.

36. Dorrell, p. 12.

37. Dorrell estimates 1,360 man-days for the walls. Add to this a conservative estimate of 140 man-days for the tower.

38. William McNeill, personal conversation, May 1991; E. Anati, *B.A.S.O.R.* 167 (1962): 25ff.

39. Mellaart, *The Neolithic of the Near East*, p. 51.

40. Dorrell, p. 12.

41. B. Bower, "Gauging the Winds of War," *Science News* 139/6 (1991): 89; C. R. Ember and M. Ember, "Resource Unpredictability, Mistrust and War," *Journal of Conflict Resolution* 36/2 (1992): 242–62.

42. Bar-Yosef, p. 158.

43. Mellaart, *The Neolithic of the Near East*, p. 71; S. Budiansky, *The Covenant of the Wild: Why Animals Chose Domestication* (New York, 1992), pp. 22–23.

44. S. Budiansky, "The Ancient Contract," *U.S. News & World Report*, 20 March 1989, p. 76.

45. Budiansky, *Covenant of the Wild*, pp. 56, 60.

46. R. White, "Husbandry and Herd Control in the Upper Paleolithic: A Critical Review of the Experience," *Current Anthropology* 30/5 (1989): 609–16.

47. Budiansky, *Covenant of the Wild*, p. 77.

48. Ibid.

49. J. Huxley, *Evolution: The Modern Synthesis* (London, 1974), pp. 532, 543; S. J. Gould, *Ever Since Darwin: Reflections in Natural History* (New York, 1977), pp. 63–66, 219–21.

50. Budiansky, *Covenant of the Wild*, pp. 77–78.

51. Ibid., p. 90.

52. Ibid. pp. 95–98.

53. S. J. Gould, "A Biological Homage to Mickey Mouse," in *The Panda's Thumb* (New York, 1980).

54. R. P. Coppinger and C. K. Smith, "The Domestication of Evolution," *Environmental Conservative* 10 (1983): 283–92; Budiansky, *Covenant of the Wild*, pp. 61, 125.

55. Mellaart, *The Neolithic of the Near East*, p. 48; Budiansky, *Covenant of the Wild*, p. 16.

56. Budiansky, *Covenant of the Wild*, p. 39; A. J. Legge and P. A. Rowley-Conwy, "Gazelle Killing in Stone Age Syria," *Scientific American* 257/2 (August 1987): 88–95.

57. S. A. Gregg, *Foragers and Farmers: Population Interaction and Agricultural Expansion in Prehistoric Europe* (Chicago, 1988), p. 118.

58. Ibid., pp. 101, 111.

59. R. Cribb, *Nomads in Archaeology* (Cambridge, Eng., 1991), p. 136; D. G. Bates, *Nomads and Farmers: A study of the Yoruk of South East Turkey* (Ann Arbor, Memoir no. 52, 1973).

60. F. Hole, *Studies in the Archaeological History of the Deh Luran Plain: The Excavation of Chaga Sefid* (Ann Arbor, Memoir no. 9, 1977), p. 14.

61. Cribb, p. 18.

62. H. E. Wright, "Natural Environment of Early Food Production North of Mesopotamia," *Science* 161 (1968): 224–39; Cribb, pp. 27–28.

63. Cribb, p. 59.

64. F. Hole, K. V. Flannery, and J. A. Neely, *Prehistory and Human Ecology of the Deh Luran Plain: An Early Village Sequence from Khuzistan, Iran* (Ann Arbor, Memoir no. 1 1969).

65. Cribb, p. 215.

66. See, for example, S. Mansfield, *The Gestalts of War: An Inquiry into Its Origins and Meaning as a Social Institution* (New York, 1982); M. Ehrenberg, *Women in Prehistory* (London, 1989); and R. Eisler, *The Chalice and the Blade* (New York, 1988?).

67. M. N. Cohen, "Prehistoric Hunter-Gathers: The Meaning of Social Complexity," in Price and Brown, p. 109.

68. P. J. Wilson, *The Domestication of the Human Species* (New Haven, 1988), p. 60.

69. Mellaart, *The Neolithic of the Near East*, p. 86.

70. Ibid., p. 89.

71. J. Mellaart, *Catal Hüyük: A Neolithic Town in Anatolia* (New York, 1967), pp. 68–69.

72. Mellaart, *The Neolithic Near East*, p. 101.

73. Ibid., pp. 101–3; R. L. O'Connell, "The Mace," *MHQ: The Quarterly Journal of Military History* 3/3 (1991): 92–93.

74. Mellaart, *Catal Huyuk*, p. 224; Mellaart, *The Neolithic Near East*, p. 100.

75. Mellaart, *The Neolithic Near East*, p. 102.

76. Mellaart, *Catal Hüyük*, chs. 4–5.

77. P. J. Wilson, pp. 99–112.

78. Cribb, pp. 28–29.

79. O. Bar-Yosef and A. Khazanov, "Introduction," in O. Bar-Yosef and A. Khazanov, eds., *Pastoralism in the Levant: Archaeological Materials in Anthropological Perspectives* (Madison, Wis, 1992), p. 5; I. Kohler-Rollefson, "A Model for the Development of Nomadic Pastoralism on the Transjordanian Plateau," in Bar-Yosef and Khazanov, p. 16.

80. Cribb, p. 20

81. R. B. Ekvall, *Fields on the Hoof* (New York, 1968).

82. Cribb, p. 24.

6 RIDE OF THE SECOND HORSEMAN

1. E. Ovada, "The Domestication of the Ass and Pack Transport by Animals: A Case of Technological Change," in O. Bar-Yesef and A. Khazanov, eds,. *Pastoralism in the Levant: Archaeological Materials in Anthropological Perspectives* (Madison, Wis., 1992), pp. 24–26.

2. J. F. Downs, "The Origin and Spread of Riding in the Near East and Central Asia," *American Anthropologist* 63 (1961): 1193–203; M. A. Littauer and J. H. Crouwel, *Wheeled Vehicles and Ridden Animals in the Ancient Near East* (Leiden, 1979), chs. 1–7; E. L. Shaughnessy, "Historical Perspectives on the Introduction of the Chariot into China," *Harvard Journal of Asiatic Studies* 48/1 (1988): 210–11.

3. E. I. Issac, *Geography of Domestication* (Englewood Cliffs, N.J., 1970), p. 92; M. Jankovich, *They Rode into Europe: The Fruitful Exchange in the Arts of Horsemanship Between East and West*, trans. A. Dent (London, 1971), pp. 25–26; A. P. Okladnikov, "Inner Asia at the Dawn of History," in Denis Sinor, ed., *The Cambridge History of Early Inner Asia* (Cambridge, Eng., 1990), pp. 81–85.

4. D. W. Anthony, D. Y. Telegin, and D. Brown, "The Origins of Horseback Riding," *Scientific American* 265 (1991): 94.

5. D. W. Anthony, "The 'Kurgan Culture,' Indo-European Origins, and the Domestication of the Horse: A Reconsideration," *Current Anthropology* 27/4 (1986): 295.

6. Anthony, Telegin, and Brown, p. 95.

7. D. W. Anthony, telephone conversation, 2 Dec. 1991.

8. D. W. Anthony and D. R. Brown, "The Origins of Horseback Riding," *Antiquity* 65 246 (1991): 23; Anthony, Telegin, and Brown, p. 96.

9. Anthony and Brown, pp. 25–31.

10. Ibid., pp. 34–35.

11. Anthony, "The 'Kurgan Culture,' " pp. 295–98.

12. D. W. Anthony, telephone conversation, August 1993.

13. J. P. Mallory, "Comments," *Current Anthropology* 27 4 (1986): 308.

14. Anthony, "The 'Kurgan Culture,' " pp. 301–3.

15. N. Merpert, "Comments on the Chronology of the Early Kurgan Tradition," *Journal of Indo-European Studies* 5 (1977): 373–78.

16. Anthony, "The 'Kurgan Culture,' " 297.

17. Ibid., p. 296.

18. Anthony, Telegin, and Brown, p. 96.

19. M. Shmagli, V. P. Dudkin, and K. V. Zin 'kovs' kii. "Prokompleksne Vivchennia Tripol 'skikh Polelen' " *Arkheologiia* (Kiev), no. 10, pp. 23–31, cited in Anthony, "The Kurgan Culture," p. 299; E. H. Minns, *Scythians and Greeks: A Survey of Ancient History and Archaeology on the North Coast of the Euxine from the Danube to the Caucasus* (Cambridge, Eng. 1913), pp. 141–42, 147.

20. R. L. Carneiro, "A Theory of the Origin of the State," *Science* 169 (1970): 733–38.

21. Adams notes: "New World agriculture . . . essentially did not involve stockbreeding. . . . Domesticated Andean camelids such as the llama were used mainly for transport. . . . Also missing in nuclear America, therefore, is the unique and powerful ambivalence of relations between herdsman and farmer, involving both symbiosis and hostility, which has shaped the social life, tinctured the history, and enriched the literature of the civilizations of the Fertile crescent" (R. M. Adams, "Early Civilizations: Subsistence and Environment," in C. H. Kraeling and R. M. Adams, eds., *City Invincible: An Oriental Institute Symposium* [Chicago, 1960], pp. 269–95).

22. R. Cribb, *Nomads in Archaeology* (Cambridge, Eng., 1991), p. 16.

23. See, for example, E. Marx, "Are There Pastoral Nomads in the Middle East?" in Bar-Yosef and Khazanov, *Pastoralism in the Levant*, pp. 255–60, and R. H. Meadow, "Inconclusive Remarks on Patoralism, Nomadism, and Other Animal-Related matters" in ibid., p. 262.

24. O. Bar Yosef and A Khazanov, "Introduction," in Bar-Yoseph and Khazanov, *Pastoralism in the Levant*, p. 5; Marx, p. 257.

25. T. E. Lavy, "Transhumance, Subsistence, and Social Evolution in the Northern Negev Desert," in Bar-Yosef and Khazanov, *Pastoralism in the Levant*, p. 66.

26. Bar-Yosef and Khazanov, "Introduction," p. 5.

27. Ibid.

28. D. W. Anthony, telephone conversation, 15 June 1994; J. N. Postgate, *Early Mesopotamia: Society and Economy at the Dawn of History* (London, 1992), p. 250.

29. Anthony, "The 'Kurgan Culture,' " p. 295; Anthony, telephone conversation, 2 Dec. 1991.

30. A. Khazanov, *Nomads and the Outside World* (Cambridge, Eng., 1984), p. 199.

31. J. Mellaart, *Excavations at Halicar* (Edinburgh, 1970); J. Mellaart, *The Neolithic of the Near East* (London, 1975), pp. 119, 167, 206-7; M Zohar, "Pastoralism and the Spread of the Semitic Language," in Bar-Yosef and Khazanov, *Pastoralism in the Levant*, p. 172.

32. M. A. Littauer and J. H. Crouwel, *Wheeled Vehicles and Ridden Animals in the Ancient Near East* (Leiden, 1979), p. 43.

33. J. Boessneck and A. von den Driesch, "Pferde in 4./3 Jahrtausend, V. Chapter in Ostanatolien," *Saugertierkundliche Mitteilungen* 31 (1976): 89-104.

34. S. Bökönyi, "Late Calcolithic Horses in Anatolia," in R. H. Meadow and H. P. Uerpmann, eds., *Equids in the Ancient World*, vol. 2 (Wiesbaden, 1991), pp. 123-31.

35. C. Grigson, "The Earliest Horses in the Middle East? New Finds from the Fourth Millennium of the Negev, *Journal of Archaeological Science*, cited in J. Clutton-Brock, *Horse Power A History of the Horse and the Donkey in Homan Societies* (Cambridge, Mass., 1992), p. 56; S. Davis (*Archaeology of Animals* [New Haven, 1987], p. 161) also cites similar finds of horse bones at Arad in the northern Negev dated slightly later (c. 2050-2650 B.C.).

36. Clutton-Brock, p. 12.

37. Postgate, p. 251.

38. I. Finkelstein and A. Perevolotsky, "Process of Sedentarization and Nomadization in the History of the Sinai and the Negev," *Bulletin of the American Schools of Oriental Research* 279 (1990): 67-88.

39. F. Hole, "Pastoral Nomadism in Western Iran," in R. A. Gould, ed., *Explorations in Ethno-Archaeology* (Albuquerque, 1978), p. 166.

40. D. W. Anthony, telephone conversation, 2 Dec. 1991; p. 37; O. Lattimore, *The Inner Asian Frontiers of China* (New York, 1951), pp. 74-75.

41. Cribb, p. 86; S. Legg, *The Barbarians of Asia* (New York, 1990), p. 42.

42. V. G. Childe, *Man Makes Himself* (London, 1936), p. 81; Cribb, p. 65.

43. G. Daniel, *The Origins and Growth of Archaeology* (New York, 1967), pp. 220-33; E. McEwen, R. L. Miller, and C. A. Bergman, "Early Bow Design and Construction," *Scientific American* 264/6 (June 1991): 80.

44. R. Hardy, *Longbow: A Social and Military History* (Cambridge, Eng., 1976), p. 30.

45. McEwen, Miller, and Bergman, pp. 80-81; C. A. Bergman, E. McEwen, and R. Miller, "Experimental Archery: Projectile Velocities and Comparison of Bow Performance," *Antiquity* 62 (1988): p. 668.

46. Legg, p. 41.

47. Sinor, p. 8.

48. D. L. Johnson, "Nomadic Organization of Space: Reflections on Patterns and Process" (1976): 25; B. Spooner, "The Cultural Ecology of Pastoral Nomads," *Addison-Wesley Module in Anthropology No. 45* (1973), pp. 35ff.

49. Cribb, p. 45; Legg, pp. 44–52.

50. D. L. Johnson, p. 38.

51. Legg, pp. 45; Denis Sinor, "The Inner Asian Warriors," *Journal of the American Oriental Society* 101/2 (1981); p.135.

52. Cribb, pp. 49–57.

53. Ibid., pp. 60–61.

54. K. C. Chang, *The Archaeology of Ancient China* (New Haven, 1968), pp. 152–54; 237; K. C. Chang, *Shang Civilization* (New Haven, 1980), p. 340; E. G. Pulleybank, "The Chinese and Their Neighbors in Prehistoric and Early Historic Times," in D. N. Keightley, ed., *The Origins of chinese Civilization* (Berkeley, 1983), pp. 457–58.

55. Cribb, pp. 57, 59.

56. A. L. Oppenheim, *Ancient Mesopotamia: Portrait of a Dead Civilization* (Chicago, 1977), p. 49; G. S. Ghurye, *Vedic India* (Bombay, 1979), pp. 9–10, 343; J. G. Macqueen, *The Hittites and Their Contemporaries in Asia Minor* (London, 1986), p. 27.

57. Herodotus, *The Histories*, trans. A. de Selincourt (Harmondsworth, Eng., 1981), 4. 125–130 (p. 312).

58. Thucydides, *The History of the Peloponnesian War*, trans. R. Crawley (Chicago, 1952), 2.96, 97.

59. J. K. Fairbank, "A Preliminary Framework" in J. K. Fairbank, ed., *The Chinese World Order: Traditional China's Foreign Relations* (Cambridge, Mass., 1968), pp. 11–12, 20–21; D. Twitchett and M. Lowe, *The Cambridge History of China*, vol. 1, *The Ch'in and Han Empires* (Cambridge, Eng., 1986), pp. 385, 390.

60. Lattimore, pp. xlvi, 240, 246.

7 URBAN IGNITION

1. Gilgamesh plays a role in several versions of epic tales. The story followed in the reconstruction is taken from the Sumerian poem "Gilgamesh and Agga," which was translated by S. N. Kramer and reconstructed from six tablets and fragments, five of which were excavated at Nippur. While the city is referred to as Erech in the text, Uruk is more common usage and is employed for the sake of clarity.

2. The Akkadian version of the myth, pieced together largely from Assyrian texts, repeatedly refers to Enkidu as a former nomad ("The barbarous fellow from the depths of the steppe," iv, 7) who is seduced and brought to Uruk by a whore ("Akkadian Myths and Epics," trans. E. A. Speiser, in J. B. Pritchard, ed., *Ancient Near Eastern Texts: Relating to the Old Testament* [Princeton, N.J., 1950], pp. 60–98.

3. "Gilgamesh and Agga," trans. S. N. Kramer, in Pritchard, *Ancient Near Eastern Texts*, line 55, p. 46.

4. Ibid., lines 103–4, 111, p. 47.

5. J. Mellaart, *Excavations at Halicar* (Edinburgh, 1970); J. Garstang, *Prehistoric Mersin*

(Oxford, 1953); J. Mellaart, *The Neolithic of the Near East* (London, 1975), pp. 119, 167, 206–7, 254–55.

6. Mellaart, *The Neolithic of the Near East*, p. 236.

7. R. M. Adams, *The Evolution of Urban Society: Early Mesopotamia and Prehispanic Mexico* (Chicago, 1966), p. 19.

8. P. Charvat, "The Kish Evidence and the Emergence of States in Mesopotamia," *Current Anthropology*, 22/6 (1981): 686–88; R. Cribb, *Nomads in Archaeology* (Cambridge, 1991), p. 159; P. J. Wilson, *The Domestication of the Human Species* (New Haven, 1988), pp. 99–116.

9. G. Johnson, "A Test of Central-Place Theory in Archaeology," in P. Ucko, R. Tringham, and G. Dimbleby, eds., *Man, Settlement, and Urbanism* (London, 1972); C. Renfrew, "Trade as Action at a Distance," in J. A. Sabloff and C. C. Lamberg-Karlowsky, eds., *Ancient Civilization and Trade* (Albuquerque, 1975).

10. Mellaart, *The Neolithic of the Near East*, pp. 278–79.

11. For a good summary of the argumentation see R. J. Wenke, *Patterns in Prehistory: Humankind's First Three Million Years* (New York, 1984), ch. 7, "The Origins of Complex Societies."

12. See, for example, E. O. Wilson, *Sociobiology: The New Synthesis* (Cambridge, Mass., 1975), p. 574; K. V. Flannery, "The Cultural Evolution of Civilizations," *Annual Review of Ecology and Systematics* 3 (1972): 399–426.

13. M. Gimbutas, "The First Wave of Eurasian Steppe Pastoralists into Copper Age Europe," *Journal of Indo-European Studies* 5 (1977): 277–337.

14. A. L. Oppenheim, *Ancient Mesopotamia: Portrait of a Dead Civilization* (Chicago, 1977), pp. 82–83.

15. R. M. Adams, "Developmental Stages in Ancient Mesopotamia," in S. Struever ed., *Prehistoric Agriculture* (Garden City, N.Y., 1971), p. 578; M. Harris, *Our Kind: Who We Are, Where We Came From, Where We Are Going* (New York, 1989), p. 394.

16. R. M. Adams, *Heartland of Cities: Surveys of Ancient Settlements and Land Use on the Central Floodplain of the Euphrates* (Chicago, 1981), p. 60.

17. Ibid., p. 244.

18. Adams, *Evolution of Urban Society*, pp. 42–43.

19. Adams, "Developmental Stages in Ancient Mesopotamia," pp. 581, 584.

20. R. Fox, *The Search for Society: Quest for a Biosocial Science and Morality* (New Brunswick, N.J., 1989), p. 214; Oppenheim, *Ancient Mesopotamia*, pp. 76–77; Adams, *Evolution of Urban Society*, pp. 82, 83, 85.

21. P. J. Wilson, *Domestication of the Human Species*, chs. 3–4.

22. W. H. McNeill, *Plagues and Peoples* (Garden City, N.Y., 1976), pp. 44–45.

23. Ibid., pp. 54–55; Adams, *Evolution of Urban Society*, pp. 70–71; R. J. Braidwood and C. A. Reed, "The Achievement and Early Consequences of Food Production: A Consideration of the Archaeological and Natural-Historical Evidence," *Cold Spring Harbor Symposium on Quantitative Biology* 22 (1957): 28–29.

24. Cited in S. N. Kramer, *History Begins at Sumer* (Philadelphia, 1981), p. 114.

25. D. Rindos, *The Origins of Agriculture: An Evolutionary Perspective* (Orlando, Fla., 1984), pp. 268–72. The flavor of the predicament is captured in a fragment of a contem-

porary poem, "The Deeds and Exploits of the God Ninurta" (Kramer, *History Begins at Sumer*, pp. 170-71).

> Famine was severe, nothing was produced,
> At the small rivers, there was no 'washing of the hands.'
> The waters rose not high.
> The fields were not watered,
> There was no digging of ditches.
> In all the lands there was no vegetation,
> Only weeds grew.

26. Adams, "Developmental Stages in Ancient Mesopotamia," pp. 611-12.

27. W. H. McNeill, "Human Migration: A Historical Overview," in W. H. McNeill and R. S. Adams, eds., *Human Migration: Patterns and Policies* (Bloomington, Ind., 1978), pp. 5-7.

28. Adams, *Evolution of Urban Society*, pp. 120-21; A. I. Tyumenev, "The Working Personnel in the Estate of the Temple of B A U in Lagas During the Period of Lugalanda and Urukagina," in I. M. Diakonoff, ed., *Ancient Mesopotamia: Socio-Economic History* (Moscow, 1969), pp. 88-89.

29. See, for example, Kramer, *History Begins at Sumer*, pp. 94-95 149-53; T. Jacobsen and S. N. Kramer, "The Myth of Inanna and Bilulu," *Journal of Near Eastern Studies* 12 (1953): 160-88; J. N. Postgate, *Early Mesopotamia: Society and Economy at the Dawn of History* (London, 1992), pp. 262-66.

30. See, for example, "Gilgamesh and the Land of the Living," "The Death of Gilgamesh," and "Inanna's Descent to the Nether World," in Pritchard, *Ancient Near Eastern Texts*, pp. 47-57.

31. Wenke, p. 211.

32. Adams, *Evolution of Urban Society*, p. 126.

33. Kramer, *History Begins at Sumer*, p. xxi; Adams, *Evolution of Urban Society*, p. 126; Postgate, p. 52.

34. Wenke, pp. 256-58; D. Diringer, *Writing* (New York, 1962).

35. I. J. Gelb, *Glossary of Old Addadian: Materials for the Assyrian Dictionary, no. 3* (Chicago, 1957); G. Roux, *Ancient Iraq* (Middlesex, Eng., 1972), p. 124.

36. Wenke, p. 258; Kramer, *History Begins at Sumer*, ch. 1.

37. Adams, *Evolution of Urban Society*, p. 126.

38. Kramer, *History Begins at Sumer*, pp. 30-35.

39. "Gilgamesh and Agga," pp. 45-46.

40. J. Lewthwaite, "Comments on Gilman," *Current Anthropology* 22/1 (1981): 14.

41. S. Andreski, *Military Organization and Society* (Berkeley: 1968), p. 95.

42. Adams, *Evolution of Urban Society*, pp. 96-109; Roux, p. 262.

43. I. M. Diakonoff, "The Rise of the Despotic State in Ancient Mesopotamia," in Diakonoff *Ancient Mesopotamia*, p. 179; I. M. Diakonoff, "The Sale of Land in Pre-Sargonic Sumer," in *Papers Presented by the Soviet Delegation at the XXIII International Congress of Orientalists: Assyriology Section* (Moskow, 1954).

44. Adams, *The Evolution of Urban Society*, p. 58.

45. Oppenheim, *Ancient Mesopotamia*, pp. 82-83.

46. Kramer, *History Begins at Sumer*, p. 118.

47. Ibid., p. 119.

48. Adams, *Evolution of Urban Society*, pp. 82, 87.

49. M. Ehrenberg, *Women in Prehistory* (London, 1989), p. 81.

50. R. Frisch and J. McArthur, "Menstrual Cycles: Fatness as a Determinant of Minimum Weight for Their Maintenance and Onset," *Science* 185 (1974): 949–51; Harris, *Cannibals and Kings: The Origins of Cultures* (New York, 1977), pp. 16–17.

51. Harris, *Our Kind*, p. 330.

52. "The Epic of Gilgamesh," "Akkadian Myths and Epics, in Pritchard, *Ancient Near Eastern Texts* iv, 13, 19, p. 75.

53. A. I. Tyumenev, "The State Economy in Ancient Sumer," in I. M. Diakonoff, ed., *Ancient Sumer: Socio-Economic History* (Moscow, 1969), pp. 71, 73.

54. Adams, *Evolution of Urban Society*, pp. 96–97.

55. Ibid., p. 102; Postgate, pp. 254–55.

56. A. I. Tyumenev, "The Working Personnel on the Estate of the Temple of BAU in Lagas During the Period of Lugalanda and Urukagina," in Diakonoff *Ancient Sumer*, p. 114 n. 115; Adams, *Evolution of Urban Society*, pp. 102–3.

57. Adams, table 1.

58. Roux, p. 121.

59. Charvat, p. 687.

60. A. L. Oppenheim, "Trade in the Ancient Near East," *International Congress of Economic History*, vol. 5 (Moscow, 1976), pp. 126–47.

61. G. Algaze, "The Uruk Expansion: Cross-Cultural Exchange in Early Mesopotamian Civilization," *Current Anthropology* 30/5 (1989): 571–73.

62. "Gilgamesh and the Land of the Living," trans. S. N. Kramer in Pritchard, *Ancient Near East Texts*, line 51, p. 48.

63. McNeill, *Plagues and Peoples*, pp. 61–62.

64. "Enmerkar and the Lord of Aratta," in Kramer, *History Begins at Sumer*, pp. 24–26.

65. Algaze, p. 587.

66. This line of reasoning follows the general approach of R. L. Carneiro, "A Theory of the Origin of the State," *Science* 169 (1970): 733–38.

67. Kramer, *History Begins at Sumer*, p. 123.

68. Ibid., pp. 122, 260; "Lamentation over the Destruction of Nippur," line 1, p. 261.

69. The best representation of the Stele of Vultures, which is in the Louvre, appears in Y. Yadin, *The Art of Warfare in Biblical Lands: In Light of Archaeological Study*, 2 vols. (New York, 1963), vol. 1, pp. 134–35.

70. V. D. Hanson, *The Western Way of War: Infantry Battle in Classical Greece* (New York, 1989), pp. 152–53.

71. Ibid., pp. 17, 31, 35, 158.

72. Cited in Roux, p. 131; Tyumenev, p. 114 n. 115.

73. Adams, *Evolution of Urban Society*, p. 149.

74. Kramer, *History Begins at Sumer*, pp. 39–43; Adams, *Evolution of Urban Society*, p. 149.

75. Kramer, *History Begins at Sumer*, pp. 39–41.

76. Ibid., p. 281.

77. "The Legend of Sargon," from Neo-Assyrian and Neo-Babylonian sources, in Pritchard, *Ancient Near Eastern Texts*, p. 119.

78. See, for example, R. M. Adams, *Land Behind Bagdad: A History of the Settlement on the Diyala Plains* (Chicago, 1965), pp. 42–45; McNeill, "Human Migration," p. 6.

79. Roux, p. 141; "Sargon of Agade," in Pritchard, *Ancient Near Eastern Texts*, i–ii I–iii–iv 44, p. 267.

80. Roux, p. 141.

81. Dates of Sargon's fifty-five-year reign vary considerably but all appear to locate it within the twenty-fourth century.

82. Oppenheim, *Ancient Mesopotamia*, p. 154.

83. "Sargon of Agade," in Pritchard, *Ancient Near Eastern Texts* v–vi 5–52, p. 268.

84. Roux, p. 142.

85. W. Hinz, *The Lost World of Elam: Re-creation of a Vanished Civilization* (New York, 1973), pp. 71–73; "Sargon of Agade," in Pritchard, *Ancient Near Eastern Texts*, v–vi 5–52, p. 268; Johannes Lehmann, *The Hittites: People of a Thousand Gods* (New York, 1975), p. 191.

86. Roux, p. 142.

87. "The Sargon Chronicle," in Pritchard, *Ancient Near Eastern Texts*, p. 266.

88. Ibid.

89. "Sargon of Agade," in Pritchard, *Ancient Near Eastern Texts*, v–vi 5–52, p. 268.

90. Roux, p. 143; Hinz, p. 73.

91. A. Goetze, "Historical Allusions in Old Babylonian Omen Texts," *Journal of Cuneiform Studies* 1/13 (1947): 256.

92. Oppenheim, *Ancient Mesopotamia*, p. 227.

93. Yadin, pp. 47–48.

94. O. Gurney, *Anatolian Studies* 5 (1955): 93–113.

95. "How the Akkadian Empire Was Hung Out to Dry," *Science* 261 (1993): 985.

96. Roux, p. 141; Adams, *Evolution of Urban Society*, p. 152.

8 ANATOMY OF THE BEAST

1. Even in so young a field there are already numerous competing models of cultural evolution. Some of the more prominent are contained in L. L. Cavalli-Sforza and M. W. Feldman, *Cultural Transmission and Evolution* (Princeton, N.J., 1981); G. P. Murdock, "How Cultures Change," in H. C. Shapiro, ed., *Man, Culture and Society* (New York, 1971); R. Alexander, *Darwinism and Human Affairs* (Seattle, 1979); C. J. Lumsden and E. O. Wilson, *Genes, Mind and Culture* (Cambridge, Mass., 1981); H. R. Pulliam and C. Dunford, *Programmed to Learn: An Essay on the Evolution of Culture* (New York, 1980); R. Boyd and P. Richerson, *Culture and the Evolutionary Process* (Chicago, 1985); and W. H. Durham, *Coevolution: Genes, Culture, and Human Diversity* (Stanford, Calif., 1991).

2. See Durham, ch. 5.

3. Boyd and Richerson, pp. 7–8.

4. This omission is truly puzzling. In the case of Durham, who emphasizes test cases,

it simply may have been too basic to mention. However, with Boyd and Richerson, who concentrate upon mechanisms of cultural transmission, it is less understandable but may also be a matter of overlooking the obvious.

5. E. O. Wilson, *Sociobiology: The New Synthesis* (Cambridge, Mass., 1975), pp. 415–37.

6. E. O. Wilson, *On Human Nature* (Cambridge, Mass., 1978), p. 162.

7. Ibid., p. 175.

8. I. Eibl-Eibesfeldt, *Human Ethology* (New York, 1989), p. 20.

9. K. Kaltenbach, M. Weintraub, and W. Fullard, "Infant Wariness Toward Strangers Reconsidered: Infants' and Mothers' Reactions to Unfamiliar Persons," *Child Development* 51 (1980): 1197–202.

10. Eibl-Eibesfeldt, *Human Ethology*, p. 175.

11. R. L. O'Connell, *Of Arms and Men: A History of War, Weapons and Aggression* (New York, 1989), pp. 23–24, 37.

12. See the discussion in chapter 3.

13. For specific examples see O'Connell, *Of Arms and Men*, chs. 2, 4, 9.

14. Ibid., pp. 100, 129–37, 141–47.

15. Eibl-Eibesfeldt, *Human Ethology*, pp. 405–6; S. Freud, *Totem and Taboo* (Leipzig, 1913), p. 330; S. Tornay, "Armed Conflict in the Lower Oma Valley: 1970–76," *Senri Ethnological Studies* (1979): 97–117.

16. For example, the Ecumenical Lateran Council in A.D. 1139 outlawed the use of the crossbow among Christians but in no way discouraged its employment against Moslems (P. Contamine, *War in the Middle Ages*, trans. M. Jones [Oxford, 1984], p. 71; T. Depuy, *The Evolution of Weapons and Warfare* [Fairfax, Va., 1984], p. 65; R. Hardy, *Longbow: A Social and Military History* [Cambridge, Eng., 1976], p. 34). Similarly, at the International Peace Conference at the Hague in 1899, Sir John Ardaugh, arguing against outlawing the dumdum bullet, explained: "Civilized man is much more susceptible to injury than savages. . . . The savage, like the tiger, is not so impressionable, and will go on fighting even when desperately wounded" (*Proceedings of the Hague Peace Conference* [London, 1920]); Pliny (the Elder), *Natural History: Loeb Classical Library*, 10 vol., trans. H. Rackham, W. H. S. Jones and D. E. Eichholz (Cambridge, Mass., 1955–66), 7.7/LB II 511.

17. S. L. A. Marshall, *Men Against Fire: The Problem of Battle Command in Future War* (Washington, D.C., 1947), pp. 42–43. Marshall has been criticized recently for a number of inconsistencies in his work. However, his central thesis on small units remains intact and was supported by the following: "Fighter I: An Analysis of Combat Fighters and Non-Fighters," *Human Resources Research Office Technical Report no. 44* (Washington, D.C., 1957); M. Van Creveld, *Fighting Power: German and U.S. Army Performance, 1939–1945* (Westport, Conn., 1982), pp. 163–64; S. A. Stouffer, *The American Soldier: Combat and its Aftermaths* vol. 2 of Social Science Research Councils, *Studies in Social psychology in World War II* (Princeton, N.J., 1949).

18. R. Fox, *The Search for Society: Quest for a Biosocial Science and Morality* (New Brunswick, N.J., 1989), p. 18; Eibl-Eibesfeldt, *Human Ethology*, p. 270; L. Tiger, *Men in Groups* (New York, 1969). For example, Greek phalanxes were composed of small pods

of well-acquainted men (V. D. Hanson, *The Western Way of War: Infantry Battle in Classical Greece* [New York, 1989], pp. 121–22).

19. See, for example, depictions of early military training in Y. Yadin, *The Art of Warfare in Biblical Lands: In Light of Archaeological Study*, 2 vols. (New York, 1963), 1:201 (Nineteenth Dynasty Egypt, thirteenth century B.C., Pritchard, ed., *The Ancient Near East in Pictures: Relating to the Old Testament* (Princeton, N.J., 1954), p. 68 (Seventh Dynasty Egypt and Early Dynastic Iraq, first half of the third millennium B.C.).

20. Flavius Vegetius Renatus, "The Military Institutions of the Romans," in T. R. Phillips, ed., *The Roots of Strategy*, (Harrisburg, Pa., 1940); R. D. Sawyer, *The Seven Military Classics of Ancient China* (Boulder, 1993).

21. W. H. McNeill, *The Pursuit of Power: Technology, Armed Force, and Society Since A.D. 1000* (Chicago, 1982), p. 131.

22. D. C. Large, "The 'Stirring Voice of the Drumme' " *MHQ: The Quarterly Journal of Military History* 2 (1992): 20–23; Marshal M. de Saxe, *Reveries or Memoirs Concerning the Art of War* (Paris, 1957).

23. B. Kraig, "Feeding the Troops, 3000 BCE," *MHQ: The Quarterly Journal of Military History* 4/4 (1992): 18–23; J. Bram, *An Analysis of Inca Militarism* (New York, 1941), p. 53.

24. Yadin, 1:103, 108.

25. Frederick II, "Testimony Politique," 1768, cited in C. Duffy, *Frederick the Great: A Military Life* (London, 1985), p. 335.

26. Herodotus, *The Histories* (New York, 1981), bk. 7, 36 (p. 457); bk. 7, 226 (p. 518); R. Sawyer, ed., *The Seven Military Classics of Ancient China* (Sun-tzu, pp. 151-52; WeiLiao-tzu, pp. 245, 263).

27. M. Harris, *Our Kind: Who We Are, Where We Came From, Where We Are Going* (New York, 1989), p. 150; K. A. R. Kennedy, "Skulls, Aryans and Flowing Drains: The Interface of Archaeology and Skeletal Biology in the Study of Harrapan Civilization," in G. L. Possehl, ed., *Harappan Civilization: A Contemporary Perspective* (Warminster, Eng., 1982), pp. 290–91; B. Kraig, "Feeding the Troops," p. 22. It is interesting to note that in societies that do not appear as highly stratified, such as early Sumer and Harappa, these differences in size appear to be less significant or even nonexistent.

28. E. O. Wilson, *Sociobiology*, pp. 300–310; B. Hölldobler and E. O. Wilson, *The Ants* (Cambridge, Mass., 1990), p. 330.

29. "The Death of Gilgamesh," in J. B. Pritchard, ed., *Ancient Near Eastern Texts: Relating to the Old Testament* (Princeton, N.J., 1950), lines 41–64 (pp. 50–51); J. A. B. van Buitenen, ed., *The Mahabharata* (Chicago, 1978), see for example pp. 481–93; Homer, *Iliad*, trans. R. Lattimore (Chicago, 1951), 21.108; O'Connell, *Of Arms and Men*, p. 47.

30. A. Ferrill, *The Origins of War: From the Stone Age to Alexander* (New York, 1985), p. 30; Kraig, p. 23.

31. O'Connell, *Of Arms and Men*, p. 47; S. Mansfield, *The Gestalts of War: An Inquiry into Its Origins and Meanings as a Social Institution* (New York, 1982), pp. 114-15, 121.

32. This is particularly true of the *Iliad*, where the results of violence are so vividly depicted that it appears at times to be a sort of Bronze Age anatomy text.

33. F. A. Kierman, "Phases and Modes of Combat in Early China," in F. A. Kierman and J. K. Fairbank, eds., *Chinese Ways in Warfare* (Cambridge, Mass., 1974), p. 62; M.

Loewe and D. Twitchett, *The Cambridge History of China*, vol. 1, *The Ch'in and Han Empires* (Cambridge, Eng., 1986), pp. 99–100; W. Conrad and A. A. Demarest, *Religion and Empire: The Dynamism of Aztec and Inca Expansionism* (Cambridge, Eng., 1984), p. 47.

34. O'Connell, *Of Arms and Men*, p. 90; A. A. Demarest, "Peace, War, and the Collapse of an Ancient American Civilization," *Proposal to the United States Institute of Peace* (Washington, D.C. 1990), pp. 1–3.

35. O'Connell, *Of Arms and Men*, p. 10.

36. N. Eldredge and S. J. Gould, "Punctuated Equilibria: An Alternative to Phyletic Gradualism," in T. J. M. Schopf, ed., *Models in Paleobiology* (San Francisco, 1972); S. J. Gould, "The Meaning of Punctuated Equilibruim and Its Role in Validating a Hierarchical Approach to Macroevolution," in R. Milkman, ed., *Perspectives on Evolution* (Sunderland, Mass., 1982), p. 83.

37. The gun appears to have arrived in Europe by 1325, however, almost a century and a half would pass before serious elaboration on its possibilities took place (A. R. Hall, "Military Technology," in C. Singer, E. J. Holmyard, and A. R. Hall, eds., *A History of Technology*, vol. 1 [Oxford, 1954], p. 726); O'Connell, *Of Arms and Men*, ch. 7.

38. R. L. O'Connell, "The Mace," *MHQ: The Quarterly Journal of Military History* 3/3 (1991): 92–93.

39. J. Mellaart, *The Neolithic of the Near East* (London, 1975), pp. 110, 113, 135–37.

40. D. Anthony, D. Telegin, and D. Brown, "The Origin of Horseback Riding," *Scientific American* 265 (1991): 96.

41. O'Connell, "The Mace," p. 93.

42. For good representations of both the Standard of Ur and the Stele of Vultures see Yadin, 1:132–35.

43. These experiments were performed by R. Gabriel and K. Metz and are cited in their unpublished manuscript entitled "Ancient Armies" (1990), p. 142.

44. Singer, Holmyard, and Hall, 1:594.

45. Yadin, 1:132–35; R. Humble, *Warfare in the Ancient World* (London, 1980), p. 17.

46. C. Oman, *The Art of War in the Middle Ages: 378–1515* (Ithaca, N.Y., 1953), p. 76.

47. See, for example, Yadin, 1:150, 192–93, 196; K. C. Chang, *Shang Civilization* (New Haven, 1980), p. 196.

48. Homer, *Iliad*, 18.478–615; O'Connell, *Of Arms and Men*, pp. 47–48.

49. R. L. O'Connell, "The Insolent Chariot," *MHQ: The Quarterly Journal of Military History* 2/3 (1990): 80.

50. Yadin, 2:286.

51. M. A. Littauer and J. H. Crouwel, *Wheeled Vehicles and Ridden Animals in the Ancient Near East* (Leiden, 1979), p. 28.

52. See fragment of relief from Ashurbanipals's palace (northwest), Yadin, 2:452.

53. Littauer and Crouwel, pp. 17, 28–29, 32.

54. "Syrian Find Shows Horses' Ancient Role," *Washington Post*," 4 Jan. 1993, p. A8; "Ancient Clay Horse Is Found in Syria," *Washington Post*, 3 Jan. 1993; "Ancient Figurine Lifts Horses' Profile," *Science News* 143 (January 1993): 22.

55. E. L. Shaughnessy, "Historical Perspectives on the Introduction of the Chariot into

China," *Harvard Journal of Asiatic Studies* 48/1 (1988): 190; *Science News*, 144 (1993): 380; D. Anthony and N. B. Vinogradov, "Bronze Age Chariot Burials in the Ural Steppes," *Archaeology* (in press).

56. H. G. Creel, "The Role of the Horse in Chinese History," *American Historical Review* 70 (1965): 650.

57. K. C. Chang Civilization, p. 240; Shaughnessy, p. 192.

58. Ibid., p. 190.

59. Littauer and Crouwel, p. 134; Yadin, 2:384–85.

60. Yadin, 2:382–85, 403.

61. Ibid., p. 456.

62. Ibid., p. 450; Littauer and Crouwel, p. 135.

63. Yadin, 1:70–71, 146–47, 158–59.

64. O'Connell, *Of Arms and Men*, p. 42; Yadin, 2:314.

65. A. T. Olmstead, *History of Assyria* (New York, 1923), p. 97. See also similar acts in Pritchard, *Ancient Near Eastern Texts*, pp. 276–80.

66. T. H. Hollingsworth, *Historical Demography* (Ithaca, N.Y., 1969), pp. 33–34.

67. R. J. Doyle and N. Lee, "Microbes, Warfare, Religion, and Human Institutions," *Canadian Journal of Microbiology* 32 (1986): 194.

68. W. H. McNeill, *Plagues and Peoples* (Garden City, N.Y., 1976), p. 73.

69. Doyle and Lee, p. 195; McNeill, *Plagues and Peoples*, pp. 192–95.

70. Professor R. T. Joy of the Uniformed Services University of the Health Sciences, telephone interview, 7 Feb. 1993.

71. E. H. Ackerknecht, *A Short History of Medicine* (Baltimore, 1982), pp. 3–6; T. A. Cockburn, "Infectious Disease in Ancient Populations," *Current Anthropology* 12/1 (1971): 48–52.

72. FM 101-10-1/2 *Staff Officers' Field Manual Organizational, Technical, and Logistical Data Planning Factors* (Washington D.C., 1987), vol. 2, pp. 1–43.

73. Dio Cassius LIII.29.3.8. *History: Loeb Library*, trans. E. Cary, vol. 6, pp. 269–71; S. Jarcho, "A Roman Experience with Heat Stroke in 24 BC," *Bulletin of the New York Academy of Medicine* 43/8 (1967): 767–68.

74. Gabriel and Metz, unpublished manuscript, p. 214.

75. A sort of cottage industry has grown up around reconstructing and testing all manner of ancient weaponry, including large siege engines and even warships. See, for example, C. A. Bergman, E. McEwen, and R. Miller, "Experimental Archery: Projectile Velocities and Comparison of Bow Performances," *Antiquity* 62/237 (1988): 658–70; E. W. Marsden, *Greek and Roman Artillery: Technical Treatises* (Oxford, 1971). J. F. Coats and J. S. Morrison conducted an increasingly complex series of experiments in an effort to determine how the oars of Greek triremes were arranged, culminating in actually building and sailing a full-scale replica of such a ship in 1987.

76. R. Gabriel and K. Metz conducted a series of experiments on the amount of force generated by a blow from a wide variety of ancient weapons. On these grounds it seems clear that all were easily capable of killing an adult human (see Gabriel and Metz, unpublished manuscript, p. 207.

77. G. Majno, *The Healing Hand: Man and Wound in the Ancient World* (Cambridge, Mass., 1975), p. 84.

78. H. Frolich, "Baracken im trojanischen Kriege," *Arch. f. path. Anat u. Physiol.* 71 (1877): 509–14.

79. Majno, p. 15.

80. Ibid., pp. 16, 191, 199–200.

81. T. Bauer, *Zeitschrift fur Assyriologie und verwandte Gebiete: Neue Folge,* (XL), 3/4 (1913), p. 234, V. A. Jakobson, "The Social Structure of the Neo-Assyrian Empire," in I. M. Diakonoff, *Ancient Mesopotamia: Socio-Economic History* (Moscow, 1969), p. 290.

82. B. G. Trigger, B. J. Kemp, D. O'Connor, and A. B. Lloyd, *Ancient Egypt: A Social History* (Cambridge, 1983), p. 175–.2.

83. D. Twitchett and M. Loewe, *The Cambridge History of China, vol. 1, The Ch'in and Han Empires* (Cambridge, Eng., 1986), pp. 63–63, 101.

84. W. T. Sanders, J. R. Parsons, R. S. Santley, *The Basin of Mexico: Ecological Processes in the Evolution of a Civilization* (New York, 1979), p. 106.

85. J. Bram, *An Analysis of Inca Militarism* (New York, 1941), pp. 38–40; G. W. Conrad and A. A. Demarest, *Religion and Empire: The Dynamism of Aztec and Inca Expansionism* (Cambridge, Engl 1984), p. 170.

86. Conrad and Demarest, p. 47.

87. B. de Sahagun, *Florentine Codex: General History of the Things of new Spain,* book 8, Kings and Lords, ed. A. J. O. Anderson and C. E. Dibble (Santa Fe, 1954), pp. 67, 75; M. Harris, *Cannibals and Kings: The Origins of Cultures* (New York, 1977), p. 110.

88. R. Hassig, *Aztec Warfare: Imperial Expansion and Political Control* (Norman, Okla., 1988), pp. 264–65; Conrad and Demarest, p. 38.

89. W. Watson, *Early Civilization in China* (New York, 1966), p. 71; J. Lee, "Migration and Expansion in Chinese History," in W. H. McNeill and R. S. Adams, eds., *Human Migration: Patterns and Policies* (Bloomington, Ind., 1978), pp. 20–26.

90. E. O. Wilson, *On Human Nature,* p. 81; O. Patterson, "Slavery," *Annual Review of Sociology* 3 (1977): 407–49; O. Patterson, "The Structural Origins of Slavery: A Critique of the Neiboer-Domar Hypothesis from a Comparative Perspective," *Annals of the New York Academy of Sciences* 292 (1977): 12–34; W. H. McNeill, "Human Migration: A Historical Overview," in McNeill and Adams, p. 12.

91. W. H. McNeill, *Population and Politics Since 1750* (Charlottesville, Va., 1990), pp. 7–8.

92. J. Keegan, *A History of Warfare* (New York, 1993), p. 60.

9 GARDEN OF OTHERWORLDLY DELIGHTS

1. Cited in B. J. Kemp, *Ancient Egypt: Anatomy of a Civilization* (London, 1989), p. 111.

2. Cited in P. Montet, *Everyday Life in Egypt: In the Days of Rameses the Great,* trans. A. R. Maxwell-Hyslop and M. S. Drower (Westport, Conn., 1974), p. 158.

3. B. J. Kemp, "Imperialism and Empire in New Kingdom Egypt," in P. D. A. Garnsey

and C. R. Whittaker, eds., *Imperialism in the Ancient World: The Cambridge University Research Seminar in Ancient History* (New York, 1978), pp. 11-12.

4. Montet, pp. 222-23.

5. Herodotus, *The Histories* (New York, 1972), quoted in Michael Rice, *Egypt's Making: The Origins of Ancient Egypt 5000-2000 BC* (London, 1990), p. 13; John Romer, *People of the Nile: Everyday Life in Ancient Egypt* (New York, 1982), p. 11.

6. K. W. Butzer, *Early Hydraulic Civilization in Egypt: A Study in Cultural Ecology* (Chicago, 1976), p. 41.

7. Ibid., p. 17.

8. Herodotus, bk. 2, 13 (p. 134).

9. L. Casson, *Ancient Egypt* (New York, 1965), p. 33.

10. A. R. David, *The Pyramid Builders of Ancient Egypt: A Modern Investigation of Pharaoh's Workforce* (London, 1986), p. 57.

11. F. Wendorf and R. Schild, "The Paleolithic of the Lower Nile Valley," in F. Wendorf and A. E. Marks, eds., *The Pleistocene Pre-history of the Southern and Eastern Mediterranean Basin* (Dallas, 1975); F. Wendorf and R. Schild, *Prehistory of the Nile Valley* (New York, 1976), pp. 289-91; M. A. Hoffman, *Egypt Before the Pharaohs: The Prehistoric Foundations of Egyptian Civilization* (New York, 1979), pp. 85-90.

12. N. Grimal, *A History of Ancient Egypt* (Oxford, 1992), pp. 21-22; K. W. Butzer, *Environment and Archaeology: An Ecological Approach to Prehistory* (Chicago, 1971), p. 591.

13. Butzer (*Early Hydraulic Civilization in Egypt*, pp. 4-10) questions the widely held assumption that plant and animal domesticates were introduced into Egypt from southwest Asia, arguing instead that this process could have taken place either indigeneously or from North African sources. However, the supersession of the so-called hair sheep by a woolly "Asiatic" breed would appear to undermine this argument. Nonetheless, it now appears that the process of neoteny could also account for the development of such a sheep in a purely African context.

14. Butzer, *Early Hydraulic Civilization in Egypt*, pp. 19, 25.

15. Ibid.

16. Hoffman, pp. 128, 148.

17. A. P. Okladnikov, "Inner Asia at the Dawn of History," in Denis Sinor, ed., *The Cambridge Ancient History of Early Inner Asia* (Cambridge, Eng., 1990), p. 100.

18. Hoffman, pp. 68, 218-19, 237, 241-42.

19. See Hoffman, "Map showing location of important prehistoric desert tradition sites," p. 218; H. A. Winkler, *Rock-Drawings of Southern Upper Egypt*, vol. 1 (London, 1938), p. 20.

20. Butzer *Early Hydraulic Civilization in Egypt*, p. 39.

21. Hoffman, p. 247; Rice, p. 36.

22. Hoffman, p. 148; Kemp, *Ancient Egypt*, pp. 33, 39.

23. Y. Yadin, *The Art of Warfare in Biblical Lands: In Light of Archaeological Study*, vol. 1 (New York, 1963), pp. 51, 122-23; B. G. Trigger, "The Rise of Egyptian Civilization," in B. G. Trigger, B. J. Kemp, D. O'Connor, and A. B. Lloyd, eds., *Ancient Egypt: A Social History* (Cambridge, Eng., 1983), p. 45.

24. J. Wilson, "Civilization Without Cities," in C. Kraeling and R. M. Adams, eds., *Cities Invincible: An Oriental Institute Symposium* (Chicago, 1960), pp. 124–64.

25. Grimal, pp. 28–29.

26. Kemp, *Ancient Egypt*, p. 47.

27. Ibid., p. 52.

28. Hermann Kees first overturned the traditional view that the south triumphed over the north in 1949 by theorizing that the unification was first accomplished under the aegis of the north, but that it broke down, only to be reaccomplished by rulers from the south (J. Vandier, *La Religion de Egyptienne* (Paris, 1949), pp. 24ff). As recently as 1985 Kaiser argued that Middle and Upper Egypt from el-Badari to Naqada were ever more influenced by northern culture (W. Kaiser, Zur Sudausdehnung der vorgeschichtlichen Deltakulturen und zur fruhen Entwicklung Oberagyptens, *Mitteilungen des Seutschen Archaologischen Instituts* [Cairo, 1985]). The weakness of these arguments is that the evidence indicating social consolidation remains all in the south.

29. Kemp, *Ancient Egypt*, p. 44.

30. Ibid., pp. 44–5; Hoffman, p. 299.

31. Rice, p. 49.

32. Butzer, *Early Hydraulic Civilization in Egypt*, pp. 20–21, 47.

33. K. Baer, "The Low Price of Land in Ancient Egypt," *Journal of American Research Center Egypt* 1 (1962): 25–42; Butzer, *Early Hydraulic Civilization in Egypt*, p. 50.

34. Butzer, *Early Hydraulic Civilization in Egypt*,

35. Ibid.

36. Ibid., p. 85.

37. J. M. Riddle, *Contraception and Abortion from the Ancient World to the Renaissance* (Cambridge, Mass., 1992), pp. 66–73.

38. Hoffman, p. 110.

39. Ibid., p. 271–79.

40. Kemp, *Ancient Egypt*, p. 55.

41. Romer, p. 65.

42. Herodotus, bk. 2, 124.

43. Kemp, *Ancient Egypt*, p. 132.

44. Ibid., pp. 112–13.

45. Trigger, p. 63.

46. B. J. Kemp, "Old Kingdom, Middle Kingdom, and Second Intermediate Period, c. 2686–1552 BC," in Trigger et al., pp. 93–94.

47. David, pp. 59, 62–98.

48. Grimal, pp. 102, 133–36.

49. W. H. McNeill, *Plagues and Peoples* (Garden City, N.Y., 1976), p. 39; Rice, p. 6.

50. C. Aldred, *Egypt: To the End of the Old Kingdom* (New York, 1965), pp. 126–29.

51. L. Manniche, *Sexual Life in Ancient Egypt* (London, 1987), pp. 43–44; B. Watterson, *Women in Ancient Egypt* (New York, 1991), p. 8; G. Robins, "Some Images of Women in New Kingdom Art and Literature," in B. S. Lesko, ed., *Women's Earliest Records: From Ancient Egypt and Western Asia* (Atlanta, 1989), pp. 105–16; R. A. McCoy, *The Golden Goddess: Ancient Egyptian Love Lyrics* (Menomonie, Wis., 1972).

52. Watterson, p. 1; H. G. Fischer, "Women in the Old Kingdom and the Heracleopolitan Period," in Lesko, pp. 5–24.

53. Romer, p. 115.

54. Kemp, *Ancient Egypt*, p. 227.

55. Yadin, 1:146–47.

56. Butzer, *Early Hydraulic Civilization in Egypt*, pp. 53–54.

57. Kemp, *Ancient Egypt*, pp. 39–41.

58. B. Bell, "The Dark Ages in Ancient History: I. The First Dark Age in Egypt," *American Journal of Archaeology* 75 (1971): 1–26.

59. Kemp, "Old, Kingdom, Middle Kingdom," p. 115.

60. Romer, p. 22.

61. Rice, pp. 7, 229.

62. Butzer, *Early Hydraulic Civilization in Egypt*, p. 29; Romer, p. 22.

63. David, pp. 31–32, 190–92; Butzer, *Early Hydraulic Civilization in Egypt*, pp. 94–96; Grimal, p. 182.

64. Grimal p. 185.

65. See "The Story of Si-nuhe," pp. 18–22; "Asiatic Campaigns under Pepi I," pp. 227–28 in J. B. Pritchard, ed., *Ancient near Eastern Texts: Relating to the Old Testament* (Princeton, N.J., 1950); Grimal, p. 186; P. H. Newby, *Warrior Pharaohs: The Rise and Fall of the Egyptian Empire* (London, 1980), pp. 19–24.

66. Kemp, "Imperialism and Empire in New Kingdom Egypt," p. 20; G. Steindorff and K. C. Seele, *When Egypt Ruled the East* (Chicago, 1963), p. 91.

67. Kemp, *Ancient Egypt*, p. 223.

68. D. O'Connor, "New Kingdom and Third Intermediate Period, 1552–664 BC," in Trigger et al., p. 192.

69. Kemp, *Ancient Egypt*, p. 229.

70. O'Connor, pp. 205–6.

71. Kemp, "Imperialism and Empire in New Kingdom Egypt," pp. 13–15.

72. O'Connor, P. 196.

10 LORDS OF EXTORTION

1. L. Waterman, *Royal Correspondence of the Assyrian Empire* (Ann Arbor, 1930–36), vol. 4, p. 213, no. 6; G. Roux, *Ancient Iraq* (Middlesex, Eng., 1966), pp. 323–24; H. W. F. Saggs, *The Might That Was Assyria* (London, 1984), pp. 279–80. The wording and spelling here follow Saggs.

2. Typical epithets used to describe Assyrian kings. See J. B. Pritchard, ed., *Ancient Near Eastern Texts: Relating to the Old Testament* (Princeton, N.J., 1950), p. 274: "Tiglath-pileser I: Expeditions to Syria, the Lebanon and the Mediterranean Sea"; p. 276: "Shalmaneser III: the Fight Against the Aramean Coalition"; or the Ashurbanipal colophon cited by Saggs, pp. 280–81.

3. Extracted from Ashurbanipal colophon cited by Saggs, p. 281.

4. Not to be confused with Sargon the Akkadian.

5. M. Cogan, "A Plaidoyer on Behalf of the Royal Scribes," in M. Cogan and I. Eph'al,

eds., *Ah, Assyria . . . Studies in Assyrian History and Ancient Near Eastern Historiography Presented to Hayim Tadmor* (Jerusalem, 1991), p. 124; Saggs, p. 114.

6. Saggs, p. 113.

7. Xenophon, *The Anabasis* (London, 1918), 3.4.10–12.

8. A. T. Olmstead, *History of Assyria* (New York, 1923), pp. 648–49.

9. Saggs. p. 247.

10. W. H. McNeill, *Plagues and Peoples* (Garden City, N.Y., 1976), p. 40.

11. Roux, p. 235.

12. M. T. Larsen, *The Old Assyrian City-State and Its Colonies* (Copenhagen, 1976), p. 124; J. Laessoe, *People of Ancient Assyria: Their Inscriptions and Correspondence* (New York, 1963), pp. 43–44; Saggs, pp. 28–30.

13. N. Joffe, "Eplaining Trade in Ancient Western Asia," *Monographs on the Ancient Near East* (Malibu, 1981), p. 2.

14. Ibid., p. 12.

15. "Middle Assyrian Laws, Tablet B," in Pritchard, *Ancient Near Eastern Texts,: Relating to the Old Testament* (Princeton, N.J., 1950), p. 186.

16. Saggs, pp. 35–36; "How the Akkadian Empire Was Hung Out to Dry," *Science* 261 (1993); 985.

17. Cited in Laessoe, p. 43.

18. Saggs, pp. 35–36.

19. P. Machinist, "Provincial Governance in Middle Assyria and Some New Texts from Yale," in K. Deller, P. Garelli, E. Porada, and C. Saporetti, eds., *Monographic Journals of the Near East: Assur* (Malibu, 1982), pp. 33–34.

20. Saggs, p. 38.

21. "Treaty Between Mattiwaz and Suppiluliumas," Rev. 8–10, E. F. Weidner, *Politische Dokumente as Kleinasian: Boghazkoy Studien* (Liepzig, 1923), vol. VIII, p. 39.

22. "Suppiluliumas Destroys the Kingdom of Mitanni," excerpt from the historical introduction to the treaty between Mattiwaz and Suppiluliumas; ibid., 6–15.

23. See J. N. Postgate in M. T. Larsen, ed., *Power and Propaganda: A Symposium on Ancient Empires* (Copenhagen, 1979), p. 202 n. 92

24. Machinist, pp. 32–33.

25. D. D. Luckenbill, *Ancient Records of Assyria and Babylonia* (Chicago, 1927), vol. 1, p. 145.

26. Saggs, pp. 48–57.

27. Ibid., p. 62.

28. Roux, p. 256.

29. Saggs, p. 61.

30. I. M. Diakonoff, "Agrarian Conditions in Middle Assyria," in I. M. Diakonoff, ed., *Ancient Mesopotamia: Socio-Economic History* (Moscow, 1969), pp. 221–22.

31. Saggs, pp. 48–49

32. Ibid., p. 137; T. H. Hollingsworth, *Historical Demography* (Ithaca, N.Y., 1969), pp. 68–69.

33. "Middle Assyrian Laws, Tablet A," in Pritchard, *Ancient Near Eastern Texts*, pp. 181, 183, 184–185.

34. Saggs, p. 145.

35. See, for example, Y. Yadin, *The Art of Warfare in Biblical Lands: In Light of Archaeological Study*, 2 vols. (New York, 1963), vol. 2, p. 413; J. B. Prichard, ed., *The Ancient Near East in Pictures: Relating to the Old Testament* (Princeton, N.J., 1954), p. 51, plate 167.

36. See P. Gay, *Freud for Historians* (New York, 1985), particularly the preface and chapter 1, for a good discussion of the topic.

37. G. W. Conrad and A. A. Demarest, *Religion and Empire: The Dynamism of Aztec and Inca Expansionism* (Cambridge, Eng., 1984), pp. 187, 214.

38. 2 Kings 18–19; Isa. 36–37; Saggs, p. 243.

39. R. L. O'Connell, *Of Arms and Men* (New York, 1989), p. 40; Yadin, 2:296.

40. Yadin, 2:295–96; Olmstead, p. 605; R. Humble, *Warfare in the Ancient World* (London, 1980), p. 24.

41. N. B. Jankowska, "Some Problems of the Economy of the Assyrian Empire," in Diakonoff *Ancient Mesopotamia*, p. 275.

42. Shalmaneser III specifically mentions crossing the Euphrates River in 845 B.C. with an army of 120,000 in the Black Obalisk, Fourteenth Year. See Pritchard, *Ancient Near Eastern Texts*, p. 280.

43. Ibid., p. 115.

44. D. G. Hogarth, *The Ancient Near East* (London, 1950), p. 25; Roux, p. 258.

45. Saggs, pp. 178–79, 246.

46. P. Garelli, "The Achievement of Tiglath-Pileser III: Novelty or Continuity?" in Cogan and Eph'al, p. 49.

47. Saggs, p. 131.

48. M. Elat, "Phoenician Overland Trade Within the Mesopotamian Empires," in Cogan and Eph'al, p. 24; Jankowska, p. 255 n. 19.

49. Laessoe, p. 106.

50. Saggs, p. 128; Roux p. 278.

51. B. Oded, *Mass Deportations and Deportees in the Neo-Assyrian Empire* (Wiesbaden, 1979); Saggs, p. 268.

52. 1 Sam. 8:11.

53. N. Na'aman, "Forced Participation in Alliances in the course of the Assyrian Campaigns to the West," in Cogan and Eph'al, pp. 80–98.

54. Ibid., p. 88.

55. Saggs, p. 80.

56. Roux, pp. 275–76.

57. A. L. Oppenheim, *Journal of Near Eastern Studies* 19 (1960): 133–47.

58. F. M. Fales, "The Account of Sargon's Eighth Campaign," in Cogan and Eph'al, p. 138.

59. A. K. G. Kristensen (*Who Were the Cimmerians, and Where Did They Come From?* [Copenhagan, 1988]) argues unconvincingly that they were actually Israelites deported after the fall of Samaria in 722 B.C. Herodotus, *The Histories* (Harmondsworth, Eng., 1981), bk. 4, 1–3, 9–13, pp. 271, 274–75.

60. Saggs, P. 97.

61. N. Kotker, "The Assyrians," *MHQ: The Quarterly Journal of Military History* 3/4

(1991): 17; A. T. Olmstead, *Western Asia in the Days of Sargon of Assyria* (Lancaster, Pa., 1906), p. 157.

62. Roux, p. 295; Saggs, p. 111.

63. Roux, p. 295.

64. Waterman, no. 327; Roux, p. 278.

11 HEAVEN'S MANDATE

1. Cited in A. Waley, *The Analects of Confucius* (London, 1956), bk. III, 7 (p. 95).

2. Cited in ibid., bk. III, 16 (p. 98); bk. V, 6 (p. 108).

3. Cited in ibid., Bk. IX, 1 (p. 138). This is the basis for the episode in the text, but it draws on details throughout *The Analects*.

4. B. I. Schwartz, *The World of Thought in Ancient China* (Cambridge, Mass., 1985), p. 58.

5. Ibid.

6. Ibid., p. 60.

7. See, for example, D. N. Keightley, "Early Civilization in China: Reflections on How It Became Chinese" in P. S. Ropp, ed., *The Heritage of China: Essays on Chinese Civilization in Comparative Perspectives* (Berkeley, 1986); D. N. Keightley, "Where Have All the Heroes Gone? Reflections on the Art and Culture in Early China and Early Greece." (paper prepared for discussion, Department of History, University of Virginia, 1990); M. H. Fried, "Tribe to State or State to Tribe in Ancient China?" in D. N. Keightley, ed., *The Origins of Chinese Civilization* (Berkeley, 1983).

8. K. C. Chang, *The Archaeology of Ancient China,* 4th ed. (New Haven, 1986), p. 412.

9. See, for example, L. Ward, "The Relative Chronology of China Through the Han Period," in R. W. Ehrich, ed., *Relative Chronologies in Old World Archeology* (Chicago, 1954), p. 130.

10. H. Ping-ti, *The Cradle of the East* (Hong Kong, 1975), p. 362; D. N. Keightley, "Ho Ping-ti and the Origin of Chinese Civilization," *Harvard Journal of Asiatic Studies* 37 (1977); 381–411; K. C. Chang, *Shang Civilization* (New Haven, 1980), p. 359.

11. K. C. Chang, *Archaeology of Ancient China,* p. 90; T. T. Chang, "The Origins and Early Cultures of the Cereal Grains and Food Legumes," in Keightley, *Origins of Chinese Civilization,* p. 73.

12. K. C. Chang, *Archaeology of Ancient China,* p. 85.

13. R. Pearson and S. C. Lo, "The Ch'ing-lien-kang Culture and the Chinese Neo-lithic," in Keightley, *Origins of Chinese Civilization,* p. 120; K. C. Chang, *Archaeology of Ancient China,* p. 143.

14. CKKH and Shen-hsi 1963: 198, cited in C. Kwong-Yue, "Recent Archaeological Evidence Relating to the Origin of Chinese Characters," in Keightley, *Origins of Chinese Civilization,* p. 364.

15. R. Pearson and S. C. Lo, p. 138.

16. K. C. Chang, *Archaeology of Ancient China,* p. 114.

17. See map detailing archaeological sites of Lung-shan in ibid., p. 245 (from Yen Wen-ming, *Wen-wu* 1981, no. 6, p. 42).

18. Ibid., pp. 274–75, 287.

19. K. C. Chang, *Archaeology of Ancient China*, p. 282.

20. Keightley, *Origins of Chinese Civilization*, xx; William Meacham, "Origins and Development of the Yueh Coastal Neolithic: A Microcosm of Culture Change on the Mainland of East Asia," in ibid., pp. 148, 170–71.

21. K. C. Chang, *Archaeology of Ancient China*, p. 270.

22. Y. Wen-ming, *Archaeology and Cultural Objects (Chinese publication)* 1982 (2). cited in ibid.

23. Herodotus, *The Histories* (New York, 1981), bk. IV, 65, p. 291.

24. See, for example, W. Eberhard, "The Growth of Chinese Civilization" in *Settlement and Social Change in Asia* (Hong Kong, 1967), p. 9 and notes; E. G. Pulleyblank, "Chinese and Indo-Europeans," *Journal of the Royal Asiatic Society* (April 1966): 31.

25. For the basis of this line of reasoning see J. F. Downs, "The Origin and Spread of Riding in the Near East and Central Asia," *American Anthropologist* 62 (December 1961): 1193–203; H. G. Creel, "The Role of the Horse in Chinese History," *American Historical Review* 62 (1965): 647–72.

26. D. W. Anthony, telephone conversation, 7 July 1994.

27. For a somewhat different perspective on the impetus behind urban intensification see P. Wheatley, *The Pivot of the Four Quarters: A Preliminary Enquiry into the Origins and Character of the Ancient Chinese City* (Chicago, 1971), chs. 1, 5; and K. C. Chang, *Archaeology of Ancient China*, pp. 361–68, epilogue.

28. Dates derived from R. D. Sawyer, *The Seven Military Classics of Ancient China* (Boulder, 1993), p. xix; K. C. Chang, *Shang Civilization*, pp. 1, 348; C. Y. Hsu and K. M. Linduff, *Western Chou Civilization* (New Haven, 1988), p. 95.

29. K. C. Chang, *Shang Civilization*, p. 2.

30. Hsu and Linduff, p. 12; K. C. Chang, *Shang Civilization*, pp. 342–45.

31. K. C. Chang, *Archaeology of Ancient China*, pp. 315–18.

32. Ibid., p. 361.

33. Ibid., p. 364; Wheatley, pp. 67–68.

34. K. C. Chang, *Shang Civilization*, p. 136.

35. K. C. Chang, *Archaeology of Ancient China*, p. 364.

36. *Tso Chuan*, under entry for 577 B.C. (the thirteenth year of Duke Ch'eng).

37. K. C. Chang, *Archaeology of Ancient China*, p. 366.

38. W. Watson, *Early Civilization in China* (New York, 1966), pp. 94–97; Wheatley, p. 71.

39. A. Waldron, *The Great Wall of China: From History to Myth* (Cambridge, Eng., 1990), p. 31.

40. D. N. Keightley, "The Late Shang State: When, Where, What?" in Keightley, *The Origins of Chinese Civilization*, pp. 526–29, 555–57.

41. Watson, p. 73.

42. E. L. Shaughnessy, "Historical Perspectives on the Introduction of the Chariot into China," *Harvard Journal of Asiatic Studies* 48 1 (1988): 193–94, 202–3.

43. Sawyer, Appendix A, p. 364.

44. K. C. Chang, *The Archaeology of Ancient China*, pp. 384–85.

45. J. Prušek, *Chinese Statelets and the Northern Barbarians in the Period 1400–300 BC* (New York, 1971), p. 39; K. C. Chang, *Shang Civilization*, p. 249.

46. Prušek, p. 40; Hsu and Linduff, pp. 25–26.

47. K. C. Chang, *Shang Civilization*, pp. 95, 114, 121–24, 331.

48. Ibid., p. 249.

49. W. W. Howells, "Origins of the Chinese People: Interpretations of the Recent Evidence," in Keightley, *The Origins of Chinese Civilization*, pp. 312–13.

50. K. C. Chang, *Shang Civilization*, p. 259.

51. *Yin pen chi*, ibid, pp. 13–14.

52. Hsu and Linuff, p. 104.

53. K. C. Chang, *Shang Civilization*, p. 249; Prusek, p. 40.

54. Hsu and Linduff, pp. 94, 111.

55. O. Lattimore, *Inner Asian Frontiers of China* (New York, 1951), p. 337; R. Gousset, *The Rise and Splendour of the Chinese Empire* (Berkeley: 1958), p. 22.

56. M. von Dewall, *Symposium in Honor of Dr. Li Chi on His Seventieth Birthday* (Taipei, 1966), pp. 1–68.

57. Hsu and Linduff, p. 109.

58. Ibid., p. 123.

59. Ibid., p. 200.

60. Ibid., p. 224.

61. Ibid., p. 158.

62. K. C. Chang, *The Archaeology of Ancient China*, p. 398.

63. Wheatley, pp. 176, 182.

64. Hsu and Linduff, pp. 249, 256.

65. Ibid., p. 225.

66. J. Legge, *The Ch'un Ts'ew with the Tso Chuen* (Hong Kong, 1970), pp. 40, 579; Lattimore, pp. 347–48.

67. Prušek, p. 150.

68. Hsu and Linduff, pp. 270–71, 279.

69. Wheatley, p. 114.

70. Schwartz, pp. 41, 57. The dates for these periods vary considerably; the ones above are taken from Sawyer, p. xix.

71. Sawyer, p. 10; D. B. Wagner, *Iron and Steel in Ancient China* (Leiden, 1993), pp. 145–48.

72. W. Eberhard, *A History of China* (London, 1977), p. 51; Lattimore, p. 374.

73. Hsu and Linduff, p. 249; Schwartz, p. 57.

74. Ibid., p. 140.

75. W. Dobson, *Mencius: A New Translation Arranged and Annotated for the General Reader* (Toronto, 1963), p. 182; Y. P. Mei, *Motse, the Neglected Rival of Confucius* (London, 1934), p. 100.

76. Schwartz, pp. 155, 157–58.

77. Waley, *The Analects of Confucius*, bk. XIV, 18 (p. 185).

78. Keightley, "Where Have All the Heroes Gone?" pp. 12–13.

79. Ibid., p. 18.

80. The military classics consist of *T'ai Kung's Six Secret Teaching, The Methods of*

Ssu-ma, Sun-tzu's Art of War, Wu-tzu, Wei Liao-tzu, Three Strategies of Huang Shih-kung, and *Questions and Replies Between T'ang T'ai-tsung and Li Wei-kung*; Sawyer, pp. 36–37, 112, 115, 149, 150, 191–92, 232.

81. V. D. Hanson, *The Western Way of War: Infantry Battle in Classical Greece* (New York, 1989), p. 9.

82. Sun-tzu in Sawyer, p. 161; See also *T'ai Kung*, in ibid., p. 53; *Wei Liao-tzu*, in ibid., p. 262, and *Questions and Replies*, in ibid., p. 360.

83. See *T'ai Kung*, in Sawyer, p. 51; *Wei Liao-tzu*, in ibid., p. 273; *Three Strategies*, in ibid., p. 306; *Ssu-ma*, in ibid., p. 126.

84. *T'ai Kung*, in Sawyer, pp. 69–70; *Ssu-ma*, in ibid., p. 141; *Sun-tzu*, in ibid., p. 155; *Wu-tzu*, in ibid., pp. 210–11; *Three Strategies*, in ibid., p. 289.

85. R. L. O'Connell, *Of Arms and Men: A History of War, Weapons, and Aggression* (New York, 1989), p. 65.

86. Sawyer, p. 387 n. 52.

87. H. G. Creel, *The Origins of Statecraft in China*, vol. 1, *The Western Chou Empire* (Chicago, 1970), pp. 256–62.

88. *Wei Liao-tzu*, in Sawyer, p. 245.

89. *T'ai Kung*, in Sawyer, p. 43; *Ssu-ma*, in ibid., p. 128; *Wu-Tzu*, in ibid., p. 223; *Wei Liao-tzu*, in ibid., p. 254.

90. *Wei Liao-tzu*, in Sawyer, p. 243.

91. F. A. Kierman, "Phases and Modes of Combat in Early China," in F. A. Kierman and J. K. Fairbank, eds., *Chinese Ways in Warfare* (Cambridge, Mass, 1974), pp. 39, 65.

92. Sawyer, pp. 10–11.

93. Prusek, pp. 176, 203.

94. Lattimore, p. 388.

95. Ibid., pp. 388, 421; Eberhard, *A History of China*, p. 60.

96. D. Twitchett and M. Loewe, eds., *The Cambridge History of China*, vol. 1, *the Ch'in and Han Empires 221 BC–AD 220* (Cambridge, Eng., 1986), p. 36; Lattimore, pp. 401, 421; Eberhard, *A History of China*, p. 60.

97. Lattimore, p. 401.

98. Waldron, p. 19.

99. Twitchett and Loewe, pp. 54–81.

100. Ibid., p. 103.

101. M. Loewe, "The Campaigns of Han Wu-ti," in Kierman and Fairbank, pp. 67–82, 104–5.

12 THE WORLD ANEW

1. This reconstruction is based primarily on Bernal Díaz del Castillo, *The Discovery and Conquest of Mexico* (New York, 1956), bk. 2, pp. 190–91; and M. Harris, *Our Kind: Who We Are, Where We Came From, Where We Are Going* (New York, 1989), p. 488.

2. G. Clark, *World Prehistory: In New Perspective* (Cambridge, Eng. 1977), p. 351; Harris, *Our Kind*, pp. 475–76.

3. "Sea's Bouty, Not the Mammals, May Have Drawn Nomads to Alaska," *New York Times*, 26 Dec. 1989, pp. 1, 7.

4. W. H. McNeill, *Plagues and Peoples* (Garden City, N.Y., 1976), ch. 1, p. 178.

5. M. E. Mosely, *The Incas and Their Ancestors: The Archaeology of Peru* (London, 1992), p. 86; C. E. Smith Jr., "Plant Remains from Guitarrero Cave," in T. F. Lynch, ed., *Guitarroro Cave: Early Man in the Andes* (New York, 1980), pp. 87–119; L. Kaplan, "Variations in Cultivated Beans," in Lynch, pp. 145–48.

6. J. W. Rick, "The Character and Context of Highland Preceramic Society," in R. W. Keatinge, ed., *Peruvian Prehistory: An Overview of Pre-Inca and Inca Society* (Cambridge, Eng., 1988), pp. 27–37; M. N. Cohen, *The Food Crisis in Prehistory* (New Haven, 1978), pp. 252, 256.

7. Mosely, pp. 68, 163–64; R. Hassig, *Aztec Warfare: Imperial Expansion and Political Control* (Norman, Okla., 1988), p. 105; R. Hassig, *War and Society in Ancient Mesoamerica* (Berkeley, 1992), pp. 57, 77.

8. Harris, *Our Kind*, p. 489; R. J. Wenke, *Patterns in Prehistory: Mankind's First Three Million Years* (New York, 1984), p. 409.

9. R. Carneiro, "A Theory of the Origin of the State," *Science* 169 (1970): 733–38; D. L. Webster, "Warfare and the Evolution of the State: A Perspective from the Maya Lowlands," University of Northern Colorado, Museum of Anthropology, Miscellaneous Series, no. 19 (1976).

10. G. R. Willey, "Maya Lowland Settlement Patterns: A Summary Review," in G. R. Willey, ed., *Essays in Maya Archaeology* (Albuquerque, 1987), pp. 114–15; Hassig, *War and Society in Ancient Mesoamerica*, p. 41; W. J. Conklin and M. E. Moseley, "The Patterns of Art and Power in the Early Intermediate Period," in Keatinge, pp. 151–52.

11. Hassig, *War and Society in Ancient Mesoamerica*, pp. 107, 130; Mosely, *The Incas and the Ancestors*, p. 245.

12. J. L. Stephens, *Incidents of Travel in Central America, Chiapas, and Yucatan* (New York, 1941), p. 105.

13. These assumptions are represented most clearly in the work of J. E. S. Thompson, *Maya Hieroglyphic Writing: Introduction* (Washington, D.C., 1950).

14. T. Proskouriakoff, *Maya History* (Austin, Tex., 1993), p. xxiii.

15. D. Stuart and S. D. Houston, "Maya Writing," *Scientific American* 261/2 (August 1989): 86.

16. G. R. Willey, "Towards a Holistic View of Ancient Maya Civilization," in Willey, *Essays in Maya Archaeology*, pp. 140–41; P. D. Harrison, "The Revolution in Ancient Maya Subsistence," in F. S. Clancy and P. D. Harrison, eds., *Vision and Revision in Maya Studies* (Albuquerque, 1990), pp. 99–100.

17. B. L. Turner, "Prehistoric Intensive Agriculture in the Mayan Lowland," *Science* 185 (1974): 118–24.

18. P. D. Harrison and B. L. Turner, *Prehistoric Maya Agriculture* (Albuquerque, 1974).

19. J. Marcus, "Ancient Maya Political Organization" (paper presented at the 1989 Dumbarton Oaks Conference).

20. Hassig, *War and Society in Ancient Mesoamerica*, pp. 73–75.

21. L. Schele, "Human Sacrifice Among the Classic Maya," in E. H. Boone, ed., *Ritual Human Sacrifice in Mesoamerica: A Conference at Dunbarton Oaks, October 13–14, 1979* (Washington, D.C., 1984), pp. 8–9.

22. Ibid., p. 44.

23. S. J. W. Wilkerson, "In Search of the Mountain of Foam: Human Sacrifice in Eastern Mesoameric," in *Ritual Human Sacrifice in Mesoamerica*, p. 116; W. L. Fash, *Scribes, Warriors and Kings: The City of Copan and the Ancient Maya* (London, 1991), pp. 85–6.

24. G. R. Willey and D. B. Shimkin, "The Maya Collapse: A Summary View," in Willey, *Essays in Maya Archaeology*, p. 51.

25. M. Harris, *Cannibals and Kings: The Origins of Cultures* (New York, 1977), p. 89; A. Demarest, "Warfare, Demography, and Tropical Ecology: Speculation on the Parameters of the Maya Collapse" (unpublished paper presented at the 89th American Anthropological Association meeting, November 1990), pp. 5–6.

26. R. M. Leventhal, "Southern Belize: An Ancient Maya Region," in Clancy and Harrison, p. 127.

27. Ibid., pp. 37, 77–78, 130.

28. Demarest, pp. 6–10; T. H. Maugh, "Mayas' Demise: A War Within?" *Los Angeles Times*, 14 Aug. 1989, II, p. 3; W. Booth, "Maya Tomb's Clues to New World Disorder: Guatemalan Dig May Explain Advanced Civilization's Violent Collapse," *Washington Post*, 26 May 1991, pp. A1, A32.

29. Willey, "The Maya Collapse," in Willey, *Essays in Maya Archaeology*, pp. 45–46; Fash, pp. 173–74.

30. Hassig, *War and Society in Ancient Mesoamerica*, p. 71.

31. W. T. Sanders, J. R. Parsons, and R. S. Santley, *The Basin of Mexico: Ecological Processes in the Evolution of a Civilization* (New York, 1979), chs. 2–6.

32. I. Clendinnen, *Aztecs: An Interpretation* (Cambridge, Eng., 1991), p. 21.

33. Sanders, Parsons, and Santley, p. 107.

34. R. Millon, B. Drewitt, and G. Cowgill, *Urbanization at Teotihuacan: The Teotihuacan Map, Part Two* (Austin, 1973).

35. Sanders, Parsons, and Santley, pp. 114–16.

36. Hassig, *War and Society in Ancient Mesoamerica*, p. 57.

37. Sanders, Parsons, and Santley, p. 134.

38. Ibid., p. 127.

39. Hassig, *War and Society in Ancient Mesoamerica*, p. 60.

40. W. T. Sanders, "Life in a Classic Village," in *Teotihuacan: Onceava Mesa Redonda* (Mexico City, 1966).

41. Hassig, *War and Society in Ancient Mesoamerica*, p. 212.

42. Sanders, Parson, and Santley, p. 137.

43. Fr. Bernadino de Sahagun, *Florentine Codex: General History of the Things of New Spain, Thirteen Volumes*, trans. A. O. Anderson and C. E. Dibble (Santa Fe, 1961), bk. 10, p. 165–66.

44. Sanders, Parsons, and Santley, pp. 144–48.

45. Ibid., p. 145; Hassig, *War and Society in Ancient Mesoamarica*, p. 115.

46. Ibid., pp. 113–14.

47. R. A. Diehl, *Tula: The Toltec Capital of Ancient Mexico* (New York, 1983), pp. 66, 94–95, 98.

48. Ibid., pp. 158–60.

49. S. D. Gillespie, *The Aztec Kings: The Construction of Rulership in Mexica History* (Tucson, 1989), pp. 3–4.

50. Ibid., pp. 68–70, 90.

51. G. W. Conrad and A. Demarest, *Religion and Empire: The Dynamism of Aztec and Inca Expansionism* (Cambridge, Eng., 1984), p. 22.

52. Ibid., p. 26.

53. J. de Acosta, *The Natural and Moral History of the Indies*, 2 vols., (New York, 1960), vol. 2, p. 480: D. Duran, *Historia de las Indias de Nueva Espana e Islas de la Tierra Firme* (Mexico City, 1967), vol. 2, pp. 75–77.

54. Cited in Conrad and Demarest, p. 35.

55. Clendinnen, p. 8.

56. Conrad and Demarest, p. 53.

57. Hassig, *War and Society in Ancient Mesoamerica*, p. 141.

58. Clendinnen, p. 116.

59. Hassig, *War and Society in Ancient Mesoamerica*, p. 141.

60. Clendinnon, p. 116; Hassig, *Aztec Warfare*, p. 101.

61. Hassig, *Aztec Warfare*, pp. 114–15; Clendinnen, pp. 115–21.

62. Hassig, *War and Society in Ancient Mesoamerica*, pp. 145–46; R. L. O'Connell, *Of Arms and Men: A History of War, Weapons, and Aggression* (New York, 1989), pp. 53, 90; Strabo 10.1.12; Polybius, 13.3.2–7.

63. Hassig, *Aztec Warfare*, p. 120.

64. Conrad and Demarest, p. 42.

65. Duran, bk. 2, ch. 44. Duran claims that the figure was actually eighty thousand, but this is based on what Clendinnen and others believe is a doubtful reading of an Aztec pictorial text (Clendinnen, p. 322. n. 7).

66. B. A. Brown, "Ochpaniztli in Historical Perspective," in *Ritual Human Sacrifice in Mesoamerica*, p. 195.

67. M. Harner, "The Ecological Basis for Aztec Sacrifice," *American Ethnologist* 4 (1977): 117–35; M. Harner, "The Enigma of Aztec Sacrifice," *Natural History* 86 (1977): 47–52.

68. Conrad and Demarest, p. 47.

69. Sanders, Parsons, and Santley, p. 176.

70. Clendinnen, p. 95; J. Keegan, *A History of Warfare* (N.Y, 1993).

71. Sahagun, *Florentine Codex*, bk. 8, *Kings and Lords*, p. 75; bk. 9, *The Merchants*, pp. 64, 67.

72. Harris, *Cannibals and Kings*, p. 110.

73. Clendinnen, p. 29; Sanders, Parsons, and Santley, p. 233–34.

74. Sahagun, *Florentine Codex*, bk. 10, *The People*, p. 80.

75. There are repeated references to terrifying characters such as "the demons of darkness (who) would descend to eat men" ibid., bk. 7, p. 27) or "the bad noblewomen . . . she terrorizes she ate people" (*Florentine Codex*, bk. 8, p. 46).

76. Ibid., bk. 9, *The Merchants*, p. 55; Conrad and Demarest, p. 59.

77. Clendinnen, pp. 46, 237–38.

78. Ibid., p. 49.

79. Ibid., p. 200.

80. Sahagun, bk. 10, pp. 37–38; Clendinnen, p. 169.

81. Ibid., pp. 53, 55.

82. Ibid., *bk. 6, Rhetoric and Moral Philosophy*, pp. 116–19, 125–26.

83. Ibid., bk. 10, p. 12.

84. C. MacLachlan, "The Eagle and the Serpent: Male over Female in Tenochtitlán," *Proceedings of the Pacific Coast Council of Latin American Studies* 5 (1976): 45–56; J. Nash, "The Aztecs and the Ideology of Male Dominance," *Signs* 4 (1978): 349–62.

85. Sahagun, bk. 6, pp. 171–72.

86. Conrad and Demarest, pp. 38–44; see also A. Demarest, "Overview: Mesoamerican Human Sacrifice in Evolutionary Perspective," in Boone, ed., *Ritual Human Sacrifice in Mesoamerica*, pp. 234–36.

87. R. M. Adams, *The Evolution of Urban Society: Early Mesopotamia and Prehispanic Mexico* (Chicago, 1965), pp. 148–49; Hassig, *Aztec Warfare*, pp. 105, 112–13.

88. Harris, *Cannibals and Kings*, p. 107.

89. Duran, vol. 2, chs. 19, 42.

90. Conrad and Demarest, pp. 172–73.

91. Hassig, *Mesoamerican Warfare*, pp. 153–54; Conrad and Demarest, p. 55.

92. B. Cobo, *Inca Religion and Customs*, trans. and ed. R. Hamilton (Austin, Tex., 1990), p. 190.

93. Mosely, *The Incas and Their Ancestors*, pp. 8, 44–46.

94. Ibid., p. 32.

95. Ibid., pp. 46–47.

96. Carneiro, "A Theory of the Origins of the State," 733–38.

97. Mosely, *The Incas and Their Ancestors*, p. 119; W. F. Allman and J. M. Schrof, "Lost Emprires of the Americas," *U.S. News & World Report*, 2 April 1990, p. 49.

98. S. Pozorski and T. Pozorski, "Early Andean Cities," *Scientific American* 270/6 (June 1994): 72.

99. Ibid., p. 68.

100. S. Pozorski and T. Pozorski, "Early Civilization in the Casma Valley, Peru," *Antiquity* 66 (1992): 867.

101. R. L. Burger, "Unity and Heterogeneity Within the Chavin Horizon," in Keatinge, p. 123.

102. Mosely, *The Incas and Their Ancestors*, p. 249.

103. Conklin and Mosely, "Art and Power in the Early Intermediate Period," in Keatinge, pp. 151–53.

104. Mosely, *The Incas and Their Ancestors*, pp. 85–92.

105. T. F. Lynch, *Guitarrero Cave: Early Man in the Andes* (New York, 1980).

106. Allman and Shrof, p. 53.

107. Mosely, *The Incas and Their Ancestors*, p. 207.

108. Ibid., p. 142; Harris, *Our Kind*, p. 489.

109. J. R. Parsons and C. M. Hastings, "The Late Intermediate Period," in Keatinge, pp. 190–91.

110. Mosely, *The Incas and Their Ancestors*, p. 207.

111. Ibid., p. 208; Keatinge, pp. 175–76.

112. Mosely, *The Incas and Their Ancestors*, pp. 219, 221.

113. W. H. Isbell, "City and State in Middle Horizon Huari," in Keatinge, pp. 164, 182.

114. M. Godelier, *Perspectives in Marxist Anthropology* (Cambridge, Eng., 1977), p. 188.

115. Mosely, *The Incas and Their Ancestors*, pp. 221–23.

116. Ibid., 223–24.

117. J. V. Murra, "El 'Control Vertical' de un Maximo de Pisos Ecologicos en la Economiai de las Sociedads Andinas," in J. V. Murra, ed., *Visita de la Provincia de Leon de Huanuco* (Huanucu, Peru, 1972), pp. 429–76.

118. See, for example, J. Bram, *An Analysis of Inca Militarism* (New York, 1941); B. C. Brundage, *Empire of the Incas* (Norman, Okla., 1963); H. L. Dobyns and P. L. Doughty, *Peru: A Cultural History* (New York, 1976), ch. 2.

119. S. A. Niles, *Callachaca: Style and Status in the Inca Community* (Iowa City, 1987), p. 7.

120. J. H. Rowe, "Inca Culture at the Time of the Spanish Conquest" in J. Steward, ed., *Handbook of South American Indians, vol. 2, the Andean Civilizations* (Washington, D.C., 1946), p. 329.

121. E. B. Dwyer, "The Early Inca Occupation of the Valley of Cuzco, Peru" (Ph.D. diss., University of California, Berkeley, 1971), pp. 145–46.

122. B. B. Bauer, *The Development of the Inca State* (Austin, 1992), p. 48.

123. Ibid., p. 99.

124. Ibid., p. 137.

125. Mosely, *The Incas and Their Ancestors*, pp. 68, 248.

126. P. de Cieza de Leon, *Chronicles of Peru, Second Part* (London, 1883), p. 49; Bram, pp. 31–33.

127. Conrad and Demarest, pp. 113–18.

128. Ibid., pp. 121–22.

129. Mosely, *The Incas and Their Ancestors*, p. 9.

130. Ibid., pp. 66, 71.

131. A. W. Crosby, *The Columbian Exchange; Biological and Cultural Consequences of 1492* (Westport, Conn., 1975), cited in "When Worlds Collide," *Newsweek*, Special Issue, Fall/Winter 1991, p. 76.

132. McNeill, *Plagues and Peoples*, pp. 180–81.

133. Ibid., p. 181.

134. Sahagun, *Florentine Codex, bk. 12, The Conquest of Mexico*, pp. 19, 40; Díaz, pp. 72, 131; Brundage, pp. 303–4.

135. Díaz, p. 59.

13 THALASSOCRACY

1. Mycenaean Greek for "Leader of the War-Host," first translated from the Linear B tablets at Pylos.

2. 1.Strom, "Aspects of Minoan Foreign Relations, LMI-LMII," in R. Hagg and Nanno Marinatos, eds., *The Minoan Thalassocracy: Myth and Reality* (Göteborg, 1984), pp. 193–94.

3. K. Branigan, *The Foundations of Palatial Crete* (London, 1970), p. 9; R. Castleden,

The Knossos Labyrinth: A New View of the "Palace of Minos" at Knossos (London, 1990), p. 39.

4. G. Dalton, "Karl Polanyi's Analysis of Long-Distance Trade and His Wider Paradigm," in J. A. Sabloff and C. C. Lamberg-Karlowsky, eds., *Ancient Civilization and Trade* (Albuquerque, 1975), p. 98.

5. M. A. Edey, *The Sea Traders* (Alexandria, Va., 1979), pp. 37–39.

6. "Final Discussion," in Hagg and Marinatos, p. 218.

7. Probably the most famous proponent of this view was the chief excavator of Knossos, Sir Arthur Evans. See A. Evans, *The Palace of Minos at Knossos*, 4 vols. (London, 1921–1936); see also J. Hawkes, *Dawn of the Gods* (London, 1968); R. F. Willetts, *The Civilization of Ancient Crete* (Berkeley, 1977), pp. 64, 112, 129.

8. C. Starr, "Minoan Flower Lovers," in Haag and Marinatos, pp. 9–12.

9. H. E. L. Mellersh, *The Destruction of Knossos: The Rise and Fall of Minoan Crete* (New York, 1970), p. 1.

10. Minos is mentioned by Homer, Hesiod, Thucydides, Herodotus, Apollodorus, Bachylides, Plutarch, Pindar, and Diodorus Siculus, frequently from differing perspectives and with some variation.

11. April 5 entry into Evans's diary, cited in A. Cotterell, *The Minoan World* (New York, 1980), p. 42.

12. Ibid., p. 43; Mellersh, p. 45.

13. Evans diary cited in *Castleden*, p. 31.

14. Cotterell, p. 165.

15. J. Alsop, *From the Silent Earth: A Report on the Greek Bronze Age* (New York, 1964), pp. 18–19.

17. Mellersh, p. 102.

18. Branigan, p. 52.

19. Ibid., p. 63.

20. Evans, *The Palace of Minos at Knossos*, 1: 4.

21. Cotterell, p. 13.

22. Castleden, p. 72; Alsop, p. 31.

23. Branigan, p. 180.

24. J. Chadwick, *The Mycenaean World* (Cambridge, Eng., 1976), pp. 188–91; Mellersh, p. 121.

25. H. van Effenterre, "Le Langage de la Thalassocratie," in *Hagg and Marinatos*, p. 57; Branigan, p. 125.

26. Ibid., p. 67.

27. Ibid., p. 85.

28. Ibid., p. 67.

29. Z. A. Stos-Gale and N. H. Gale, "The Minoan Thalassocracy and the Aegean Metal Trade," in *Hagg and Marinatos*, p. 59.

30. Branigan, p. 65; P. Warren, "The Place of Crete in the Thalassocracy of Minos," in *Hagg and Marinatos*, p. 41.

31. R. F. Willetts, *Everyday Life in Ancient Crete* (Amsterdam: 1988), p. 102.

32. Branigan, pp. 115–16.

33. Warren, "The Place of Crete in the Thalassocracy of Minos," p. 39.

34. A. Bingham, "A Minoan Maginot Line," *History Today* 41 (October 1991): 3-4.

35. J. L. Davis, "Minos and Dexithea: Crete and the Cyclades in the Later Bronze Age," in J. L. Davis and J. F. Cherry, eds., *Papers in Cycladic Prehistory* (Los Angeles, 1979), pp. 143-57.

36. W-D Niemeir, "The End of the Minoan Thalassocracy," in Hagg and Marinatos, pp. 207-8; the dissenting view is represented by C. G. Starr, "The Myth of the Minoan Thalassocracy," *Historia* 3 (1954-55: 282-91.

37. L. Cohen, "Evidence for the Ram in the Minoan Period," *American Journal of Archaeology* (1938); 486-94. For indications that Cretan vessels did employ rams see Willetts, *Everyday Life in Ancient Crete*, fig. 49, p. 100; R. Laffineur, "Mycenaeans at Thera: Further Evidence?" in Hagg and Marinatos, fig. 8, p. 138.

38. S. Marinatos, *Excavations at Thera VI* (Athens, 1974), col. pl. 7.

39. J. A. MacGillivray, "Cycladic Jars from Middle Minoan III Contexts at Knossos," in Hagg and Marinatos, p. 157.

40. E. Schofield, "Coming to Terms with Minoan Colonists," in Hagg and Marinatos, pp. 45-47.

41. M. H. Wiener, "Crete and the Cyclades in LM I: The Tale of the Conical Cups," in Hagg and Marinatos, p. 17; S. Hiller, "Pax Minoica Versus Minoan Thalassocracy: Military Aspects to Minoan Culture," in Hagg and Marinatos, p. 27.

42. Willetts, *Everyday Life in Ancient Crete*, fig. 87 ("Captain of the Blacks"), p. 133; Laffineur, pp. 134-37.

43. Laffineur, p. 133.

44. "General Discussion," comment of J. C. Poursat, in Hagg and Marinatos, p. 113; A. R. David, *The Pyramid Builders: A Modern Investigation of Pharaoh's Work Force* (Boston, 1986), p. 185.

45. E. Sakellarakis and Y. Sakellarakis, "Drama of Death in a Minoan Temple," *National Geographic* 159 (1981). 205-22.

46. P. Warren, "Archaeological Report," *Stratigraphical Museum Excavations, 1978-1981* (Knossos, 1980-81), pp. 73-92; Warren interviewed by M. Billings for *The Minoans and Their Gods*, BBC Radio, 1987.

47. Castleden, pp. 140-41; Cotterell, p. 91.

48. Alsop, pp. 161-62.

49. J. V. Luce, *The End of Atlantis* (London, 1969).

50. Chadwick, p. 11.

51. Castleden, pp. 145-46.

52. Cotterell, pp. 96, 98.

53. Castleden, p. 154, fig. 50, p. 168; Mellersh, p. 132.

54. Alsop, pp. 18-19, 27-28.

55. Ibid., p. 30.

56. Strom, "Aspects of Minoan Foreign Relations," p. 191.

57. Ibid., p. 192 n. 9; F. Schachermeyr, *Die Minoische Kultur des Alten Kreta* (Stuttgart, 1964); Castleden, p. 196.

58. Chadwick, pp. 2, 4, 85, 115.

59. Mellersh, pp. 109, 117.

60. Alsop, p. 95; Mellersh, pp. 147-62; Chadwick, pp. 178-79.

61. S. F. Bondi, "The Course of History," in M. Andreose, ed., *The Phoenicians* (New York, 1988), p. 38.

62. Plutarch, *Praecepta gerendae reipublicae*, iii, 6 (*Moralia*, Didot, II, p. 967).

63. S. Moscati, "Territory and Settlement," in Andreose, p. 28.

64. D. Harden, *The Phoenicians* (Harmondsworth, Eng., 1980), p. 23.

65. Edey, p. 17.

66. Harden, pp. 46, 149; S. Moscati, *The World of the Phoenicians* (London, 1968), p. 9.

67. S. Moscati, "Who Were the Phoenicians?" in Andreose, p. 25.

68. Ezek. 27:1-25; "The Trade Network of Tyre According to Ezek. 27," M. Cogan and I. Eph'al, eds. *Ah, Assyria . . . Studies in Assyrian History and Ancient Near Eastern Historiography Presented to Hayim Tadmor* (Jerusalem, 1991), p. 74.

69. Edey, p. 61.

70. 1 Kings 5, 6, 7; 1 Kings 9, 10.

71. Harden, p. 171.

72. Ibid., p. 105.

73. S. F. Bondi, "The Origins in the East," in Andreose, p. 35.

74. C. Singer, E. J. Holmyard, and A. R. Hall, eds., *A History of Technology, vol. 1* (Oxford, 1954), pp. 762ff.; W. F. Albright, *Journal of the American Oriental Society* 57 (1947): 153ff.

75. F. Mazza, "The Phoenicians as Seen by the Ancient World," in Andreose, p. 548.

76. Moscati, "Territory and Settlement," p. 26.

77. S. Ribichini, "Beliefs and Religious Life," in Andreose, p. 120.

78. J. B. Hennessy, "Thirteenth Century B.C. Temple of Human Sacrifice at Amman," in *Studia Phoenicia III: Phoenicia and Its Neighbors: Proceedings of the Colloquium . . . December 1983.* (Louvain, 1985), pp. 85-104.

79. Ribichini, pp. 120-21.

80. L. E. Stager, "The Rite of Child Sacrifice at Carthage," in J. G. Pedley, ed., *New Light on Ancient Carthage* (Ann Arbor: 1980), p. 9.

81. S. Lancel, G. Robine, J.-P. Thuillier, "Town Planning and Domestic Architecture of the Early Second Century B.C. on the Byrsa, Carthage," in Pedley, pp. 13-23.

82. Edey, p. 90.

83. S. F. Bondi, "Political and Administrative Organization," in Andreose, pp. 126-28.

84. Ibid., p. 128.

85. Bondi, "The Course of History," pp. 41-43.

86. Harden, fig. 11, p. 50.

87. M. Elat, "Phoenician Overland Trade Within the Mesopotamian Empires," in M. Cogan and I. Eph'al, eds., *Ah, Assyria*, p. 21.

88. A. L. Oppenheim, *Journal of Near Eastern Studies* 19 (1960): 146; J. N. Postgate, *Mesopotamia* 7 (1978): 205-6.

89. Elat, p. 24.

90. Harden, plate 51.

91. S. Moscati, "Colonization of the Mediterranean," in Andreose, pp. 47-48.

92. Harden, pp. 32-33.

93. E. Lipinski, "Products and Brokers of Tyre According to Ezekiel 27," *Studia Phoenicia III*, p. 220.

94. Herodotus, *The Histories*, trans. A. de Selincourt (Harmondsworth, Eng. 1981), 4:42 (pp. 283–84); Harden, pp. 162–63.

95. Harden, p. 170.

96. P. Bartoloni, "Ships and Navigation," in Andreose, p. 74.

97. Ibid., p. 75.

98. L. Basch, "Phoenician Oared Ships," *Mariner's Mirror*, 55/2 (1969): 141.

99. Ibid., p. 150.

100. Bartoloni, pp. 72–74.

101. C. R. Whittaker, "Carthaginian Imperialism in the Fifth and Fourth Centuries," in P. D. A. Garnsey and C. R. Whittaker, eds., *Imperialism in the Ancient World: The Cambridge Research Seminar in Ancient History* (New York, 1978), p. 59.

102. Ibid., pp. 66, 71.

103. Ibid., p. 60.

104. Ibid., pp. 88–89.

105. C. Renfrew, "Trade as Action at a Distance: Questions of Integration and Communication," in Sabloff and Lamberg-Karlowsky, p. 25.

106. S. Biswas, "Dholavira Harappan Treasure Trove," *India Today*, 31 Aug. 1993, pp. 158–61; D. P. Agrawat and R. K. Sood, "Ecological Factors and the Harappan Civilization," in G. L. Possehl, ed., *Harappan Civilization: A Contemporary Perspective* (Warminster, Eng., 1982), p. 223.

107. John Marshall, *Mohenjo-Daro and the Indus Civilization: Being an Official Account of Archaeological Excavations at Mohenjo-daro Carried Out by the Government of India Between the Years 1922 and 1927*, 3 vols. (Delhi, 1972); J. G. Shaffer, "Harappan Culture: A Reconsideration," in Possehl, p. 41; M. Wheeler, *The Indus Civilization: Supplementary Volume to the Cambridge History of India* (Cambridge, Eng., 1968), pp. 4–22.

108. K. C. Jain, *Prehistory and Protohistory of India* (New Delhi, 1979), pp. 111–18; S. Asthana, *Pre Harappan Cultures of India and the Borderlands* (New Delhi, 1985), pp. 1–3, 82, 241.

109. Marshall, 1:9; G. F. Dales, "Mohenjdaro Miscellany," in Possehl, pp. 103–4.

110. M. Wheeler, *The Indus Civilization*, p. 68; B. Allchin, "Substitute Stones," in Possehl, p. 236.

111. Marshall, 1:48; M. Wheeler, *The Indus Civilization*, p. 15.

112. Marshall, 1:vi.

113. K. A. R. Kennedy, "Skulls, Aryans, and Flowing Drains: The Interface of Archaeology and Skeletal Biology in the Study of Harappan Civilization," in Possehl, pp. 290–91.

114. Marshall, 1:p. 91.

115. Vishnu-Mittre, "The Harappan Civilization and the Need for a New Approach," in Possehl, p. 31.

116. Jain, p. 148.

117. Shaffer, p. 49.

118. Ibid., p. 44.

119. Ibid., p. 45.

120. Ibid., pp. 46–47.

121. M. Wheeler, *The Indus Civilization*, p. 80.

122. Y. M. Chitawala, "Harappan Settlements in the Kutch-Saurashtra Region: Patterns of Distribution and Routes of Cummunication," in Possehl, p. 198.

123. I. Mahadevan, "Terminal Ideograms in the Indus Script," in Possehl, p. 311.

124. F. R. Allchin, "The Legacy of the Indus Civilization," in Possehl, p. 328; M. Wheeler, *The Indus Civilization*, p. 132.

125. Agrawat and Sood, p. 225; Kennedy, p. 292.

126. P. C. Chakravarti, *The Art of War in Ancient India* (Delhi, 1972), pp. 131–33.

14 CONCLUSION: THE HORSEMAN'S FALL

1. G. Brook-Shepherd, *The Storm Birds: Soviet Postwar Defectors* (New York, 1989), p. 332.

2. Ibid., p. 330.

3. R. L. Garthoff, *Detente and Confrontation: American-Soviet Relations from Nixon to Reagan* (Washington, D.C., 1985), p. 883.

4. "Speech of General Secretary of the Communist Party Comrade Yu.V. Andropov," *Kommunist*, no. 9 (June 1983): 5.

5. "Statement by Yu.V. Andropov, General Secretary of the Communist Party of the USSR, September 29, 1983," in Garthoff, p. 1017.

6. M. Marder, "Defector Told of Soviet Alert," *Washington Post*, 8 Aug. 1986, p. A22.

7. Brook-Shepherd, p. 329.

8. Ibid.

9. Marder, p. 1.

10. Brook-Shepherd, p. 330.

11. J. Mueller, *Retreat from Doomsday: The Obsolescence of Major War* (New York, 1989), p. 13.

12. W. H. McNeill, *Population and Politics Since 1850* (Charlottesville, Va., 1990), pp. 2, 10–11.

13. Ibid., p. 13.

14. Ibid., p. 4.

15. W. H. McNeill, *Plagues and Peoples* (Garden City, N.Y., 1976), p. 243.

16. Anthropologist Helen E. Fisher cogently discusses these issues in "Back to the Future," *In These Times*, 31 May 1993, pp. 16–20, and in her recent book, *Anatomy of Love: The Natural History of Monogamy, Adultery, and Divorce* (New York, 1992), chap. 16.

17. F. Fukuyama, *The End of History and the Last Man* (New York, 1991), pp. 51–54.

18. R. Dawkins, *The Selfish Gene* (New York, 1976).

19. Compte de Saint-Germain, *Memoirs* (Paris, n.d.), p. 178.

20. R. L. O'Connell, *Of Arms and Men: A History of War, Weapons, and Aggression* (New York, 1989), chs. 7–9; J. Ellis, *The Social History of the Machinegun* (New York, 1975).

21. I. S. Bloch, *The Future of War in Its Technical, Economic, and Political Relations* abridged (Boston, 1902); H. G. Wells, *The World Set Free* (London, 1926).

22. W. H. McNeill, *The Pursuit of Power: Technology, Armed Force, and Society Since AD 1000* (Chicago, 1982), pp. 253–54; E. J. Leed, *No Man's Land: Combat and Identity in World War I* (Cambridge, Eng., 1979), pp. 16–17, 51–55.

23. A. du Picq, *Battle Studies: Ancient and Modern Battles* (New York, 1921), pp. 101–2, 113, 181, 229.

24. N. Angell, *The Great Illusion: A Study of the Relation of Military Power to National Advantage* (London, 1914); B. P. Gooch, *History of Our Time, 1885–1911* (London, 1911), pp. 248–49.

25. W. Pfaff, "The Fallen Hero," *The New Yorker*, 8 May 1989, p. 105.

26. O'Connell, *Of Arms and Men*, pp. 272–80.

27. Flavius Vegetius Renatus, *De Re Militari*, III, prologue, in Major T. R. Phillips, ed., *The Roots of Strategy* (Harrisburg, Pa., 1940).

28. Martin van Creveld (*The Transformation of War* [New York, 1991], pp. 3–10) points out that war planning also went on within the United States defense community, with an equivalent lack of success.

29. It is generally assumed that these fears of China's entry were unwarranted, if still real enough at the time. According to Vietnam era historian John Newman, however, the Chinese did seriously debate the possibility of intervention (interview, 1 April 1994).

30. K. N. Waltz, "The Emerging Structure of International Politics," *International Security* 18/2 (Fall 1993): 51–54.

31. *Handbook of Economic Statistics* (Washington, D.C., 1992), p. 15.

INDEX